Chen Tàijíquán

The Theory and Practice of a Daoist Internal Martial Art:

Chen-style 太极拳 Tàijíquán

Volume 1 – Basics and Short Form

First Published in Great Britain 2016 by Mirador Publishing

Copyright © 2016 by Thomas Hayes and Wang Hai Jun

All rights reserved. No part of this publication may be reproduced or transmitted, in any form or by any means, without permission of the publishers or author excepting brief quotes used in reviews.

First edition: 2016

A copy of this work is available through the British Library.

ISBN : 978-1-911473-48-0

Mirador Publishing

10 Greenbrook Terrace

Taunton

Somerset

TA1 1UT

Chen Tàijíquán

The Theory and Practice of a Daoist Internal Martial Art:

Chen-style 太极拳 Tàijíquán

Volume 1 – Basics and Short Form

By

Thomas Hayes and Wang Hai Jun

Table of Contents

Acknowledgements ii
Disclaimer iii
Foreword by GrandMaster Chen Zheng Lei – In Chinese 1
Foreword by GrandMaster Chen Zheng Lei – in English 4
About the Authors 8
Introduction 17
Chapter 1 – Principles and theory of Daoism 39
Chapter 2 – Internal Alchemy 100
Chapter 3 – The Origins of Chen-style 太极拳 tàijíquán 178
Chapter 4 – 站桩 zhàn zhuāng 202
Chapter 5 – Principles of 太极拳 tàijíquán 233
Chapter 6 – Warm-Up Exercises 288
Chapter 7 – 缠丝劲 Chán sī jìn 328
Chapter 8 – 18 movement short form 368

Acknowledgements

A special thank you is offered to those who have assisted in producing this book. It has truly been a team effort. In no particular order of merit, the following people have provided their skills and are warmly thanked:

Garth Williams-Hulbert, who was responsible for the photography, and the very gifted Zhao Xiang Hua, for the calligraphy and professional contribution to this book. The authors are profoundly grateful to both of them.

Special acknowledgement is also afforded to Grandmaster Chen Zheng Lei, who has both maintained and successfully promoted his family's unique martial art. He has tirelessly brought countless joy to his students for several decades through his kindness and awesome skills. Without his dedication to the art, this book would not have been possible.

Wang Jun You and Jia Ling assisted with translations with untiring effort and no complaint.

Niall Ó Floinn undertook a thorough and comprehensive review of the book based on his knowledge of the subject. His incisive comments were a major contribution and have helped immensely.

Éibhir Mulqueen performed unstinting work in editing and his dedication and enhancements to the book are greatly appreciated.

We are particularly grateful to Sarah and her team at Mirador Publishing who have made unstinted efforts in handling this complex tome, have been flexible and tenacious, acting cheerfully and with good grace at all times.

Lara Kroll in her capacity as our agent has been more than supportive, understanding and unstinting at all times and her efforts cannot be overstated. We truly thank her for her efforts.

Finally, it would be remiss of us not to mention the contribution of the Daoists, whose heritage and deep insights and knowledge of The Way have influenced the authors on an ongoing basis.

Disclaimer

The content provided in this book is designed to provide information to the reader on the subject of Chen-style 太极拳 tàijíquán. The book is not intended, nor should it be used, to diagnose or treat any medical or heath condition. For diagnosis or treatment of any medical problem, please consult your physician. Neither the publisher nor the authors are responsible for any specific health or allergy needs that may require medical supervision and are not liable for any damages or negative consequences from any treatment, action, application or preparation to any person reading or following the information in this book.

No warranties or guarantees are expressed or implied by the publisher's choice of including any of the content in this volume. Neither the publisher nor the individual authors shall be liable for any physical, psychological, emotional, financial or commercial damages, including, but not limited to, special, incidental, consequential or other damages. Our views and rights are the same: You are responsible for your own choices, actions and results.

Foreword by Grandmaster Chen Zheng Lei – in Chinese

良师益友 弘扬太极

为英国汤姆斯、王海军著作面世而序

我和汤姆斯认识是 1997 年夏天他随太极拳访问学习团，来到河南郑州学习太极拳。因我在河南省体委武术馆任副馆长兼教练。

他们一行十几位在郑州学习了太极拳和太极枪。练得非常刻苦认真，汗流浃背从不叫苦！训练结束，带他们到太极拳发源地——河南温县陈家沟，拜祖朝圣、参观旅游。

第二次是 1998 年夏天，汤姆斯再次随太极拳学习旅游团到中国。他们也是先到郑州学习后到陈家沟、少林寺参观完后，他们邀请我们全家随团去湖南张家界旅游。（当时我们家 5 口人，师母、娟、斌、媛）。在张家界大家玩的很开心。逛国家森林公园、坐汽船漂流，早晚打太极拳。

那时汤姆斯还没结婚，他非常喜欢中国传统文化如：太极拳、易经、道德经、气功和中医等。他很想找一个中国姑娘作为终身伴侣，我们大家都在为他物色好姑娘。黎明就在湖南湘潭老家找了一位年轻漂亮的姑娘。说来很巧，就在旅行将要结束的时候，汤姆斯有病了，不能随团走了。黎明就在当地让汤姆斯住院治疗，让这个姑娘来帮助照顾。这真是"万里姻缘一线牵，一百棒槌打不开。"就这样一个星期的照顾，两个人加深了认识和了解，产生了一种说不清的爱慕之情。汤姆斯病愈回到英国没几个月就又返回湖南，将这个姑娘带回了英国。这个姑娘就是现在在唐人街开一家大的"湘菜馆"酒店老板——叫 韩丽。建立一个幸福美满的家庭。

1999 年，我的儿子陈斌到英国曼彻斯特上学，得到了汤姆斯多方面的照顾。英语学习、生活关照、各地旅游都详细备至。为了鼓励陈斌学习上进、对家传太极更感兴趣，每周带陈斌一边玩，一边让陈斌教他太极拳和中文，还给陈斌报酬。他真是用心良苦。

我今年 2013 年 5 月又到伦敦讲学。汤姆斯听海军说后，他正好在伦敦附近开会，专程到车站接我和海军，中午在一起吃饭，下午逛街，还要带去大英博物馆，因天要下雨了，才恋恋不舍地分手了。我和汤姆斯认识至今已有 17 年了！给我的影响他是一个有文化、有知识、有技术、品德优秀的英国绅士。他是电脑专家。对中国的传统文化倍感兴趣。我认识他十几年来，他始终坚持练习太极拳、气功。学习易经、道德经、佛学、中医等。对中国的传统文化有较深的理解。

就在这次接触过程中，汤姆斯提出说："我和海军最近合作写一本书，是体现中国的传统文化太极拳、气功、中医、易经等方面的内容。我是站在西方文化的基础上来理解体悟东方文化的。请您给我们写个序言好吗？"我看他对中国传统文化那么虔诚，多年的接触人品又那么好，又是给海军合作，我就欣然答应了。说老实话，我的时间很紧，去年排今年的时间今年排明年的时间，满满的一个星期走一个州，在欧洲一个星期一个国家的转，每到一个地方除了讲课，就是应酬，会朋友、记者采访，忙的焦头烂额。我就只好挤时间了。这一点小文章我就分三、四次才写完。劝君千万别当名人，没有自由，没有自己的时间和空间。（开个玩笑）

谈起海军我更了解，1972年出生在郑州的一个小伙子。我和他爸是好朋友，1983年，海军12岁。他爸送他到陈家沟住在我家里，和我的三个孩子一起生活长大，上学、练拳。师母路丽丽把他当做自己的孩子一样，给他做饭、洗衣服、看他上学学习、练拳。照顾的无微不至。直到1988年，省体委领导调我到平顶山市体委工作，帮助开展太极拳健身活动。和武汉体育学院联合办起了一所"太极少林武术研习院"。本来我是让海军去上学的，院领导看他太极拳练得那么好，就让他边学习边代课当教练。就这样王海军从此走上了太极拳的专业道路。1989年17岁的王海军开始打比赛，一直到1998年10年间，海军在河南省、全国太极拳、剑、推手比赛中曾多次获得单项冠军和全能冠军。并在武术院兼任教练期间，带出了一大批优秀学员，也在省、全国、国际比赛中取得冠军等优异成绩。

1993年至2000年我曾多次带海军到国外表演讲学。2001年，汤姆斯想在英国更好的推广陈氏太极拳，想找一个好的教练。这时我儿子陈斌正好在英国上学，就向汤姆斯推荐了海军（他俩从小一起长大、学习、练拳。对他这个大哥哥太了解了）。先办了工作签证，后来全家都到了英国。现在有房子、有车，有一个幸福美满的家庭。太太贾玲也是武汉体育学院武术专科毕业，两个女儿，大的今年就要上大学了，小的还在念中学。我和太太路丽丽每次到英国曼彻斯特，住在海军家里，看到他们和谐幸福的生活，我们也从内心感到高兴。每次贾玲都对我们亲切地照顾，端茶倒水送水果，我们都很感谢！她总是说：没有您们的培养，哪有海军的今天。我们孝敬二老是应该的！

所以，通过海军的成长过程的事例，我们明白一个道理。人的一生你没有付出就没有回报！至今我练拳50多年，教拳40多年。直接教过的国内外学生也有十几万人次，收徒400余人，直接与间接影响到的太极拳粉丝也有数百万人。在40多个国家设有百十个教学点。每到一处，看到的都是鲜花笑脸。看到那么多受益者讲述他（她）们的感受，讲他（她）的感激之情，我们几十年的奔波劳顿、辛苦之情顿消失在九霄云外。所以我认为：我的弟子、学生都不是普通的拳师，不是普通的教练，我们是优秀民族文化的传播者，是向全人类送健康的使者。我们的太极事业是一项造福人类的伟大事业！只要我们团结一致，不懈努力，我们的太极事业会更加辉煌！

最后，我诚挚的祝愿：汤姆斯、海军的著作面世后，会对太极拳的发展起到良好的作用，对陈家沟、陈氏太极拳及中国传统文化的推广普及起到积极地作用，为太极拳爱好者提供一本珍贵的参考资料！让更多的外国人民认识了解中国的传统文化！让更多的人受益！

Figure 1-1: Grandmaster Chen Zheng Lei and Grandmaster Wang Hai Jun

Foreword by Grandmaster Chen Zheng Lei – translated into English

To a Mentor and Good Friend,

Thomas Hayes

I met Thomas in the summer of 1997. He was a member of one of the taijiquan groups that came to Henan to study.

The group consisted of more than ten people learning taijiquan. They practised really hard and conscientiously. They sweated a lot but didn't complain! When their training finished, they travelled to Chenjiagou village, the birthplace of taijiquan to pay homage to the ancestors.

In the summer of 1998, Thomas visited China with a taijiquan group that consisted of more than 20 people. They went to Chénjiāgōu to study taijiquan and also visited the Shaolin temple and invited my family to Zhang Jia Jie National Park. My family consisted of my wife, my younger and elder daughters and my son. They had a really great time in Zhang Jia Jie. The National Park is amazing. We also undertook some white water rafting during the day, and practised taijiquan in the mornings and evenings.

Thomas loves all things regarding Chinese culture: taijiquan, the Yì jīng, Dao de Jing, qigong, traditional Chinese medicine, etc. He also met a woman from Xiang Tan, who was really pretty and young. He met her in hospital, they got engaged and he brought her to England. They now have a Hunan cuisine restaurant in Chinatown, Manchester. In 1999, my son Chen Bin went to Manchester to study, Thomas helped him a lot in his studies, in his life and with his travel. He encouraged Chen Bin to study hard and took him around the UK, teaching him English, and, in exchange, he was taught Chinese and taijiquan. He spared no effort to learn from Chen Bin. They became really good friends and both of them learned a lot from each other.

In 2013, I came to London to teach. Thomas heard about my visit and came to meet me at the train station. He took us for lunch and to do some sightseeing, taking in the British Museum as it was raining. I have known Thomas for 17 years. In my mind, he is a man with a lot of knowledge and high moral standards and is an English gentleman. He is an expert in computers and has an enormous interest in Chinese culture. He never stops learning and practising taijiquan, qigong, Dao de Jing and Daoism. I think that he has a deep understanding of traditional Chinese culture,

When we were in London, Thomas mentioned that he and Wang Hai Jun wanted to write a book about taijiquan and Daoism. This would be based on the Western perspective of understanding Eastern culture. He asked me if it would be okay for me to write this foreword. I was touched by his devotion to Chinese culture and I have always known him to be a person of high moral standing. He was writing this book with Wang Hai Jun, and so I was happy to agree without any hesitation.

As it was, my schedule was so tight at that time: all of the events had been booked a year in advance and I had to complete my European commitments within a week. I had meetings and demonstrations and scheduled appointments. I was extremely busy and could only squeeze in a little time for this foreword. Do not think that famous people can manage their time! So, I wrote this over three breaks.

Wang Hai Jun

Wang Hai Jun was born in Zhengzhou in 1972. His father was my best friend. In 1983, when Wang Hai Jun was 12 years old, his father sent him to live with me. I was his guardian and he lived with me and grew up with my three children, studying and practising together. My wife treated him like her own son, cooking, washing his clothes and looking after him. In 1988, I was transferred from Henan Sports Department to Pingdingshan Sports Department. The purpose was to promote taijiquan in association with Wuhan Sports Department; together, we promoted taijiquan and Shaolin martial arts.

I took Wang Hai Jun with me because the government leader also recognised that he was good at taijiquan, so he wanted him to be a coach also. That was Grandmaster Wang's first step in becoming a taijiquan coach. In 1989, at the age of 17, until 1998, he continued to take part in competitions. In these 10 years, he won numerous championships in Henan, and all China levels in form, sword and Push Hands. During his tenure as a coach, he has also had a lot of students who won provincial, national and international competitions.

Between 1993 and 2000, I took Wang Hai Jun with me abroad many times to demonstrate and teach. In 2001, Thomas wanted to promote Chen-style taijiquan in the UK. He wanted a good coach and at that time my son was studying in England. He introduced Wang Hai Jun to Thomas. Wang Hai Jun was like a big brother to my son as they had lived, studied and practised together. So Thomas obtained a work permit for Wang Hai Jun. and his family emigrated to the UK. They now enjoy a good family life. Wang Hai Jun's wife Jia Ling is a graduate from Wuhan Sports University. They have two daughters. The eldest will soon be ready for university; the youngest is at primary school.

Every time my wife and I come to Manchester, we stay at Wang Hai Jun's house. They look after us really well. It feels like our second home. We feel really happy for them and Jia Ling takes really good care of us. We really appreciate the hospitality. Jia Ling says that if I had not been so kind as to have raised Wang Hai Jun, they would not be in the position they are so it is their duty to look after us.

So, through growing up with Wang Hai Jun, we all know one thing. In life, there is no taking without giving. I have now practised taijiquan for more than 50 years and taught for more than 40 years. I have taught more than 100,000 students and have more than 400 indoor students and perhaps influenced several million people. We have an association in more than 40 countries and in 100 locations throughout the world.

Every time I see smiling faces and fresh flowers and big welcomes, people tell me of the benefits that come from their own practice and how grateful they are to me. I then feel that all of my hard work and trials and tribulations were worth the effort and are paying dividends. I think that my students and all of the taijiquan culture and benefits are spreading out widely to the people. If we all work together and unstintingly concentrate our efforts, we will make taijiquan truly resplendent.

Finally, I want to give my sincerest good wishes to Thomas and Wang Hai Jun. I believe that this book will make taijiquan expand even further and will have positive benefits. It will become a precious reference book for Chen-style taijiquan for Westerners who wish to understand traditional Chinese culture and will bring benefits to many people.

About the authors

Grandmaster Wang Hai Jun

Grandmaster Wang Hai Jun was born on 10th January 1972 in 郑州 Zhèngzhōu, 河南 Hénán province, China. At an early age, he decided that his life's goal was to study martial arts. Fortunately for him, circumstances were at that time ripe to make this feasible. Due to his being a resident in 郑州 Zhèngzhōu and because his father's friend knew Grandmaster Chen Zheng Lei, it was possible for him to become a martial arts student. So aged eleven, he commenced his studies at his new teacher's house whilst at the same time undertaking his academic education, attending school in 陈家沟 Chénjiāgōu village. He was the first non-Chen family student in modern times to be traditionally trained in Chen Village in 河南 Hénán.

To begin with, he was taught in the traditional way, learning 老架一路, lǎo jiā yī lù, Old Frame, First Part. At that time, 缠丝劲 Chán sī jìn, Silk Reeling Energy was not taught as a separate skill, but was incorporated into the forms sets. Due to his diligence practising for many hours every day, he was able to make significant progress. Grandmaster Chen Zheng Lei provided personal guidance and made subtle and nuanced adjustments to Grandmaster Wang's posture, slowly but surely passing on his vast wealth of knowledge and skill.

Grandmaster Wang's initial practice routines were focused on improving his posture, without any emphasis placed on the breathing methods and the sinking of his 气 qì. This is known as moulding like a statue or 架子 niē zi, the third stage of the six stages in the advancement of the 太极拳 tàijíquán Chen style 太极拳 tàijíquán mastery. He initially found some of the practice routines somewhat frustrating. As a young boy, he considered the method of having to learn slow movements as counter to his desire to release his energy in an overtly yang fashion, for example, by running around. This is natural for all young people. His practice seemed like an interminable routine of repeat sets of the form and, at times, he was tempted by the 少林 Shàolín martial arts that were taught nearby and that seemed very athletic and energetic. Regardless of these early reservations and impatience, he persevered with his practising.

For the first three years of training, he found the practice very hard and his legs were always tired; from the time when he woke up in the mornings through to when he retired at night, his muscles constantly ached.

When Japanese students first started to arrive in 郑州 Zhèngzhōu, he assisted Grandmaster Chen Zheng Lei in teaching, travelling throughout China. A breakthrough in his practice occurred when he was 17, and he felt for the first time the harmony of total co-ordination and mind power. This was a feeling of effortless effort, a relaxed awareness whilst performing the movements.

He continued to practise, and sometimes this was with the movements performed so slowly that the form would take two hours to complete. As the training and dedicated teaching began to bear fruit, his level became so high that he started to take part in competitions, winning titles at provincial and national competitions.

In the autumn, 1988, he was accepted into the 武汉 Wǔhàn Physical Culture University, one of the top universities of its kind in China. He competed successfully for his university in many competitions. After graduation he was assigned to the post of coach of 平顶山 Píng dǐng shān 武术 Wǔshù Research and Study College, which is near 郑州 Zhèngzhōu in 河南 Hénán province. He was appointed as a senior state 武术 Wǔshù referee, the president and head coach of 郑州 Zhèngzhōu 武术 Wǔshù Research and Study College, and coach of the 河南 Hénán Chen Zheng Lei 太极拳 Tàijíquán Culture Company Ltd. He is now recognised as an official lineage holder of Chen-style 太极拳 tàijíquán 12th generation Chen master.

He has an extensive list of accomplishments in competitions. Between 1988 and 1991, he won 15 gold medals in the form, push hands and sword categories of all-styles martial arts competitions in 河南 Hénán. In 1992, he won three gold medals in the all-China national competition, and the championship gold medal in push-hands at the National 太极拳 Tàijíquán Boxing, Sword and Push-hands Competition. In 1994, he won two gold medals in 太极拳 tàijíquán boxing and sword at the International 温县 Wēnxiàn 太极拳 Tàijíquán Championships and the 80-kilo championship in push-hands at the National 武术 Wǔshù Championships. For three consecutive years – 1996 to 1998 – Grandmaster Wang won gold medals in form, sword and push-hands as well as the all-around champion's gold medal at the all-styles martial arts Chinese National Championships – the highest level of competition in China. There are very few 太极拳 tàijíquán or martial arts practitioners with his extensive level of skills, pedigree and competition achievements.

Since 1990, Grandmaster Wang has been taking his students to top-level competitions in 河南 Hénán province and to the Chinese National Championships. His expertise as a teacher is demonstrated by the fact that his students have won more than 30 gold medals at these competitions. Students of his who have won gold medals at the Chinese National Championships include Fu Nubbin (push hands, 56 kg class), Fu Lihue (push-hands, 52 kg class), Yang Lei (push-hands, 65 kg class), Zhao Zhuang (push-hands, 63 kg class) and Shi Sherwin (push-hands, 52 kg class).

In 2001 Grandmaster Wang was invited by Thomas Hayes to come to Manchester to provide authentic and the highest level of Chen style 太极拳 tàijíquán teaching. In addition to throughout the UK he also teaches in; Bulgaria, France, Ireland, Poland, Spain, and the USA. He also taught and has students in Australia, China and Polynesia.

His philosophy for success can be summarised as follows:

- Take one day at a time
- Do not think about the end result or goal

- Find the right teacher
- Use your intuition
- Practise hard
- For Westerners – loosen the joints which tend to be stiff
- Pay attention to the 五心 wǔ xīn Five Xin:

- 敬心 jìn xīn – calm mind
- 耐心 nài xīn – patience
- 决心 jué xīn – determination
- 信心 xìn xīn – confidence/faith
- 恒心和 héng xīn – perseverance

Figure 1-2: Grandmaster Wang Hai Jun

Thomas Hayes

Thomas Hayes has studied and practised the Daoist Arts for 30 years. In addition to Chen-style 太极拳 tàijíquán, he has studied nèi gōng 內功, the 易经 Yìjīng and metaphysics with several Chinese masters of different traditions. He has also studied and practised in other Buddhist and Yoga traditions. He met Grandmaster Chen Zheng Lei in 1997 and has been a friend of the family since. In 2001, he went to China to offer Grandmaster Wang Hai Jun the opportunity to come to Manchester, England, to teach authentic Chen-style 太极拳 tàijíquán. He works in information management and is a Fellow of the British Computer Society. He lives in Manchester. He never ceases to learn and practise, and has a lifelong devotion to self-cultivation.

Figure 1-4: Thomas Hayes in Chenjiagou in 1997 – back row fifth from left. Grandmaster Chen Zheng Lei is fourth from the right on the middle row and Grandmaster Wang Hai Jun is second from the right.

Figure I–5: Thomas Hayes at the back row, third from the left. Grandmaster Chen Zheng Lei is situated in the centre. Other teachers also appear here: Davidine Sim and Janet Murray are pictured second and third from the left in the front row. David Gaffney is fourth from the left in the back row and Ian Murray is seventh on the back row. Master Yue Liming to the left of Grandmaster Chen Zheng Lei and Master Kongjie Gou is immediately to his right.

Figure 1-6: Grandmaster Wang Hai Jun

Figure 1-7: A youthful Grandmaster Wang Hai Jun winning a gold medal

Figure 1-8: Grandmaster Wang Hai Jun practising sword

Figure 1-9: Grandmaster Wang Hai Jun in full flight

Introduction

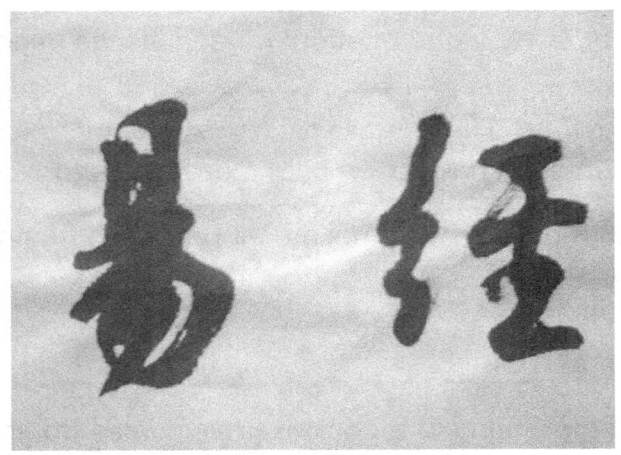

The purpose of this book

This book has been written with a two-fold purpose. At the outset, the intention was to provide some tangible benefits to those readers who were already interested in the subject matter of 太极拳 tàijíquán and who wish to further develop their understanding.

Additionally, it is aimed at those readers for whom the subject of 太极拳 tàijíquán is new, and therefore our aim in this regard is to provide some guidance and explanation of the fundamental principles.

太极拳 tàijíquán or, as it is often known in the West, T'ai Chi, is often only superficially understood. This is due to the following reasons. It is grounded in Daoist metaphysics. and people have difficulties in fully comprehending that subject. There is also the sublimity of its martial arts component, seemingly so easy to perform, yet so difficult to master. Then there is the seeming paradox of the simultaneous inclusion of stillness and meditation in the movements, the interplay of yin and yang and the perplexing aspect of why there is an associated health benefit.

These factors are all difficult to appreciate. Our aim, therefore, is to contribute to providing an explanation of the subject in enough detail to provide some level of understanding. In so doing, the art may then be appreciated and hopefully properly practised so that its benefits may be fully realised.

The book is the culmination of many years of effort. Indeed, it was understood from the outset that it would be a daunting task. Like the practice of 太极拳 tàijíquán, progress during the writing and subsequent editing and presentation have been both measured and progressive.

It has not always been smooth progress. Sometimes it has been frustrating and it has continuously required focus and concentration.

The difficulty that arises with the writing of a book like this comes from the need to provide an easily understandable narrative. There is a balance to be struck. This is between, on the one hand, explaining as simply as possible the theoretical concepts of the ancient philosophy and metaphysics upon which 太极拳 tàijíquán is founded. And, on the other, that of providing practical guidance on the associated practices to be followed in the training and practice.

To arrive at the correct balance between the ancient and the modern requires a nuanced approach. This is particularly so when undertaking the translation from the historical and esoteric language used in the source materials and the rendition of this into something understandable to a readership for Western and Chinese cultures in the 21st century.

Two essential components need to be brought into the methodology that we adopt to help explain 太极拳 tàijíquán to the reader. The first relates to the presentation of the detail and choosing the best approach for providing the narrative and descriptions. Secondly, underpinning this is the requirement to use the most appropriate language in order to be clear and understood. Hence, we need to establish the most suitable terminology for the terms used in the book.

The use of language

We have hitherto introduced two non-English words, 太极拳 tàijíquán and Tai Chi,[1] which are etymologically derived from Chinese. Also, as you would expect, most of the words and terms used in the book are also derived from Chinese. As the book is written in English and is primarily intended for a Western readership,[2] it leaves the author with a question. That is, how should these Chinese words best be represented to the non-Chinese reader without recourse to a crash course in Chinese?

One solution would be to avoid the use of any Chinese words or terminology at all and simply use the English translation. This approach we believe would suffer from the limitation that arises when attempting to precisely translate a subject into English from Chinese that uses as a primary source very esoteric and ancient words. Depending on the context in which the words are used, they may have more than one meaning, which may change what is meant to be said. Anyone who has read translations of the 道德经 Dàodéjīng (Tao Te Ching) will realise that the English is different in each translation – in some cases, a lot so.

We have therefore concluded that solely adopting the use of the English language has little merit, losing as it would much of the vitality to be derived from the use of the Chinese language. This approach is therefore rejected.

Another solution would be to use a joint linguistic approach, presenting the Chinese words together with their English translation. This is the method that we have chosen as it offers the best of both worlds. We feel that this will provide good communication and, hence, ease of comprehension to you, the reader.

---

[1] T'ai and Chi are recognised as two separate words for the purpose of this explanation, but this is sometimes written as one word, Taichi.
[2] There will be subsequent translations of the book into other languages, including Chinese, in the future.

So, let us now examine the use of the Chinese language factor in the literal equation. We appreciate that the reader is not undertaking a course in Chinese Please bear with us and you will find that this introduction serves a purpose in analysing this linguistic topic. First, it needs to be appreciated that the Chinese words that are used and that look like the English (Latin-based) are not the same as the characters that are written by the Chinese in their everyday language. This sounds like an odd statement to make but further explanation will enlighten you as to precisely what we mean.

Many of the transliterated Chinese words that we are familiar with, such as the word *tàijíquán*, are derived from the Pinyin system.[3] This is a means of converting Chinese words and sounds into the Romanisation[4] system used extensively in the West. Pinyin is used solely as a method to help non-Chinese speakers with their word pronunciation. The characters themselves resemble pictograms.[5] For example, 太极拳 is the Chinese character representation for the Pinyin word, tàijíquán.

We will encounter more examples of pinyin in subsequent chapters so it is useful at this stage to review some examples of four words that predominate in the subject matter that we shall cover.

There are two main systems using Romanisation to represent the sounds of Chinese words: Pinyin, referred to above, and the Wade-Giles system.[6] Let us illustrate the differences by choosing a relevant example. You may have seen written the word *Tai Chi Chuan* rather than 太极拳 tàijíquán. The former is from the Wades-Giles system. This was the original system used prior to the introduction of Pinyin to provide the phonetics for non-Chinese speakers using Chinese words. The word 太极拳 tàijíquán is the later Pinyin version.

Frequently used Pinyin words and their Wades-Giles' equivalents that you are likely to encounter are:

---

[3] Pinyin is currently the most commonly used Romanisation system for Standard Mandarin. It means 'phonetics', or more literally, 'spelling sound' or 'spelled sound'.
[4] Romanisation is the representation of a written word or spoken speech with the Roman (Latin) alphabet. Each Romanisation has its own set of rules for pronouncing the Romanised words.
[5] A pictograph is an ideogram that conveys its meaning through pictorial resemblance to a physical object. They were first used by the ancient Chinese around 5000 BC and were at that time drawings used to describe the words.
[6] The Wade-Giles system was developed by an Englishman, Thomas Francis Wade, who was British ambassador to China and a Chinese scholar. The system was refined in 1912 by Herbert Allen Giles, a British diplomat in China. and his son, Lionel Giles, a curator at the British Museum

气 qì – Chi or Ch'i

气功 qìgong – Chi Kung

丹田 dān tián – Tan Tien

老子 Lǎozǐ – Lao Tzu

The problem with the Wade-Giles system is that when it was introduced, it did not set an international standard and each country adopted its own way of representing the words. For example, the English representation of a Chinese word would be different to the French one.

Later on, the Pinyin system was introduced. Chinese linguists devised this in order to provide a standard way of representing the common mainland Chinese language known as Mandarin Chinese.[7] Yes, there are different Chinese languages as well!

To expand on this theme, let us take another common example. Peking is the Wade-Giles name for the capital of China. Older readers will remember this. In Pinyin, it is translated as Beijing, the modern and now most common name. It is still very common, however, to see the continued use of the Wade-Giles system in some books and other media. Where reference is made to books or tracts that were written using the Wade-Giles system, however, we retain the authenticity of the reference by adopting the original system.

Another confusion that arises for English speakers is that some Pinyin words are written in an exactly identical fashion to others, and to all intents and purposes, they seem to be exactly the same but, in fact, have different meanings. These are known as homophonic words, and there are more in use in the Chinese language than in any other.

To add further precision to the Pinyin method, and in order to reduce ambiguity, the Chinese

---

[7]This system is based on the dialect spoken in northern China, geographically based around the Beijing area. Mandarin is an attempt to standardise a common sound where previously none existed and where there are still seven major dialects (and many further sub-branches!)

language uses four pitched tones[8] and a 'toneless' one. These four tones represent the phonetic sounds made when verbalising the Pinyin words by adding the extra dimension of pitch. The toneless one sounds as the name suggests, that is, without tone, although there are not many examples of this type of word in use. To understand the reason for using these four tones, you need to appreciate that the Chinese language has a relatively small number of syllables. There are around 400 in use. English, in comparison, has about 12,000.

Tones were adopted to extend the range of use of the relatively small number of syllables. This has the effect of reducing the number of synonyms, but does not however, eliminate them entirely. There still remains a number of identical Pinyin words that have different meanings.

- First tone ā – a long and sustained sound of even pitch
- Second tone á – a rising tone from a low voice then rising
- Third tone ǎ – a low and curved tone downwards and then rising
- Fourth tone à – a downwards tone lowering from a raised voice
- Neutral tone a – a neutral tone that is short and light

Only the Chinese characters used at the time that they were written provide their precise and unique meaning. We will provide clarity where any confusion is likely to arise for the reader. We certainly do not expect every reader to have fluency in Chinese; although we would not discourage this. The sounds of Chinese consonants and vowels are shown in Table 1 below.

For the sake of completeness and where appropriate, we also show the original Chinese characters for Chinese names or terms. There are two major systems in use throughout the Chinese-speaking world for writing characters.[9] There is the modern system of writing that was codified and modernised in the 1950s and 1960s by the Chinese government.[10] This is now used on mainland China and will be used for this book. It is known as Simplified Chinese characters.

---

[8] In addition to mainland China, Hong Kong, Singapore and Taiwan use the Chinese language.
[9] In addition to mainland China, Hong Kong, Singapore and Taiwan use the Chinese language.
[10] Chinese characters, which date back many thousands of years, are based on pictures.

The original system used before the reforms that led to Simplified Chinese being implemented on mainland China has been retained in other Chinese populated places, for example, Hong Kong and Singapore. This is known as the system of Traditional Chinese characters. Although this does not differ radically from Simplified Chinese, some variations do exist.

In addition to the prevailing modern and previously used methods of character writing, there also exists an ancient form that many of the Daoist texts referred to are written in. This is the form of ancient Chinese characters. In reference and in analysing these, we stick to the use of the modern Simplified Chinese characters. This will remove for the English reader a further level of complexity that would ensue from bringing into play the characters used from this ancient Chinese system.

What also now needs to be added into this etymological mix is that in Chinese, the family name precedes the forename. That is, the Chinese method that is used to represent people's names is different to that used in the West. For example, if we take the full Chinese name Hai Jun. Wang is the family name or surname. Hai and Jun are the man's two forenames. The Chinese method of appellation will be used consistently throughout the book for Chinese (although not for English) names.

Hopefully all of this has not confused you too much. As we have only the use of language and that of the written word as the medium for our communication with you, we feel that it is useful for you to be aware of these facts as way of helping you to understand how the terms we use are derived.

Martial arts heritage in China

This book is intended to be circulated to existing Chen-style 太极拳 tàijíquán students, ranging from those at the beginner level to the more advanced practitioner. Equally, it is aimed at those new to the art who wish to gain an understanding of its theory and application. Furthermore, it will also be of interest to those who are followers of Daoism or any of its arts as the fundamental principles of 太极拳 tàijíquán are derived from those sources and we will go into some detail in covering these topics.

Additionally, some of the techniques and information that we include will be of value to practitioners of other styles of 太极拳 tàijíquán, or indeed those with an interest in 气功 qìgong and the health benefits that accrue from its practice. Just as we have explained the use of language in its literary sense, we also provide a brief explanation of the terminology used when describing the martial arts in both the Chinese and English languages.

As we all know, China is famous for its martial arts heritage. There are many numbers of styles that have developed over the centuries. These styles can be classified as family 家, jiā, sect 派, pài, or school 門 mén depending on their individual derivative roots and history.[11] Those styles that focus on 气 qì manipulation or are primarily based on the use of mind and energy are known as 內家拳, Nèijiā quán, and are described as belonging to the internal category. They are also sometimes referred to as soft martial arts. This is in contrast to those types that concentrate on external body strengthening and cardiovascular fitness that are known as 外家拳 wàijiāquán, or external or hard styles.

Nèi jiā 內家, or internal family style is used to categorise those martial arts that practise nèi jìng 內劲, internal strength. Nèi jìng is developed by the practice of nèigōng 內功, or internal exercises, as opposed to that of wài gōng 外功, which uses external exercises to develop strength. Soft or internal styles are regarded as those that emphasise spiritual, mental and 气 qì related aspects.

---

[11] Please note that this sequence will be used throughout the book; character, Pinyin then English.

We will elaborate further on the meaning and use of 气 qì later in the book. It is important to state at this point that these are all relative descriptions and that there are no completely hard or soft or internal and external styles.

These internal and external distinctions date back to the 17th century, and are partially based on a book written by 黃宗羲 Huáng Zōngxī in 1669. This is known as *An Epitaph for Wang Zhengnan*.[12] The distinction between internal and external martial arts was predicated upon what was regarded as being indigenous to the Chinese, that is Daoist martial arts and those that were foreign imports, such as the Buddhist ones – most notably Shǎolin. Shortly afterwards. Huang Zongxi's son. Huang Baijia, further developed on his father's work with his own book Nèijiā quánfǎ, *The Method of Internal Boxing*.

The modern definitions are based on a classification of how the martial art is learned and the training methods applied. Internal martial arts are regarded as being initially learned by the adoption of the practice of internal 气 qì development. This practice develops as a journey of refinement from the internal to the external, utilising and developing 气 qì. External martial arts take the opposite route, placing an initial emphasis on the development of external muscular strength, and then latterly building internal power. This distinction, whilst useful as a shorthand, is however a simplistic approach to the subject and perfectly illustrates that there are always shades of grey rather than the black and white of total opposites. This sets our theme for later analysis of yin and yang principles, where we shall see that nothing is ever totally yin or completely yang.

Beyond this level of classification that distinguishes internal and external styles, there is a more sophisticated method that we can use, one that is better suited to our aim of increasing our knowledge and understanding. That is, to adopt an analysis that defines a paradigm of relativity that avoids using absolutes and thereby to refine our approach.

---

[12] *An Epitaph for Wang Zhengnan* was written in 1669 by Huang Zongxi (1610-1695 A.D.)

These categorisations of soft and hard, internal and external can only be relative descriptions, as neither can be absolutes because each type is what it must be, a reflection and incorporation of shades of the other. The distinction between internal and external styles is better expressed by an understanding of the different types of emphasis and approach used by each style of martial art. The three most famous internal martial arts of China are 太极拳 tàijíquán,[13] 八卦掌 Bāguàzhǎng and 形意拳 Xíng yì quán.[14]

There is a saying in Chinese that is useful to mention at this point to further illustrate the relativity of the descriptions:

内外相合，
外重手眼身法步
内修心神意气力.

This translates as:

Train both the internal and the external;
External training includes the hands, the eyes, the body and stances;
Internal training includes the heart, the spirit, the mind, breathing and strength.

Chen-style 太极拳 tàijíquán belongs to the internal school of Chinese martial arts. The emphasis is on the development and movement of 气 qì and the relaxation of the body. As with all of the internal martial arts, it incorporates the use of both yin and yang energies. Indeed, it has been described as a yang or hard style in comparison to the other 太极拳 tàijíquán styles, hence illustrating our point about the use of relative descriptions.

---

[13] Bāguàzhǎng or eight trigram palm models its movements on the trigrams of the Yì jīng.
[14] 形意拳; Xíng yì quán. It translates as 'Form/Intention Boxing'. It is characterised by linear movements and explosive power. Several different styles exist.

The Chinese word for martial arts is most commonly translated as Kung Fu in the English language, and this is the term most familiar to us all in the West. In Chinese however, the word 武术 wǔshù or military arts is used. The word wǔshù is formed from the word 武 wǔ, which means martial, and 术 shù, that translates as skill or discipline.

In modern parlance, however, the term 武术wǔshù is also used to describe a form of martial arts that is much more akin to a sport. This is performed in a choreographed way and judged by technical criteria, rather than promoted as a fighting skill in its own right.

For ideological reasons, the Chinese government promoted 武术 wǔshù as something new. This arose from their aim to be wholly independent from the control of the traditional martial arts societies, many of which were ruthlessly persecuted during the the Communist government of Mao Ze Dong and his successors.[15]

The Chinese word that is translated into English as Kung Fu come from 功夫; gōngfu, which has a meaning that is only tangentially related to martial arts, describing a skill achieved through persistent efforts. This term can equally apply to activities other than those of the martial arts. For clarity, then, we use the less ambiguous English term of martial arts and thus avoid falling foul of any possible confusion.

On a final note with regard to language, and with respect to our friends in the US in mind, please note – if you have not already done so – that this book is written in the British English (or at least our version of it), not in American English.

---

[15] He was the leader of the Communist Party in China at that time and a dictator. He was the founding father of the People's Republic of China (PRC) in 1949, and held authoritarian control over the nation until his death in 1976. He suppressed many ancient Chinese religions and cultural practices and was responsible for widespread persecution and deaths on an unsurpassed scale.

This Volume - The Theory of Chen-style 太极拳 Tàijíquán

This book is the first volume of what will form a subsequent series. Further volumes will cover the whole spectrum of Chen-style 太极拳 tàijíquán, with a different topic covered in each individual book. We anticipate that we will require several volumes to cover the whole subject matter.

Volume One covers both the theory and the practice of Chen-style tàijíquán. The theory is the yin and the practice, the yang aspects of the art. Each is relative to the other, and both parts are required for completeness. More explanation of these yin and yang concepts will be provided in subsequent chapters.

So, at this point it is sufficient to submit to you the proposition that both the theory and the practice are relatively opposite yet complementary aspects of the one whole. We can leave this as a statement to rest in the reader's mind to await further explanation. Emphasising the yin theory over the yang practice creates an imbalance, and would be detrimental to the aim of providing a complete approach to our subject. Thus, we will ensure that the true principles of tàijíquán from both aspects are presented.

It is important to understand the theory as it helps with the practice, and writing a book of this kind is the most appropriate way to proceed. It should be stated clearly, however, that in practising 太极拳 tàijíquán, you will gain a lot of the benefits we describe, whether the theory is fully understood or not. Having an appreciation of the theory provides you with a common language to discourse with your teachers and fellow students. This leads to an understanding that allows you to advance more swiftly and to recognise the steps that you are progressing through. It also helps you to grasp the psychology and mind-set of the many experts who developed the art.

Hence, in this book we have sought to combine both the theory and practice of 太极拳 tàijíquán by blending both together to produce a composite that has the benefit of providing completeness for you, the reader. Volume One covers the theory in some detail and is thus relevant to all of the applications of 太极拳 tàijíquán, regardless of the individual form. Subsequent volumes will focus on the practical applications of the other forms, weapons and push hands.

In order to reach out to both categories of completely, and relatively, new students to the subject, we will go wide as well as deep and in so doing cover a range of topics. The fundamental aspects will be examined in detail. This may cause the more experienced student to wonder whether by starting from the basics, we are 'teaching them to suck eggs' in repeating these fundamental principles.

We consider that the benefits that are derived from adopting a wide and deep approach to the subject from the level of basic principles upwards will be beneficial to the reader. This is due to the presentation of a wide range of knowledge from this volume that acts as the foundation for further intellectual and practical development. Although the internet has opened up access to a range of source material on Chen-style tàijíquán, there are in our opinion, still very few, if any, single all-encompassing sources of material on the subject.

Tàijíquán 太极拳 translates as Supreme Ultimate Fist, although this is just one descriptive term for it. Further explanation will be included in the following chapters. Let us first analyse the etymological components of the term. 太 means great or supreme. This is a superlative for the word da 大 that means big. Ji 极 literally means ridgepole or the earth's pole, a pertinent term then, but now a little used term.

The Chen-style tàijíquán form that this book covers originated in the Chen village of 陈家沟 Chénjiāgōu, situated in 河南 Hénán province, China. There are many other styles or forms of 太极拳 tàijíquán, mostly offshoots of the Chen family system, as will be discussed in Chapter 1. Styles, however, such as 杨氏 Yang [6] or 武当 Wǔdāng,[17] together with those of other schools, share the same Daoist origins and philosophy that underpin the art of tàijíquán.

Most styles of 太极拳 tàijíquán taught in the West today, and even in China, are unlikely to be Chen style. The most widely practised is the Yang style. The fact that the Chen style is not as universally widespread as the others is due to the fact that the Chen family kept its secrets hidden from outsiders and it was only taught within the local village for a long period of time. In fact, it is only within the last 20 years or so that it has been fully taught in the West. Nowadays, however, its popularity is growing rapidly.

There are also many other styles that are derived from, or are closely related to, Chen. Some styles now place much less emphasis on the martial arts side and focus more on health and meditation. They thus teach 太极 tàijí, omitting the word 拳 quán, or fist, to reflect this. We shall explain the health benefits of Chen-style 太极拳 tàijíquán and explain how it is possible to incorporate both aspects of the martial art and meditation to produce a doubly beneficial exercise.

The intention of the authors is to produce a map of the journey from the roots of 太极拳 tàijíquán in Daoist philosophy and metaphysics through to its practical application as a martial art and, in doing so, describe its associated health benefits. These benefits are achieved by adhering to the Daoist principles of balance and the harmony of mind, body and spirit.

We believe that there are many good books on the theory of 太极拳 tàijíquán and this is equally so with regard to those that refer to the practice. As we have stated, our intention in writing this book is to achieve a good blend of both theory and practice within this so that the two aspects complement each other and the Daoist aim of achieving the harmony of yin and yang is accomplished.

This book, Volume One, is therefore the first step in a journey that starts by providing the reader with an explanation of the foundation and basics of Chen-style 太极拳 tàijíquán. The book concentrates on the basic building blocks as these are the most important elements for the whole system. These basic practices are incorporated throughout the whole range of Chen-style 太极拳 tàijíquán practices. Without learning these basics, more advanced progress cannot be made. These basic principles are embedded in all of the practical applications of 太极拳 tàijíquán and are equally appropriate and applicable to the other forms.

Perspectives

A further aim of ours is to produce a book that covers 太极拳 tàijíquán from two perspectives. One is from that of the master who, through perseverance, provision of good teaching and much practice. has achieved a high level of skill. From this perspective, there is much valuable information to be imparted to those who are yet to reach this stage.

The other perspective is from that of the student who is unfamiliar with the theory, struggles with the practice and may seem daunted by the long journey ahead. The intention is to integrate both of these aspects to best help the reader in their studies and practice. For those who wish to further develop their skills and obtain a visual representation, there is a DVD that accompanies this book.[16]

We have taken a decision just to cover those subjects and topics that are within our own range of knowledge and experience. Those styles of 太极拳 tàijíquán of which we have little or no knowledge, and of which many others have accrued great expertise, we shall neither discuss nor even proffer an opinion. To do otherwise would be unfair to the followers of those styles and to our own readers. In quoting from the classic texts, we reference either the original source or the work of authors who have translated them into English so that you may be able to further extend your knowledge.

---

[16] For further details, please see www. wanghaijun.com

Contents

In this volume, we cover the silk reeling exercises or 缠丝劲 chán sī jìn [17], some 气功 qìgōng,[18] the warm up exercises and the 18 movement short form. As will be discussed in Chapter 10, 缠丝劲 chánsījìng is not only a fundamental part of 太极拳 tàijíquán, it is also a very effective form of 气功 qìgong in its own right. We will also introduce the practice of 站桩 zhàn zhuāng,[19] which is also known as Standing Pole or 无极 wú jí stance.

Again, this is a very powerful form of 气功 qìgong in its own right. Traditionally taught Chen-style 太极拳 tàijíquán focuses on the forms and holds the postures statically as a form of 站桩 zhàn zhuāng. For example, Single Whip or Lazy about Tying Coat could be used. Warm-up exercises are included as they form a very important part of preparation for the form, including the loosening of the joints of the body.

Fundamental to the explanation of 太极拳 tàijíquán is the concept of qì 气功. This is the basis for a number of Chinese martial arts, of both internal and external styles, and underpins Chinese metaphysics and its practical application in medicine and martial arts. 气 qì can best be described in English as internal energy or life force that circulates and is stored within living beings.

There are many different types of qì, as will be explained in subsequent chapters. Working with 气 qì is known as 气功: qìgōng, the word 功 gōng being the same as that mentioned in 功夫 gōngfu above. 太极拳 tàijíquán incorporates stand-alone qìgong exercises as well as 气 qì movement embedded within the form itself.

The 18 movement form that we describe in detail and have chosen for Volume One as the best

---

[17] The name is derived from the metaphor of 'reeling the silk from a silk worm's cocoon'.
[18] The practice of aligning breath, movement, concentration, meditation for health, spiritual development or martial arts
[19] A form of standing 气功 qigong, where the body remains stationary and energy flows internally. Also sometimes known as Standing like a Tree.

introduction to Chen-style 太极拳 tàijíquán is a very recent one. Grandmaster Chen Zheng Lei created this within the last 20 years. Grandmaster Chen Zheng Lei is Grandmaster Wang Hai Jun's teacher, and further details on him are included in the book.

This 18 movement form is an ideal way for beginners to learn 太极拳 tàijíquán. It covers the basic movements and eliminates the repeat sequences that exist within the original longer forms. Some students find learning the long form straightaway (it consists of 74 or 81 movements, or even more in some styles) is too big a challenge. The 18 movements are easier and form a good starting point from which to progress further if you so desire. The 18 movement form is also valuable for practising even you do not wish to pursue any other forms as it will benefit your health and wellbeing.

We have included a brief and somewhat potted history of the Chen family with respect to notable individuals and their development of the art. This will provide the reader with a general historical perspective.

In order to fully understand how Chen-style 太极拳 tàijíquán developed, it is necessary to explain some Daoist concepts and philosophy. These range from the basic to the more esoteric. The most appropriate elements in the mapping of Daoist theory on 太极拳 tàijíquán principles are discussed in this book. To do full justification to all of the philosophy and tenets of Daoism would require hundreds if not thousands of books, and a Daoist canon exists that is available for that purpose.[20] It is also worth noting that there are many schools and different approaches within the entire Daoist range. We have therefore struck what we consider to be a good balance in our selection of the most appropriate texts to explain the theory and principles that appertain to the practice of 太极拳 tàijíquán.

In keeping a historical perspective on the background, we have included a chronological list of the Chinese dynasties. This will perhaps be of interest to the more scholarly amongst the readership, and is shown in Table I-1.

---

[20] 道藏; Dào Zàng; meaning Treasury of Dao or Daoist Canon, consists of almost 5,000 individual texts. These were collected circa C.E. 400, which is one time after the 道德经 Dào dé jīng and 莊子 Zhuāngzǐ, the core Daoist texts.

Please note that with regard to the coverage of the origins of Chen-style 太极拳 tàijíquán, the historical picture that we have drawn is somewhat sketchy. This is due to the reign of terror that was unleashed by the Chinese government during the 1950s to early 1970s, when many historical artefacts and manuscripts were destroyed. This is a source of regret, not just within the context of this book in being unable to provide a complete picture, but also with regard to the wider consequences.

In the near future, our aim in producing further volumes will be to cover the full range of Chen-style 太极拳 tàijíquán, including the long forms, competition forms, weapons and push hands.

Using photographs and words to describe the movements is not an easy way to pick up the form. The accompanying DVD will be useful in combination or, if you have access to a teacher, the book will help to explain what you have been taught in class. Either way, we hope that you enjoy reading this book and find it of use. We invite readers to contact us with any feedback or questions so that both subsequent reprints of this and additional volumes can incorporate any improvements suggested.

Sounds of Pinyin words

Consonants:

There are 24 consonants in Pinyin that are pronounced a lot like in English:

b, p, m, f, d, t, n, l, g, k, ng, h, j, q, x, zh, ch, sh, r, z, c, s, y, w

Pronunciation of Consonants

b as in boy

p as in pine

m as in mother

f as in food

d as in dip

t as in talk

n as in none

l as in loud

g as in good

k as in kind

ng as in song

h as in hot

j as in jeep

q like 'ch' in cheat

x like a sound between the 's' in see and the 'sh' in she

zh like 'dg' in nudge

ch as in children

sh as in shade

r as in rat

z like 'ds' in words

c like 'ts' in eats

s as in song

y as in Y before the word 'how' yhow

w as in we

Vowels:

There are six simple vowels.

a, o, e, i, u, ü

Pronunciation of Vowels

- a as in man
- o as in drop
- e as in early
- I as in sit
- U as in look
- ü like the u in the French rue

Table I-1

The historical dynasties of China				
Dynasty	Chinese	Pinyin	Dates	Duration
Three Sovereigns and the Five Emperors	三皇五帝	sān huáng wǔ dì	Before 2070 BC	628+
Xià Dynasty	夏	xià	2070 BC–1600 BC	470
Shang Dynasty	商	shāng	1600 BC–1046 BC	554
Western Zhou Dynasty	西周	xī zhōu	1046 BC–771 BC	275
Eastern Zhou Dynasty Traditionally divided into the Spring and Autumn period Warring States Period	東周 春秋 战国	dōng zhōu chūn qiū zhàn guó	770 BC–256 BC 722 BC–476 BC 475 BC–221 BC	514 246 254
Qìn Dynasty	秦	Qín	221 BC–206 BC	15
Western Han Dynasty	西汉	xī hàn	206 BC–AD 9	215
Xin Dynasty	新	Xīn	9–23	14
Eastern Han Dynasty	東汉	dōng hàn	25–220	195
Three Kingdoms	三国	sān guó	220–265	45
Western Jin Dynasty	西晋	xī jìn	265–317	52
Eastern Jin Dynasty	東晋	dōng jìn	317–420	103
Southern and Northern Dynasties	南北朝	nán běi cháo	420–589	169
Sui Dynasty	隋	Suí	581–618	37
Tang Dynasty	唐	Tang	618–907	289
Five Dynasties and Ten Kingdoms	五代十国	wǔ dài shí guó	907–960	53
Northern Song Dynasty	北宋	běi sòng	960–1127	167
Southern Song Dynasty	南宋	nán sòng	1127–1279	152
Liao Dynasty	辽	Liáo	916–1125	209
Jin Dynasty	金	Jīn	1115–1234	119
Yuan Dynasty	元	Yuan	1271–1368	97
Ming Dynasty	明	Míng	1368–1644	276
Shun Dynasty	顺	Shun	1644	<1
qìng Dynasty	清	Qīng	1644–1911	268
Empire of China			1912–1916	

Chapter 1 – Principles and theory of Daoism

乾 qián

The Dao has reality and evidence, but no action and no form. It may be transmitted but cannot be received. It may be attained but cannot be seen. It exists by and through itself. It existed before heaven and earth, and indeed for all eternity

Joseph Needham – Science and Civilization in China[21]

---

[21] Joseph Needham (1900–1995) was a famous British Sinologist, as well as a scientist and historian.

Earliest History

In order to fully understand both the principles and the theory behind 太极拳 tàijíquán, first we should proceed with a brief but appropriately detailed explanation of Daoism. This Daoist philosophy is also known as 道教 dào jiào, and the term covers the teachings of the 道 Dào together with its many tenets, ranging from the purely philosophical to the mostly liturgical and religious.

We start therefore with the roots of Daoism and develop our understanding of how it expands into its associated offspring, such as traditional Chinese medicine, 风水 fēng shuǐ, astrology, spirituality, 气功 qìgong, internal alchemy, yoga and meditation, as well as the martial arts. Finally, we focus on the Daoist principles that have most directly influenced 太极拳 tàijíquán. If the reader grasps these principles, they will make great strides in their progress, as well as acquire a well-grounded knowledge of this seemingly paradoxical subject that is often only vaguely understood.

Daoism is also known in English as Taoism, and is often written or pronounced in this way, as well as sometimes being uttered with other comical sounds from the mouths of Westerners struggling with the use of hybrid phonetics. The word Taoism derives from the use of the Wade-Giles system pronunciation, the method explained in the Introduction.

Daoism is distinctly Chinese in both its origin and its unique expression through the use of philosophical and allegorical descriptions. It is an ancient and mystical teaching method transmitting an original body of knowledge of Man and the Universe, having a history that is at least 5,000 years old. 道 Dào is translated as 'The Way' or 'The Path'.

Today, we can look around to see just how fragmented we are as a species by viewing the rampant global ideological, political and religious warfare and abuse of the Earth that occurs. There is nothing holy about destroying the present to achieve a perceived utopia of the future. The word 'holy' is derived from the word 'whole'. This integration is not possible unless one merges with the Dao, in other words with what is.

One must first of all start with oneself, one's body, thoughts and emotions. If one rejects rather than understands 'what is', one then moves away from reality into a world of ideas based on fear, greed, competition and ignorance. How can we understand that which is unlimited, immense, unknown if we cannot even understand ourselves?

The images that we have of the so-called 'spiritual' are mostly projections of our own limited understanding based on our cultural and religious conditioning. If one starts with belief, one will end up in doubt. If one starts with innocence, enquiry and without psychological authorities, one is free to look at and may be able to understand one's conditioning. And when one sees this as such, this conditioning is immediately cut away without effort or discipline. What is left is the unknown, which cannot be described: it has no path leading to it and is ever changing.

The mind must be totally free to merge with this so that there is no longer an observer and an observed but simply 'observing'. As Jiddhu Krishnamurti so clearly put it, 'Observation without evaluation is the highest form of intelligence'.

This Supreme State of Being is something that is attainable to each and every one of us. It is not to be sought externally, although people inevitably do so. Neither is it something, moreover, that should be regarded as holy or otherworldly. It is something that can be the experience of each and every one of us. It is already there, integral to the human being. It is just that it is not observed by ordinary consciousness. As Alan Watts[22] said, 'It may reign but it does not rule. It is the pattern of things but not the enforced'.[23]

---

[2] Alan Watts (1915–1973) was a contemporary of John Blofeld. Also British, he spent many years in the US. He was known as a philosopher, writer and speaker. He was most famous during the 1960s and 70s, and was one of the pioneers of the time who brought Buddhism, Daoism and Eastern philosophy to Western audiences.
[23] Alan Watts, *The Watercourse Way.*

As a philosophy or way of life, Daoism places much importance on being natural, or living in accordance with nature or the natural way. The Universe is understood to be integrated as a whole and human beings are a representation thereof, existing as a microcosm in relation to the macrocosm. Hence, we are not born into the world as is often stated, but quite the reverse. The world is born in us.

This view is not just an Eastern or Chinese one but is one also held by many Western esoteric traditions.[24] An important aspect of this natural state is that of acting in a state of 无为 wú wéi or effortless effort. This is a much misunderstood precept. When it is misconstrued, and, alas, most of the time this is indeed the case, and thereafter taken to its wrong conclusion, it may lead one to an idle or inert state or lifestyle. This is the very opposite of 太极拳 tàijíquán training! This misconception happened for some people in the UK and US in the 1960s. They adopted 无为 wú wéi as a sort of alternative to society norms or as the then hippy lifestyle. It is, in fact, a spontaneous act.

It is that of acting in the moment without abstract thought hindering the natural action. It is uncontrived and appropriate to the situation, unencumbered by pre-conceived notions of what is right or wrong.

The Dao is that which arises both spontaneously and effortlessly. It is there naturally, existing from a sense of connection to all things. To follow it requires the absence of any idea of separateness to others or to nature. All is one. Only the discriminating mind sees objects as being entirely separate, as, of course, individuals so perceive themselves. As a principle, it is difficult to grasp, but it may be experienced. This may be by the practice of 太极拳 tàijíquán or 气功 qìgong or meditation. It arises from the sense and feelings that a calm mind, effortless movement and abundant energy are created.

Chinese metaphysicians have in all eras described the mind as akin to that of the restless monkey,

---

[24] For example, see Franz Bardon, *Initiation into Hermetics*, page 31.

jumping around, chattering ceaselessly and moving from one concept to another, rather than experiencing directly. This mind or state of mind is the cause of the feeling of separateness from the Dao, and this is why esoteric practices put great emphasis on meditation to still the mind.

This description of the inter-connectedness between all people and between each person and the Universe is not unique to Daoism. The great challenge of metaphysics and esoteric religions is to resolve the paradox of being both an individual and connected at the same time. It is how to perceive and then apperceive beyond what the five senses display to the mind and how the mind discriminates between subject and object and discerns differences. It is more accurate to say that it is not to resolve the paradox of self and the world but to acknowledge that this situation exists and thereby to transcend the dualistic tendencies of discriminative mental activity.

Underpinning our attempt to understand what the Dao is, and hence the Daoist schools of thought, are the following principles. The first is that of the state of Manifestation from the Unmanifest. The Unmanifest is the Dao in the state of rest. For this state of rest, the creation of 阴 yīn and 阳 yáng manifests all that is observable in the Universe. These are all created through the interaction of 阴 yīn and 阳 yáng and are relative parts of the whole. They appear as opposites but are, in fact, just different by degree. One cannot exist without the other because their expression and our classification of either one can only be accomplished relatively, one through the other. For example, if no males existed, the existence of and how we classify a female could not be attempted. The Manifest world moves through cycles: spring through to winter, night to day, and birth to death. Everything is in motion; nothing is at rest even if it so appears. It is just how the one totally connected and interacting Dao expresses itself.

The Dao is the Substantial Reality that is behind, and moves, the phenomenal world. This is described in Daoist terminology through the energy that pervades all, the expression of the Dao. Hence, Daoists describe these things through a paradigm of energy. Although as individual human beings we are separate (or seem to be so), at the level of the phenomenal world, we are connected energetically as expressions of the Dao, the Manifest expressing itself in the phenomenal world.

As above, so below is a principle that may also be expressed as the relationship of the microcosm to the macrocosm. That of the person being a smaller scale version of the Universe. The aim of Daoists is therefore to try to increase their energetic vibrations so as to merge with the Unmanifest that is Dao. Therefore, the understanding of this should be internally directed. This cannot be achieved by the intellect. In fact, the Dao cannot really be explained by the intellect. It can only be experienced. Even the Buddha did not speculate as to such things that cannot be known. Why does the Dao choose to express itself? If we knew, we would not have written and nor would you be reading this book!

Daoism, however, unlike some Eastern religions and philosophies, does not abjure the body but works with it to achieve harmony with the Dao. Mind and body are inter-connected, and each affects the other. Moreover, the distinction is superficial. They are just different rates of energy vibration and classifications. Ultimately all is one. And if you wish to worship anything, why would you not start with your own body? Your mind and body are not separate things. Those extreme schools that abjure the body and tell you that the mind is all and the body is unimportant then proceed to tell you that the body must be in the right position in order to meditate! This is merely a dualistic approach whilst attempting to be non-dualistic. Please beware of the risk of getting lost in these doctrinaire obfuscations. Daoism adopts a practical and holistic approach.

From its antecedents over many thousands of years, various branches and techniques of Daoism have developed, each emphasising different aspects of Daoist teachings, including those that we are most interested in here: 太极拳 tàijíquán. In particular, we need to examine those theories and techniques that have influenced 太极拳 tàijíquán, 气功 qìgong and health. Daoism has also influenced Zen Buddhism. Zen is a Japanese word derived from the Chinese word, Chán, which, in turn, is derived from the Sanskrit word, dharma.[25] The origins of the Chan sect are based in the Shǎolin temple that, interestingly, is located very close to Chénjiāgōu village in Henan. Chan Buddhism or Zen is a synthesis of Daoism and Mahayana Buddhism.[26]

The approach of some authors in writing about Daoism is to explain the subject through the prism of a chronology of historical events, categorised into fixed periods of time. This is an excellent abstract exercise but may mislead the reader into assuming that we are dealing with the linear progression of a homogenised system. Like all attempts to describe the profound, the act of writing anything at all about Daoism risks straying from the essence. We shall persevere, nevertheless, through the medium of this book, and at least point to the general direction of the moon. There are so many schools of Daoism that the variation from one extreme to the other may lead you to assume that they are entirely different. These differences range from their philosophical to their highly religious and ceremonial aspects, as well as in the amount of influence they derive from Buddhism and Confucianism.

For our own analytical purposes, we prefer to adopt the approach of examining key figures and works of Daoism for illustration purposes and for the representation of key philosophical concepts. However, for ease of reference, we will describe these within the parameters of a loose chronological sequence. Please be aware that these historical periods are referenced retrospectively and, thus, do not suggest any linear or progressive theme, such as would be the case, for example, in describing an aspect of Western science, such as the development of medicine. Our referencing also underlines that time is not, in fact, linear anyway. But that discussion is for another time and will not be indulged within the context of this volume.

Daoism has a recorded history of at least 5,000 years, and quite probably dates back 8,000 years, depending on how the criterion is applied for the calculation of the starting point. Within this period,

---

[25] Chán is a school of Mahāyāna Buddhism introduced during the 6th Century CE during the Song dynasty. The Chinese word is derived from the Sanskrit word, dhyāna, meaning 'meditation'.
[26] Mahāyāna Buddhism emerged as the 'Greater Vehicle' (literally, the 'Greater Ox-Cart') school of Buddhism and comprises many different traditions.

some important people emerge who have profound messages for us to reflect upon. It is easy for us to connect with people and events. Hence, historical or even mythical people as enunciators of Daoism will be presented, together with their writings. We include them on the basis of the importance of their profound doctrines, rather than focussing too much on their historical or mythical personalities.

King 伏羲 Fú Xī

We shall now choose what we consider to be the most appropriate starting point. This may be regarded as a sort of beginning, although it is, in fact, just a snapshot that we have chosen from a series of events within a timeless sequence. Like some other countries, China has shamanic roots. One of the historical Chinese shamans, 伏羲 King Fú Xī, who lived circa 2,800 BCE, developed a system that he used to represent the underlying structure of the Universe. This was in a time that was very different to ours. No sophisticated means of literal communication in the form of words existed, let alone Twitter or Facebook. So 伏羲 Fú Xī chose a method to describe what he understood about the Universe and reality. All of the tenets that we earlier introduced to describe the Dao are incorporated, using diagrams rather than words.

Hence, 伏羲 Fú Xī developed a system to represent the Universe and that of nature and Man within it. This was based on a system that used eight 八卦 bā gùa. These are sets of three horizontal lines, stacked one upon the other, which are known as trigrams. Each trigram is represented by a solid line to represent 阳 yáng and a broken line to represent 阴 yīn. Thus, each line can only represent 阴 yīn or 阳 yáng, the two primary states of the Universe. Legend states that he obtained these trigrams from his observation of the scales of the back of a tortoise as it emerged from the water.

So, what is the significance of the number eight that equates to the number of trigrams in total? This number is arrived at by a calculation of the number of permutations of trigrams. That is, it is the sum total of all of the possible combinations of 阴 yīn and 阳 yáng. This equates to two choices - 阴 yīn or 阳 yáng, multiplied by the number of lines, which is three. Hence, 2x2x2 = 8. For mathematicians, this is known as 2^3 (2 to the power of 3). This was King 伏羲 Fú Xī's representation of the movement of the Dao as the Universe expressed itself from the unchanging and Unmanifest to the ever-changing and expressed Manifest state.

As seen above, the Unmanifest state is eternal and limitless. It is devoid of any boundaries, yet it is a state that does not take up any space. It is not part of time, which is of the Manifest sphere and which is a process that covers the separation of one instant from another, the naming and individuation. To name the Unmanifest is impossible, and it can only be described in a limited fashion as it is beyond relative terms. If we were able to express it, we would be excluding its completeness when we talk about the 先天八卦 xiān tiān bā guà or 后天 八卦 hòu tiān bā guà, pre and post heaven arrangements.

Others may define this Unmanifest state as God or the Void, sometimes as That, or many other words used to express the inexpressible. You may even refer to it as Bill or Jill. Names do not matter and, in any case, whatever appellation used would be insufficient to even attempt to describe it. It is the state to which religions aspire to but often are clueless as to how to attain. This theme will be developed further on in the chapter. The 易经 Yì jìng is a very early binary system, and this system was, of course, later used to develop a machine code that was used on the earliest computers. Indeed, computers are nothing more than on or off switches, or states of 阴 yīn and 阳 yáng but all have been created by a creative principle in the minds of people. And how many manifestations have resulted from this expansion outwards of the binary principle in the use of the computer? So maybe there is some Twitter and Facebook in there after all.

Some of you will now realise that 阴 yīn and 阳 yáng and the number 8 resonate with 太极拳 tàijíquán and other internal energy practices. But do not worry if you do not. We will develop this theme later on. The book that developed from 伏羲 Fú Xī is known as the 易经 Yì Jīng, or the *Book of Changes*. It is also more usually known as the *I Ching*, using the Wade-Giles phonetic system.

The 易经 Yì jīng or the *Book of Changes*

Background

The importance of the 易经 Yì Jīng on Chinese culture is immense, reflecting as it does the expression of all things. That is, Dao, together with its contribution to the metaphysical and spiritual development of the world, is difficult to exaggerate. It is like a Chinese Bible, but describing a divine creation and a unified Universe, rather than parables and people. The 易经 Yì Jīng is best known in the West as a system of divination, and this is true enough, but whilst it is legitimate to use it for such purposes, it provides us with just the husk of its profound offerings.

The 易经 Yì Jīng is a book much admired by Buddhists and Confucians alike. Indeed, Lama Anagarika Govinda,[27] in his introduction to the book *The I Ching & the Genetic Code - The Hidden Key to Life*[28] states:

> ...the philosophical principles on which the spiritual attitude of the 'Book of Changes' is based, and at the same time the results of most modern scientific research, are largely identical with the basic ideas of Buddhism and more particularly with the principles of Tantric Buddhism of the Tibetan kind.

In this book, author Dr. Martin Schonberger analyses the correlation between the 64 hexagrams and the DNA code. Other references to the 易经 Yì Jīng were made by the scientist Nils Bohr,[29] who related it to atomic science and used 阴 yīn and 阳 yáng symbols in his calculations.

---

[27] Lama Anagarika Govinda (1898–1985), also known as Anagarika Govinda, was a writer on Tibetan Buddhism, Buddhist meditation and other Buddhist themes.
[28] ISBN 10: 094335837X / ISBN 13: 9780943358376. Author, Martin Schonberger.
[29] Niels Henrik David Bohr (1885–1962) was a Danish scientist who won a Nobel Prize in 1922.

The Daoist Immortal, Lǚ Dòngbīn[30]呂洞宾, who is written about many times in Daoist literature and, in particular, in the book *Eight Immortals Achieving the Tao* 八仙得道搏Bā Xiān dé dào tuán wrote:

> Although the words are very clear, yet they are also very vague. The shallow may take the 易经 Yì Jīng to be a book of divination, but the profound consider it the secret of the celestial mechanism.[31]

The 易经 Yì Jīng represents the interrelationships between the levels of heaven, earth, and man, the relationships within themselves and the different phenomena that are manifested. It uses its system of lines, which, as we have noted, are based on the binary system and are represented by trigrams to map the movement of energy through the various stages of manifestation. Prior to the Han dynasty, the 易经 Yì Jīng was also known as The 周易 Zhōu yì, having been named after the Zhou dynasty.

It also makes use of natural elements as symbols to describe the energies, such as lake and thunder, and human family members, such as father, son and eldest daughter to describe the attributes of the energies as they are moving through these cycles. This symbolism is not intended to be taken as absolutes or fixed tangible elements per se, but is merely used as a description of the relative characteristics. They are thus the characteristics of the energies together with the interrelationships within the context of the whole. Indeed, it should be noted that the imagery and use of metaphor represents both the environment and culture at that time of writing. These natural elements are depicted in Table 1-1 below. The 易经 Yì Jīng describes how the Manifest moves and creates as energy, and it is this description that permeates through Daoism and leads into our practice of 太极拳 tàijíquán.

In ancient China, as a literal definition, a trigram was the name given to an instrument that was used to measure a sphere. A pole would be attached to a rope or cloth, and this was then used to survey

---

[30] Lǚ Dòngbīn was born in Jingzhao Prefecture (京兆府 Jīngzhào Fǔ) around 796 CE during the Tang Dynasty.
[31] *The Taoist I Ching* by Thomas Cleary.

the movement of the sun and the direction of the wind. 爻 yáo was the term used to describe one of these lines. The literal definition is two intersections. In a year, one intersection represented the summer solstice, which occurs between June 21st and June 22nd, and the other represented the winter solstice, that occurs between December 22nd and December 23r

The systems of trigrams used in the 易经 Yì Jīng therefore represent the movement of energy within the Manifest state, and is determined within the cycle of time, describing the changing of the seasons and the time of the day. It is also used to describe the Unmanifest, as we will describe later. As we have stated, the trigrams consist of three lines. There are solid ones that represent the 阳 yáng or male. These are known as 阳爻 yáng yáo. There are also broken ones that represent the 阴 yīn or the female 阴爻 yīn yáo.

A famous poem written by 伏羲 Fú Xī goes as follows:

无极生有极、有极生太极、

太极生兩儀、即阴阳；

兩一儀生四象:即少阴、太阴、少阳、太阳、

四象演八卦、八八六十四卦

This is translated as:

The Wuji produces the delimited,

And this is the taiji.

The taiji produces two forms, named yin and yang

The two forms produce four phenomena, named lesser yin, great yin, lesser yang, great yang.

These four phenomena, the lesser 阴 yīn, great 阴 yīn, lesser 阳 yáng and great 阳 yáng act upon the eight trigrams. The word 无极 wú jí translates as Limitless. This term is prominent in 太极拳

tàijíquán as will be seen later, and is used when you are stationary, just before the movement of the form commences. 太极 tàijí is known as the Absolute. 太极 tàijí depicts the movement of the two forces of 阴 yīn and 阳 yáng and the different types of energy that Manifest themselves through this interplay of the two forces. Great 阴 yīn is also known as tài yīn 太阴, which also means the Moon, and Great 阳 yáng is 太阳 tài yang, which also means the Sun. Hence all things can be described in terms of their constituent components of 阴 yīn and 阳 yáng.

We can see that the way that movement occurs through the Universe is represented as either 阳 yáng or 阴 yīn and that these are constantly acting against and within each other as a dynamic. Energy is created and moves through this interplay of 阴 yīn and 阳 yáng, each one being both dependent and non-existent without the other.

As Daoism also holds to the belief that the human being is a microcosm of the Universe, the movement of energy within the body logically follows the same pattern. This is the part that the 易经 Yì Jīng plays in 太极拳 tàijíquán as the principles are applied to both the human body and the movement of energy or 气 qì, within.

周文王, King Wen or Zhōu Wén Wáng 1099–1055BCE, who lived during the Shang dynasty, further developed 伏羲 Fú Xī's system based on an expanded number of 64 hexagrams known as 六十四卦 liù shí sì gùa that forms the modern 易经 Yì Jīng.

The hexagrams are comprised of a series of figures of six lines, rather than the three that make up the trigrams. With six lines, the possible permutations are 2x2x2x2x2x2 or 2^6 (2 to the power of 6), which aggregates to 64. As there are six lines or 爻 yáo to each of the 64 hexagrams, the total number of 爻 yáo in this arrangement consists of 384 爻 yáo or lines. Please refer to Table 4 for details of the eight trigrams.

阴 yīn and 阳 yáng are opposite yet dependent forces that must always be in balance. The reason for this is simple. 阴 yīn and 阳 yáng are merely the expressions of relative energies within the context of the unified whole. These opposites (relatives) of 阴 yīn and 阳 yáng flow in a natural cycle, one always replacing the other. The trigrams of 伏羲 Fú Xī and the hexagrams of 周文王 Zhōu Wén Wáng represent the phases of this movement. 阴 yīn and 阳 yáng move in cycles from active to passive, and from dark to light, just as the seasons move through cycles and create different patterns of hot (summer) through to cold (winter). As we have seen, the theory of 阴 yīn and 阳 yáng evolved from a belief of mutually dependent opposites. One cannot exist in the absence of the other, and this is represented by the familiar 阴 yīn - 阳 yáng symbol illustrated below.

Figure 1-1: The 太极拳 tàijíquán 阴 yīn - 阳 yáng symbol

In Figure 1-1, the 太极 tàijí symbol is shown. This is another way of representing 阴 yīn and 阳 yáng energies. It is a very widely known symbol, often being depicted in both the cultures of the East and of the West. 阳 yáng is depicted as white and 阴 yīn as black. Within 阳 yáng there is a small element of 阴 yīn, and within 阴 yīn, there is a small element of 阳 yáng. Hence, there is never a situation where there is total 阳 yáng or total 阴 yīn. The figure illustrates the inter-dependency of the male and female energies and their mutual dependence. Only by viewing this from a discriminating perspective does any dualism appear.

People often become confused about the concepts of energy manifestations and relativities versus absolutes, especially when they are represented in diagrammatic form. It is important to realise that these are neither static nor concrete elements in the way depictions of edifices on a building plan are. Nothing could be further from the truth. 阴 yīn and 阳 yáng are representations of movements, of relative states, one to the other that are used to describe energy. The words are meaningless inasmuch that you could label the two black and white or Henry and Matilda. This is the conundrum faced when trying to express and explain in words something that needs to be experienced.

阴 yīn and 阳 yáng ebbs and flows constantly, and the process is never static. Each energises the other and contributes to the endless continuum that is the expression of the Dao. This situation of the static in terms of a diagram to represent the fluid is akin to that of one taking a photograph of someone practicing 太极拳 tàijíquán, like the ones shown in the later chapters of this book. In reality, there was movement at the point at which the photographs were taken, but the outcome is that all appears to be static.

Going back to our historical figures and their interesting times, 周文王 Zhōu Wén Wáng was imprisoned by the last Emperor of the 商朝 Shāng cháo or 殷代 Yīn dà dynasty. He made the most of the opportunity of his imprisonment (if being locked up can be described as such) by using the time to write explanations to accompany the judgements and images of the 易经 Yì Jīng. It would appear that he was not imprisoned in the worst of prison environments that existed during those times if he was able to undertake such a task. It is likely that few prisoners in those times would have had access to writing materials or (any hands left) or even had the luxury of the space to do so.

His son, 周公 Zhōu Gōng, the Duke of Zhou, who later went on to found the 周朝 Zhōu Cháo dynasty, further developed the understanding of the 易经 Yì Jīng. His brother is said to have created the 爻辞 yáo cí, the Explanation of Horizontal Lines. This clarified the significance of each horizontal line within each of the hexagrams. Only then was the 易经 Yì Jīng more fully understood. The philosophy and metaphysics contained therein heavily influenced the contemporary literature of the government administration of the 周朝 Zhōu Cháo, Zhou Dynasty (1122 BCE–256 BCE).

Many versions of the 易经 Yì Jīng have been written, translated and commented on over the millennia in China and, latterly, in the West. The version that first became a standard one originated in the ancient text 古文经 Gǔ Wén Jīng, transmitted by 费直 Fèi Zhí, who lived between 50 BCE and 10 AD during the 漢朝 Hàn Cháo, Han Dynasty. At this time during the Han Dynasty, this version competed with the bowdlerised new 今文经 Jīn Wén Jīng version transmitted by the Duke Tai of Tian Qi 田齊太公 Tián Qí Tài Gōng, also known as 田和 Tián Hé at the beginning of the Western Han dynasty. By the onset of the 唐朝 Táng cháo – Tang Dynasty, however, the ancient text version, which survived the Emperor Qin's book burning by being secretly preserved by the peasantry, became the accepted standard by Chinese scholars.

The earliest extant version of the text, which was written on bamboo slips, is the 楚简周易 Chu Jiǎn Zhōu yì, and dates to the latter half of the 战国时代 Zhànguó Shídài, Warring States[32] period, between the mid-4th to the early 3rd century BCE. It is similar to the standard text, apart from a few significant differences.

During the 楚简周易 Chu Jiǎn Zhōu yì Warring States period, the text was re-interpreted as a system of cosmology and philosophy that subsequently became intrinsic to Chinese culture. It was based on the principle of the dynamic balance of opposites. The evolution of events from this interplay of these opposite forces was stated as an ongoing process, and together with this came the logical acceptance of the inevitability of change. The practice of 风水 Fēng Shuǐ, literally translated as Wind and Water, and now also popular in the West uses these same principles.

As we have previously explained, the expansion in the number of permutations of 阴 yīn and 阳 yáng, representing the energy movements of all manifestations, increases in the range of 2, 4, 8, 16, 32 and 64.

This is the binary system, using what is now known as octal numbers. And so the maximum number

---

[32] The Warring States period ran from about 475–221 BC

of combinations was concluded at 64. But this does not mean that this maximum cannot be exceeded or that everything stops. This is merely the convention used of stopping at that number. There is no ceiling; it is infinite. Do not forget that this derived number is considered to be the optimum one to be used for explaining the combinations of the Manifest.

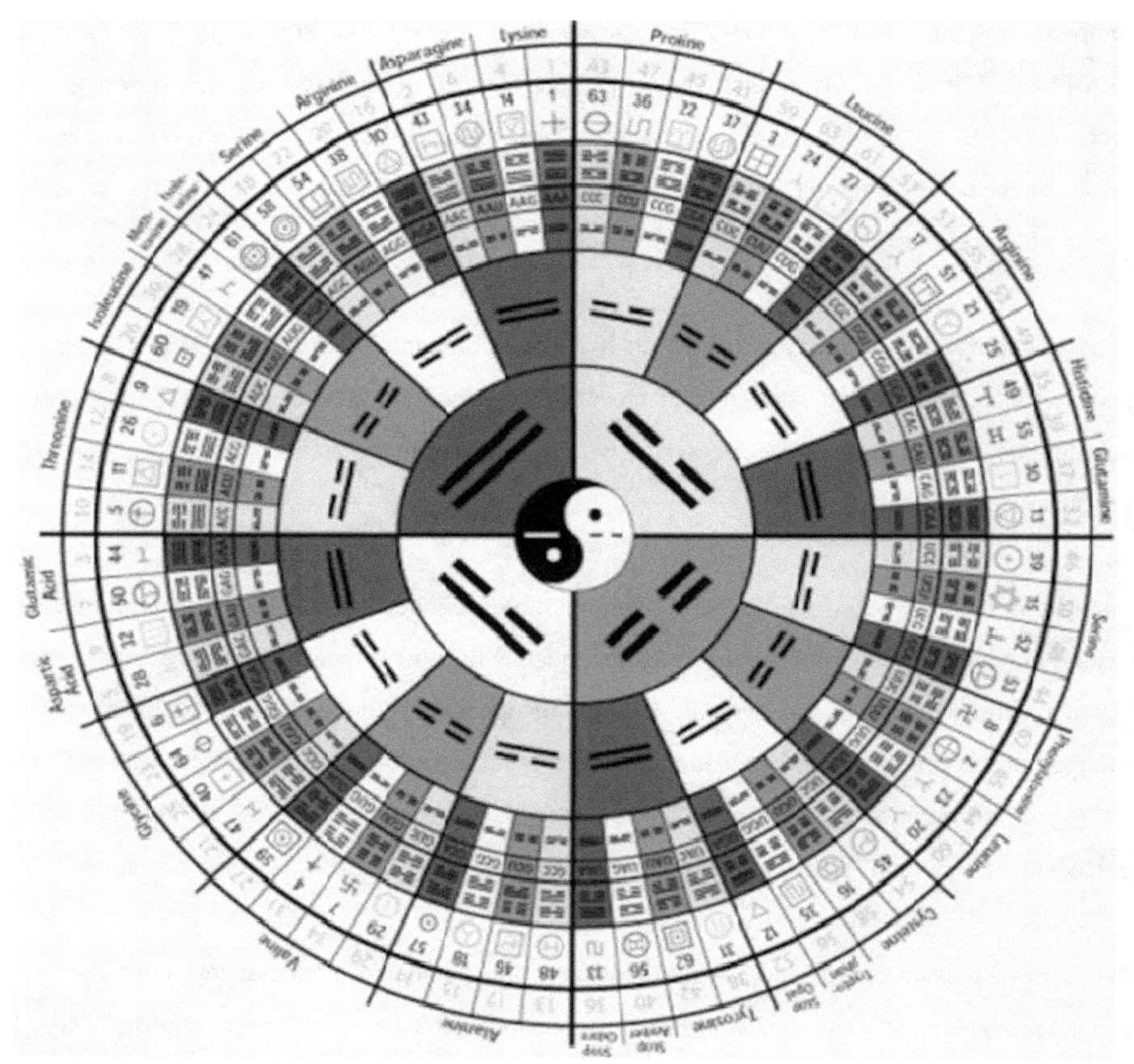

Figure 1-2: The 64 hexagrams in pictorial form

Each of the 64 possible hexagrams has its own associated meaning. These are expounded as descriptions in what is known as the Commentaries. During the time of the spring and autumn dynasty, Kǒng Zǐ; 孔子 Confucius[33] is said to have added a set of the 十翼 Shí Yì or Ten Wings, or what was often called 易传 Yì Zhùan or the Commentary on the Yì jīng. This, together with the 易经 Yì Jīng, comprised what is known as the 周易 Zhōu Yì, or the Changes of Zhou. All later texts about the 周易 Zhōu yì were explanations only.

By the time of 汉武帝 Hàn Wǔ Dì of the Western Han Dynasty, around 200 BCE, the 十翼 Shi Yi was often called 易传 yì Zhùan, or Commentary on the 易经 Yì jīng. In combination this became known as the 周易 zhōu yì, Changes of Zhou. Later versions of the 周易 zhōu yì were ongoing explanations of the 易经 Yì jīng. Hence, the 易经 Yì Jīng was added to and expanded from its original version.

Using the 易经 Yì Jīng for prediction

The 易经 Yì Jīng is a much used tool for prediction. This use is based on the application of the higher mind and connection to the Dao. This method bypasses the intellect and thereby shows what the energies are and how they will manifest themselves.

This is a coherence of what is often seen as a dichotomy – that of free will and fate. They are intertwined, as the amount of fate suffered varies proportionately to the more robotic a person is, captured by the wanderings of the monkey mind. Hence it can be said that thoughts create actions, and actions cause behaviour. Behaviour leads to traits. Traits form the personality. Personality leads to karma and karma leads to fate.

In order to assess a situation, it is necessary to form hexagrams. These hexagrams will depict the situation you wish to explore. This will ensure that the commentaries that accompany the hexagram can be read and that the question posed can be answered.

The two traditional methods of constructing a hexagram are the Yarrow Stalk Method and the Coin Method. The Yarrow Stalk Method consists of using 50 stalks and selecting individual ones from

---

[33] Confucius, who lived between 551–479 BCE, is very famous in China. As philosopher, he had as great an influence on Chinese thought as Socrates had in the West.

stacks laid out. The Coin Method consists of using three traditional Chinese coins that are metallic but have a square hole in their middle. This method involves tossing or spinning the coins to produce the hexagram results.

Two hexagrams are formed, a primary and a secondary one. A secondary hexagram is derived from the primary hexagram in certain situations where there are specific combinations of 阴 yīn and yáng. These are known as 'moving lines'. Each moving line in the primary hexagram has a special comment written by the 周公 Zhōu Gōng, the Duke of Zhou.

The secondary hexagram is formed as the moving lines change from 阳 yáng into 阴 yīn. The moving 阴 yīn lines transform into 阳 yáng lines. The secondary hexagram has its own value and meaning that must be considered within the context of the primary hexagram and its moving lines.

Much more could be written about the various methods used to construct the hexagrams, but this would lead us into the subject area of divination. This is a fascinating topic to explore but would lead us away from our subject area of 太极拳 tàijíquán and its roots in Daoism. We recommend, however, that you immerse yourself in this fathomless ocean of sublime metaphysics as an adjunct to the physical practices.

Analysis of the 易经 Yì Jīng

As we have explained, the 八卦 bā gùa eight symbols are used in Daoist cosmology to represent the fundamental reality of the Universe, based on the energy forces that are both Unmanifest and manifest. There are, therefore, two sequences of the 易经 Yì Jīng arrangement of the hexagrams. There is the Unmanifest or the primordial 先天八卦 Xiān Tiān Bā Guà, which is also known as Earlier Heaven and ascribed to 伏羲 Fú Xī. This is thus known as the Fú Xī bā gùa. The manifested or post-heaven arrangement 后天八卦, Hòu Tiān Bā Guà, is ascribed to 周文王 Zhōu Wén Wáng.

The 易经 Yì Jīng states the following principles in three laws.[34]

The following is taken from a wonderful and very scholarly piece of work called An Anthology of the I Ching by W.A.Sherrill and W.K.Chu.

---

[34] *An Anthology of the I Ching* by W.A. Sherrill and W.K. Chu

The Law of Evolution

The Dao is posited in the formation and development of the Yijing.

The Dao is nameless yet exists everywhere. (This was later expressed in the 道德经 Dào dé jīng, and will be covered later in this chapter).

The Dao possess both Thought and Force.

The Dao possesses Matter, differentiated, and undifferentiated. As differentiated, it is known as consciousness. As undifferentiated, there are atoms, possessing neither positive nor negative consciousness.

Atoms move from a quiescent state to many forms and states of development, and attract similar atoms.

Atoms must be given consciousness and experience. They start at the lowest level and work ever higher. For atoms to have consciousness, there must be activity. This activity is the interaction of yin and yang.

The Dao is nameless because it cannot be named. There are no words that can be used to describe that which is beyond description. As soon as a word is used, the confines of that meaning cannot be sufficient to explain something that cannot be grasped by the intellect. The intellect deals with concepts and the abstract. Words are simply labels that cannot be absorbed by anything other than thoughts.

The Dao can only be experienced and is all encompassing as nothing fits outside of it. Everything seen and felt or heard is merely an expression of it. Matter is a form of energy and within that, there is a consciousness that rises from the lowest levels to the highest and where the vibration increases accordingly. Hence, the quest of mankind to cultivate and purify this energy so as to resonate at the highest levels with the Dao.

The Law of Change

Everything is progressing or retrogressing. Standstill is but an appearance. Thought, Force and Matter formulate the Universe. Force is vibratory.

The expression of the Dao is that of constant motion at various levels of vibration. Nothing is static even where it appears to be so. Your own body is constantly changing, even as you sit there reading this book. Your cells are constantly being replenished. This is not noticeable in the present. But glance at some of your old photos, and you will grimace at the changes that have taken place – that is, if you were not already grimacing when the photograph was taken.

Law of Enantiodromia

This is also known as the Law of Reversal in Extremis, deriving from the Greek word Enantiodromia that is composed of two words: enantios, which means opposite, and dromos, which means running course.

This principle was made famous by the psychiatrist, Carl Jung,[35] in the last century. He stated that the extreme abundance of any force would inevitably produce its opposite. This is equivalent to the principle of equilibrium in the natural world, in that any extreme is opposed by the system in order to restore balance. The more extreme the force is, the greater is the reaction. Hence, when you meditate and your body is in a still situation then, internally, energy will move.

---

[35] Carl Jung (1875-1961) a famous western psychiatrist, who was interested in Eastern thought and psychology.

This can be visualised by the movement of a pendulum, where the range of swing is determined by the amount of stretch and eventual equilibrium. As human beings, we can help ourselves by not exaggerating these swings. These cycles of good and bad fortune can be exaggerated and, hence, unnaturally extended by either increasing the energy and emotions applied and accentuating the effects. With a calm mind, we can surf the waves of good and bad fortune like Californians surfers and still remain afloat. (no disrespect to other surfers who are no doubt equally skilful).

Between 1972 and 1974, some archaeologists in China excavated a site in 马王堆, Mǎwángduī, the name of which is translated literally as Horse King Mound, at a location near Changsha, the capital of 湖南 Húnán province. There they found intact some Han dynasty-era tombs. The tombs belonged to the first Marquis of Dai, his wife, and a male who is believed to have been their son.

One of the tombs contained the 马王堆Mǎwángduī Silk Texts, a second century BCE new text version of the 易经 Yì jīng, the 道德经 Dào dé jīng, and some pictographs of 導引dào yīn exercises and other works. 導引 Dào yīn consists of movement and breathing exercises co-ordinated together for health preservation. The version of the 易经 Yì Jīng discovered in textual form suggests that it was prepared from an old text version for the use of a contemporary Han patron.

When the world began, there was Heaven and Earth
Heaven mated with the Earth and gave birth to everything in the world
Heaven is qìan-gua, and the Earth is Kun-gua
The remaining six gua are their sons and daughters

周文王 Zhōu Wén Wáng

Pre and Post Heaven Arrangements

As we mentioned earlier, the 易经 Yì Jīng represents both the Manifest and the Unmanifest. These two states are depicted as the Pre-Heaven (unmanifest) or Post-Heaven (manifest) arrangements. In actual fact, both states exist simultaneously. It is just by means of perception that there appears to be two states. Nevertheless, the ancient Daoists were looking to explain phenomena and associated characteristics through a paradigm of energetic movements, and through the prism of what could now be described as a mathematical model. Please always remember that ultimately *all* is *one*, and that these methods of attempting to describe what can only really be experienced are the best that we can do. At least if the intellect can grasp these concepts, it becomes easier to experience and transcend to higher levels of consciousness.

Hence, when you see an arrangement of the 易经 Yì Jīng displayed, it will either be the Pre-Heaven (Unmanifest) or the Post-Heaven (Manifest) configurations. This concept of a Manifest and an Unmanifest is very profound and is used by other spiritual traditions. It is also used as a model in meditation in order to attain the supreme state and to differentiate between that which is transitory and that which is permanent. That is to discover, the real 'you', which is the Unmanifest, and that which is impermanent, which is the Manifest.

Figure 1-3: 先天八卦 Xiān Tiān Bā gua or Earlier Heaven Arrangement

In the Daoist description of man's relationship to his environment - wedged between heaven and earth, the pre-heaven or Unmanifest arrangement represents heavenly influences, and the post-heaven or Manifest arrangement represents the earthly influences. The trigrams that are used to represent both are the same, but their arrangements or relative positions to each other differ. The full list of trigrams used for both arrangements are shown in Tables 1-3 and 1-4 below.

According to the precepts of Daoism, our aim as human beings is to align ourselves with the principles revealed in the 易经 Yì Jīng and, by the adoption of practices such as 风水 Fēng Shuǐ, 气功 qìgōng and 太极拳 tàijíquán, to derive the greatest benefit from heavenly and earthly influences. This is a cultivation undertaken in order to merge with those principles that lie behind the expressions of the Dao and, in our example, through our practice of 太极拳 tàijíquán, to harmonise with the energy movements manifested.

In the 先天八卦 Xiān Tiān Bā gua, or pre-heaven arrangement, the trigram 乾 qián, heaven, is in the top part and represents south in direction. This is the opposite direction to that used in the West, and this is due to the fact that this is how it was displayed in Chinese maps in those early days.

坤 Kūn, the earth, is in the bottom part, representing the north. 離 Lí, fire, and 坎 Kǎn, water, are on the left and on the right-hand side, forming a matching pair. 震 Zhèn, thunder, and 巽 Xùn, wind, form another pair, one facing opposite the other. 艮 Gèn, mountain, and 兑 Duì, lake, form another pair, one facing opposite the other. These opposites are in balance and harmony. The adjustment of the trigrams is symmetrical by forming exact contrary pairs. They symbolise the opposite forces of 阴 yīn and 阳 yáng and represent an ideal state when everything is in balance.

To re-cap, these eight forces as represented by trigrams are not intended to be tangible, physical entities, but are merely decryptions to explain the relative energetic transformations and manifestations. They were terms that had a relative meaning to a society that was wedded to the natural environment. They are meant to be used as a metaphor and to be understood as types of energy, each relative to the other, to describe changes in their arrangement over time.

Figures 1-3 and 1-4 show the attributes ascribed to the trigrams, which can be placed within the following categories:

- An Image in Nature
- The Cardinal Direction
- The Family Relationship
- The Body Part
- The Stage/State
- The Animal

There are other attributes ascribed to the trigrams based on the development of 风水 Fēng Shuǐ. Each is meant to describe a characteristic relative to each of the others and within the context of the whole.

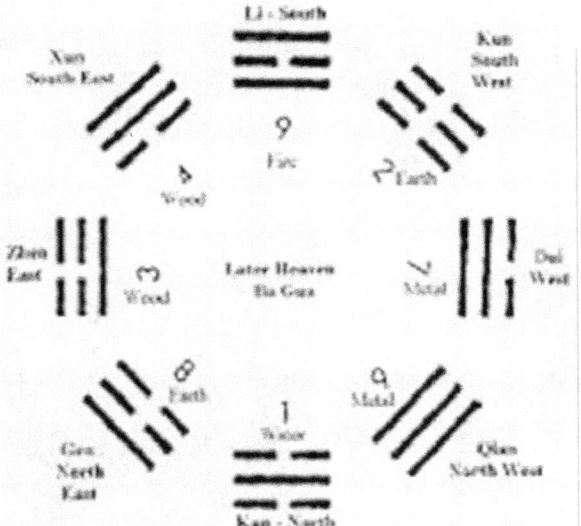

Figure 1-4: 后 天八卦 Hòu Tiān Bā Guà or Later Heaven Bagua

The 易传 Yì zhuān, also called the 'Great Commentary' and known in full as the 易大传 yì dà zhuān, was produced in order to distinguish it from those subsequent commentaries by later students of the

周易 Zhōu yì. It consists of seven parts. These are divided into two sections, the first, of three, and the second, of seven, so that they are called the 'ten wings' 十翼 shí yì. These seven parts are explained below:

彖传 Tuàn zhuàn. This means Structure. The commentary is written as an explanation of each hexagram and its lines. It applies to the first and second of the ten 十翼 shí yì Commentaries.

象传 Xiàng zhuàn. This is Appearance. It is divided into the lesser 小象 Xiǎo Xiàng and the greater 大象 Dà Xiàng. It is written as an explanation of the wordings of the classic. It applies to the third and fourth of the ten 十翼 Shi Yì Commentaries.

繫辞传 Xī cí zhuàn. This concerns the relationship of the hexagrams. The Xī cí commentary provides an overview of both the position and the meaning of the 易经 Yì Jīng in terms of the order of the world and the place of human beings. Due to the nature of its overall meaning, it is also known as dà zhuàn 大传, the Great Commentary. It applies to the fifth and sixth of the 10 Shi YI Commentaries.

文言传 Wén yán zhuàn. This is concerned with the characters. This commentary provides an explanation of the overall meaning of the first two hexagrams, 乾 qìan and 坤 kun, that represent heaven and earth.

说卦传 Shuō guà zhuàn. This provides a more granular explanation of each of the hexagrams. It explains how each hexagram may change into another. Furthermore, it explains how this is related to the realms of heaven, earth and man. It also provides an explanation of the relationship of the hexagrams with other objects.

序卦传 Xù guà zhuàn. This is the order of the hexagrams. It is an aide memoire for the sequence of the hexagrams. The last Commentary identifies similar or opposite hexagrams and highlights their relationship.

杂卦 **Zá guà zhuàn miscellaneous hexagrams**. This identifies similar or opposite hexagrams and explains their relationship.

Table 1-1: 八卦 Bā Guà – the eight trigrams

No.	Trigram Figure	Name	Translation:	Image in Nature	Direction	Family Relationship	Body Part	Stage/ State	Animal
1	☰	乾 qián	Creative, force	Heaven, sky 天	Northwest	Father	Head	Creative	Dragon
2	☱	兌 duì	Joyous, open	Swamp, marsh 泽	West	Third daughter	Mouth	Tranquil (complete devotion)	Sheep
3	☲	離 lí	Clinging, radiance	Fire 火	South	Second daughter	Eye	Clinging, clarity, adaptable	Pheasant
4	☳	震 zhèn	Arousing, Shake	Thunder 雷	East	Eldest son	Foot	Initiative	Horse
5	☴	巽 xùn	Gentle, Ground	Wind 风	Southeast	Youngest daughter	Thigh	Gentle entrance	Fowl
6	☵	坎 kǎn	Abysmal, Gorge	Water 水	North	Second son	Ear	In motion	Pig
7	☶	艮 gèn	Keeping Still, Bound	Mountain 山	Northeast	Third son	Hand	Completion	Wolf, Dog
8	☷	坤 kūn	Receptive, Field	Earth 地	Southwest	Mother	Belly	Receptive	Cow

后天八卦, Hòu Tiān Bā Guà or Later Heaven Bagua

The sequence of the trigrams in 后天八卦, Hòu Tiān Bā Guà, also known as the Bagua of 周文王 Zhōu Wén Wáng, or the Post-Heaven Arrangement, describes the patterns of the energy changes as relating to the manifestation on the physical level, the so-called 10,000 things that came from one, two and three, as expounded by 老子 Lǎozǐ, 坎Kǎn is placed downwards and 離Lí at the top, 震Zhèn in the East and 兑 Duì in the West. This will be covered in more detail in the chapter.

Figure 1-5 – UnManifest Pre-Heaven Bagua with numbers

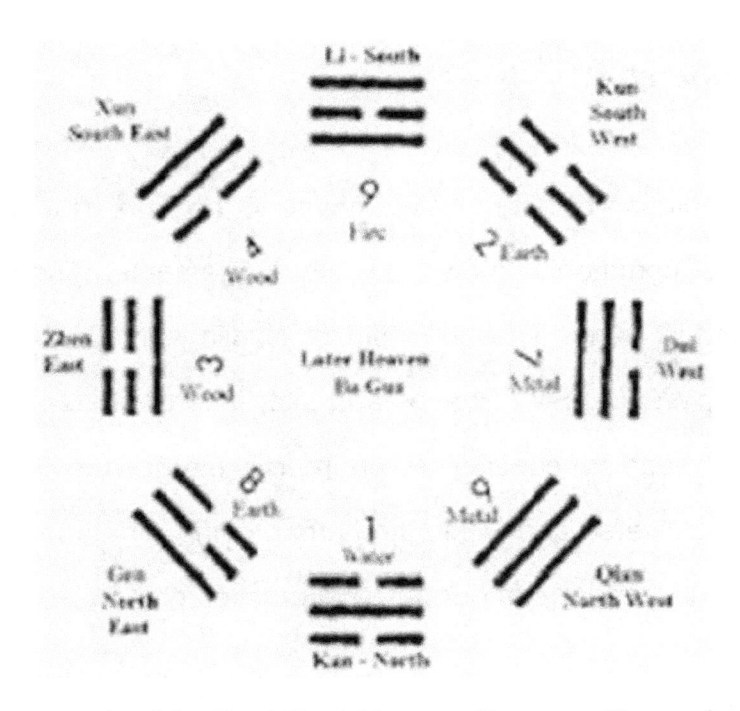

Figure 1-6 – Manifest Post-Heaven Bagua with numbers

Contrary to the Unmanifest state, that of the quiescent Earlier Heaven Bagua, this one is a dynamic Bagua, where energies and the aspects of each trigram flow towards the following one. It is the sequence used by the 罗盘 luó pán compass that is used in 风水 Fēng Shuǐ to analyse the movement of the qì that affects us in our existence and our inter-relationship with external and internal energy manifestations

The 后天八卦, Hòu Tiān Bā Guà thus takes into account the movement of the opposing energies of 阴 yīn and 阳 yáng, the movement of 气 qì and the time of the day and of the relevant seasons. All of the elements and energy movements are in a constant state of flux.

The important point to emphasise is that the Manifest and the Unmanifest exist simultaneously, and only a change in consciousness allows us to recognise there is a whole, indivisible inter-connectedness between all things. This is accepted by many esoteric and spiritual traditions, leading some to take an extreme view that all is unreal and the individual is an illusion, that the body does not matter. The Chinese have a practical philosophy and take the ordinary as extraordinary.

The Five Phases – 五行 Wǔxíng

Connecting to the representation of energy as it moves in the 易经 Yì Jīng is that of the system of Five Phases or, as it is commonly known, the Five Elements. These Five Phases are the interactions of 阴 yīn and 阳 yáng. The concepts of 阴 yīn and 阳 yáng and the trigrams were developed before the 五行 Wǔxíng Five Phases and later incorporated into them. To state it again, do not get too hung up on these concepts. They are merely methods to illustrate the expression and unfolding of energy in the Universe and within the human being. The analysis is useful, however, as you will see the patterns and language of our physical practice predicated on these concepts.

The latter term, 'elements' implies something fixed and tangible or solid, but this is due to the translation from the Chinese into English of the term 五行 Wǔxíng as Five Elements. The initial translation was made as Five Elements, and this term has now become the de facto standard translation. This misleads the reader into thinking that these are fixed and real concrete phenomena.

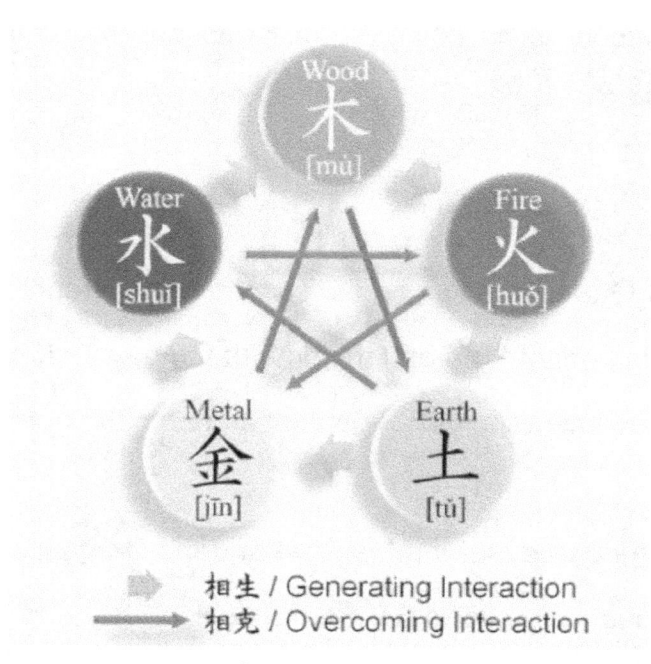

Figure 1.7. –五行 Wǔxíng The Five Phases

But this is not the case: the Five Elements show the unfolding of energy movements through cycles. Remember this – the photographs in the book are static snapshots of something that is actually dynamic and moving. Hence, the external problem arises when trying to describe movement with nouns that imply a fixed reality. In fact, nothing is static and that which appears to be solid is arrangements of moving molecules.

The 五行 Wǔxíng Five Elements can be described as follows: the four directions with the centre as shown below.

- Wood – Spring
- Fire – Summer
- Earth – Change of seasons (Every third month)
- Metal – Autumn
- Water – Winter

The five elements with the Chinese characters and Pinyin are:

- Fire: 火, huǒ
- Earth: 土, tǔ
- Metal: 金, jīn
- Water 水, shuǐ
- Wood 木, mù

There are two cycles through which these five phases of energy move. These are known as the Generating Cycle and the Destroying Cycle.

The Generating Cycle or 生, the shēng cycle, moves as follows:

- Wood feeds Fire;
- Fire creates Earth;
- Earth bears Metal;
- Metal carries Water;
- Water nourishes Wood.

Hence, there is a mutual dependency between each of the elements as the one morphs into the other. Cause and effect are mutual correspondences that turn into an endless dance of 阴 yīn and 阳 yang

The Destroying 克, kè Cycle moves in the opposite direction as follows:

- Wood displaces Earth;
- Earth absorbs Water;
- Water quenches Fire;
- Fire melts Metal;
- Metal chops Wood.

The relationship of the 八卦 bā gua to the eight trigrams is shown in Table 1-5 below:

Table 1-2 – the 五行 Wǔxíng Five Phases and the 八卦 bā gua

5 Phases:	Metal	Earth	Wood	Wood	Water	Fire	Earth	Metal
易经 Yì jīng	Heaven	Earth	Thunder	Wind	Water	Fire	Mountain	Lake
Trigrams	☰ 乾 qián	☷ 坤 kūn	☳ 震 zhèn	☴ 巽 xùn	☵ 坎 kǎn	☲ 離 lí	☶ 艮 Gèn	☱ 兌 duì

Table 1-3 – the 五行 Wǔxíng Five Phases and the 天干 tiāngān Heavenly Stems

5 Phases	Wood	Fire	Earth	Metal	Water
Heavenly Stem	jiǎ 甲 yǐ 乙	bǐng 丙 dīng 丁	wù 戊 jǐ 己	gēng 庚 xīn 辛	rén 壬 guǐ 癸
Year ends with	4, 5	6, 7	8, 9	0, 1	2, 3

A further analysis of the 五行 Wǔxíng Five Phases is show in Table 1-4 below.

The 五行 Wǔxíng Five Phases and Chinese Medicine

Early Chinese doctors developed the inter-relationship between the 五行 wǔxíng Five Phases and the body's organs. Please note that these organs that are known as 脏腑 zàng fǔ are energy characteristics and not solid physical organs as described in Western medicine, although they are linked energetically.

The relationship of the organs to one another and their mutual interactions are described by the 五行 wǔxíng Five Phases theory. Each organ is associated with one of the five phases. When the person is healthy, the organ acts in harmony with respect to that element. Hence, the pattern is that of the Generating Cycle mentioned previously.

The 阴 yīn and 阳 yáng organs relate to each other primarily in terms of resonating energies. For example, the kidneys and the bladder are the 阴 yīn and 阳 yáng water organs respectively, and resonate with each other. One does not really regulate the other; they work together.

Table 1-4 – Attributes of the 五行 Wǔxíng Five Phases

Element Analysis	Wood	Fire	Earth	Metal	Water
Season	Spring	Summer	Change of seasons (Every third month)	Autumn	Winter
Direction	East	South	Centre	West	North
Colour	Green	Red	Yellow	Black	Blue
Phase	New Yang	Full Yang	Yin/Yang balance	New Yin	Full Yin
Climate	Windy	Hot	Damp	Dry	Cold
Planet	Jupiter	Mars	Saturn	Venus	Mercury
Heavenly creature	Azure Dragon 青龙	Vermilion Bird 朱雀	Yellow Dragon 黄龙	White Tiger 白虎	Black Tortoise 玄武
Heavenly Stems	甲, 乙	丙, 丁	戊, 己	庚, 辛	壬, 癸

If energy flows through the human body, as it should, that is, without any blockage, then there will be harmony and corresponding good health for the individual. Hence, ensuring that the interchange of 阴 yīn and 阳 yáng energies occurs unimpeded is the recipe for health and longevity.

The five elements are associated energetically with the following 脏腑 Zàng-Fu organs. 木 mù, Wood, and 肝 gān, liver, is associated with the 魂 hún, the cloud-soul, and 魂 pò, the Ethereal or Corporeal Soul, and is paired with the 胆 dǎn gall bladder. The 魂 pò relates to the anima and the 魂 hún, to the animus. During one's life it is essential to harmonise these two, otherwise there is a conflict, with each striving for mastery. At death, they separate and the 魂 pò descends to earth. The 魂 hún ascends and becomes a 神 Shén, a spirit or god, to ultimately return to the Dao.

火 huǒ Fire Heart 心 xīn is associated with the 神 Shén, or Aggregate Soul paired with the small intestine 小肠 xiǎo cháng and, secondarily, the 三焦 sān jiāo or triple burner and 心包 xīn bāo, the pericardium.

土 tǔ Earth and Spleen 脾, pí, are associated with the 意 yì, Intellect, paired with the stomach 胃 wèi.

金 jīn Metal, Lung 肺 is the home of the 魄 pò, Corporeal Soul, and is paired with the 大肠 dà cháng, large intestine.

水 shuǐ Water and Kidney 肾 shèn is the home of the 志 zhì, Will, and is paired with the 膀胱 páng guāng, bladder.

脏 Zàng is the term that is used to denote the six 阴 yīn organs. These are the heart, liver, spleen, lungs, kidneys and pericardium.

腑 Fǔ is the term that is used to denote the six 阳 yáng organs. These are the small intestine, large intestine, gall bladder, urinary bladder, stomach and the San Jiao

This model takes into account the generating cycle that we depicted previously. For example, the liver is the Wood Phase, and is regarded as the generator or controller of the heart that is the Fire Phase. The kidneys are the Water Phase and the generator or controller of the liver.

Traditional Chinese medicine and its model of the body and its functions did not rely totally on the purely physical analysis that were arrived at by conducting anatomical studies, as was the case with Western medicine. On the contrary, emphasis was placed on the function of 气 qì flow through the body and the networks through which it was carried, known as the meridians. This analysis was underpinned by the model of the energetic transformations of the Dao as it expresses itself.

The functions of the organs are also connected to Shén, which can be translated as mind, attention, focus or spirit, Jing essence, 血 xuě, blood. This theme will be developed further with regard to the practise of 太极拳 tàijíquán.

An understanding of these concepts of the organs within a Western context can cause confusion where a literal translation is attempted. Two of the organs in the Chinese model that do not have a specific location in the body are the pericardium, and the Triple Burner. In western medicine, the pericardium is a literal sac of membrane that encases the heart. In the Zàng Fu system, the purpose of the pericardium is to protect the heart. So, translators assumed Chinese physicians were talking about the sac when in fact they were not. The triple burner has no correlation in Western medicine and is often described as having a global immune function.

Table 1-5 Expansion of attributes of the Five Phases

Movement	Wood	Fire	Earth	Metal	Water
Planet	Jupiter	Mars	Saturn	Venus	Mercury
Zàng - 阴 yīn organs)	Liver	Heart/pericardium	Spleen/pancreas	Lung	Kidney
Fu - 阳 yáng organs)	Gall bladder	Small intestine/San Jiao	Stomach	Large intestine	Urinary bladder
Emotion	Anger	Happiness	Love	Grief, sadness	Fear, scare
Sensory organ	Eye	Tongue	Mouth	Nose	Ears
Mental Quality	Sensitivity	Creativity	Clarity	Intuition	Spontaneity
Body Part	Tendons	Pulse	Muscle	Skin	Bones
Colour	Green	Red	Yellow	White	Black(Blue)
Body Fluid	Tears	Sweat	Saliva	Mucus	Urine
Finger	Index finger	Middle finger	Thumb	Ring finger	Little finger
Sense	Sight	Speech	Taste	Smell	Hearing
Taste	Sour	Bitter	Sweet	Pungent	Salty
Smell	Rancid	Scorched	Fragrant	Rotten	Putrid
Life	Birth	Youth	Adulthood	Old age	Death

As we have previously stated, Daoist teachings are underpinned by the belief that Man is a microcosm of the Universe, which is the macrocosm. Hence, the flow of energy in humans follows the same patterns as that of the Universe.

This is how we arrived at the transition from the 易经 Yì Jīng to the 五行 Wǔ Xíng Five Elements and through to the body energies/organs, as described above. This theme will be developed further when we explain the inner aspects of 太极拳 tàijíquán.

道德经 Dào dé jīng and other great Daoist works

Our analysis of the 易经 Yì jīng, 阴 yīn and 阳 yáng energetic transformations and associated paradigms and models reflect the earliest treatises on Daoism. We will now concentrate on a later, more philosophical method of description and discuss the works of 老子 Lǎozǐ. 老子 Lǎozǐ wrote one of the most famous texts of Daoism, the 道德经 Dào dé jīng. This was written around 2,500 years ago. 老子 Lǎozǐ is known as one of the main influences on both philosophical and religious Daoism. Daoism at the time of 老子 Lǎozǐ was not known by that contemporaneous or indeed by any other name. Long before this time, the teachings were, however, already long established as a metaphysical set of teachings.

老子 Lǎozǐ provided a philosophical rendition to the traditional teachings. What is often not appreciated, however, is that the teachings that may appear on the surface as a discourse on how to live a harmonious life are also intended to reach deeper levels than is first gleaned from just a superficial reading of the words. On one level, a commentary is provided on the role of society and the behaviour of its citizens. At a deeper level, however, there is also an energetic and alchemical aspect that is expounded.

Hence, this great treatise, the 道德经 Dào dé jīng, that may be translated as *The Classic of the Dao and the Virtue,* is the most famous of his attributed works. The 道德经 Dào dé jīng is a compilation of mysticism about the Dao, and its power, 德 dé. 经 Jīng is the Chinese word for "classic". It is not tangential or separate from or to other Daoist works, but is a continuum and an expression from the same rich soil. The fact is that it is merely couched in a different, more philosophical way in describing the Dao from both that which preceded it, the 易经 Yì Jīng, and from those works that followed. These latter works also use what to us may seem strange ways of describing Daoist teachings. We will discuss this in later chapters when we talk about Inner Alchemy.

The historical perspective of 老子 Lǎozǐ is that he grew up in the state of Ch'u and went by the name of Li Erh Ch'u. It is important to point out that all great metaphysicians (in fact all famous writers) speak within the context and background of the times that they live in. Hence, during the turbulent times that existed for 老子 Lǎozǐ, there was an emphasis on a return to natural values, such as the understanding of, and harmonising with, nature.

One of the earliest recorded references about 老子 Lǎozǐ was that made by the historian 司马迁 Sīmǎ Qiān[36] who lived from around the time of 145–86 BCE. He mentions 老子 Lǎozǐ in his book, known as The Records of the Grand Historian, Shiji. This work combines three stories. In the first, 老子, Lǎozǐ is recorded as being a contemporary of 孔夫子 Kǒng fū zǐ (Confucius), who as we have mentioned, lived from 551 to 479 BCE. His surname was Li 李 Plum, and his personal name was 耳 ěr, ear or 聃 dān long ear. Sima identifies 老子 Lǎozǐ as an archivist named Li Er or Li Dan at the Zhou dynasty court (1046–221 BCE).

Chinese history abounds with legends regarding 老子 Lǎozǐ. It is written in folklore that he was immaculately conceived by a shooting star and carried in his mother's womb for 82 years to be born as a full grown wise old man. 老子 Lǎozǐ is said to have been the keeper of the royal archives but that, tiring of his court life, he decided to retire. He journeyed west into the mountains and sought to leave the country at the Hankao Pass. It is assumed that he was traveling to Tibet. There he was

---

[36] Sīmǎ Qiān lived during the Han Dynasty. He is a very famous figure due to his work, Records of the Grand Historian, 史记), that covers more than 2,000 years, ranging from the era of the Yellow Emperor to that of Emperor 汉武帝 Han Wudi.

recognised as a sage by the guard at the gate, who persuaded him not to leave until he had committed to writing down his wisdom and insights. For three days, he worked on the 81 chapters of his 道德经 Dào dé jīng, and, upon their completion, handed them to the guard and was never seen again.

The 道德经 Dào dé Jing represents the first philosophical work of Daoism and is nowadays one of the most translated books in the world. It is also one of the most influential books in modern Chinese literature. It has been the subject of more than a thousand commentaries. It has also has been translated into English countless numbers of times.

So let us explore some of the words of the 道德经 Dào dé jīng.

The Dao is undifferentiated and formless
Unified, empty, and silent.
It precedes the existence of Heaven and Earth.
It is inexhaustible; and moves ceaselessly.

There was something undifferentiated and all-embracing which existed before.
It is so ancient, and so fundamentally different from all other beings that it does not have a name and cannot be described in ordinary language. However, in order to describe it, we cannot but give it a name, so it is called 'Dao' (道 Dao) or 'Great' (大 Da).

This is expounded in Chapter Twenty-Five.[37] It is a clear definition of the Unmanifest state expressing itself, the interplay of the nuomenom and the phenomenon. It also records the impossibility for people to be able to describe something that is fathomless.

---

[37] Wang Keping: The Classic of the Dao: A New Investigation, Beijing: Foreign Language Press, 1998, p. 230.

Figure 1-7: Lǎozǐ

The 道德经 Dào dé jīng teaches that one should endeavour not to strive to do anything in order to accomplish something. This is written in Chinese as 无为 wú wéi, which we encountered earlier. This does not mean to literally do nothing, like would be the case of a person in a coma who would be rendered incapable of acting. What it really means is that one should follow the natural forces of which one is actually a part of. That is to be assimilated with and, at the same time, is a force of nature. This is to be both the influencer and in turn, influenced by nature, rather than acting against it. This adoption of 无为 wú wéi is fundamental to the principles and practice of 太极拳 tàijíquán.

The Dao is described as the mystical source of all existence. It cannot be seen, but is the root of all things. Human beings have free will and the ability to cultivate themselves and change their own nature. Acting naturally, or in harmony with the Dao is the aim of cultivation in order to achieve a return to the natural state. Hence, this position squares the circle of the often-debated question, 'Is there Free Will or is Everything Fated?' The answer is that Man influences and is influenced by the Dao such that both parts of the question are correct (and false!).

The opening lines of the 道德经 Dào dé jīng have a resonance with the modern world. A common interpretation is similar to Korzybski's observation that 'the map is not the territory'. This has latterly been adopted by the creators of Neuro Linguistic Programming (NLP).

The opening lines of the 道德经 Dào dé jīng are:

道可道，非常道
名可名，非常名

The Way that can be described is not the true Way.
The Name that can be named is not the constant Name.

The standard edition is divided into 81 chapters and is comprised of two parts. Part one consists of chapters one through to thirty-seven. This is known as the 道经 Dào jīng or Scripture of Dao. Part Two consists of chapters thirty-eight through to eighty-one, and is known as the 德经 Dé jīng or Scripture of Virtue. In its entirety, it contains some 5,000 Chinese characters. The discoveries at 马王堆 Mǎwángduī, mentioned earlier, have unearthed early versions on silk strips.

The 道经 Dào jìng was not originally divided into chapters, but a scholar named Hu San Gong later divided the book into 81 chapters, and into two parts. The Book of Dao is represented by the first 37 chapters and the Book of De by the remaining 44.

The Dao is described as the formless and ineffable way. It forms the source of all creative power in a Universe that is in a state of constant transformation. The Dao is the mother of Heaven and Earth and is both a spontaneous and self-generating creator of the Universe. All things in the Universe have their own virtue or power, 德 dé that if permitted to be in harmony with the Dao, will bring an equilibrium and natural order to the Universe.

Human beings, however, have the capacity, known as free will in western thinking, to act in discordance with the Dao. The natural order is upset by deviance from the harmonious relationship between themselves and the Universe. Feeling that they are individuals with a power over this natural state, they impose their will upon the world, not realising where the source of real power is derived from. The ego separates a person from not just the rest of humanity but to nature as well. This is down to the lack of realisation that the body is created from the genes of parents, that there is a dependency on food for survival and that, in turn, there is also an equal dependency on the rest of the ecosystem. That breathing depends on oxygen is also a part of this whole unified Universe.
And so, it is only when we acknowledge and experience our inter-dependence as part of the whole and act as the wave in the ocean, seemingly separate but adding to and a part of the ocean, may we truly be in harmony with our existence.

The 道德经 Dào dé jīng expounds on how when people abandon the natural way, they develop hierarchical and elitist societies governed by draconian rulers at war with each other.

The enlightened ruler or sage can, however, establish order and harmony within a small community by cultivating spontaneity within himself and by following a path of 无为 wú wéi, or non-aggressive action. By virtue of the ruler's own charismatic power, order will naturally arise within human societies, because in the feudal hierarchy of ancient China, he was seen as the foundation and embodiment of the whole people. This is also a metaphor for the individual and how the harmony of Mind, Body and Spirit can be adversely affected by imbalances arising from inner conflict and how cultivation can achieve the natural balance.

The book has several levels of meaning, and although it appears and is resonant with the way of rulers and society, it also applies to the human being as a mirror of the Universe and being subject to the same forces.

Figure 1-8: Lǎozǐ

Figure 1-9: 老子 Lǎozǐ with Yin Xi, the guardian of the Gate of Tibet

To illustrate how the words of the 道德经 Dào dé jīng relate to the human body and energies and hence are a signpost for the practice of internal alchemy, let us ponder Chapter Ten[38], which has a very alchemical perspective, as follows:

Can you unite Hun and Po[39]
And embrace them as one
Rather than allowing them
To divide your life
In the pursuit of extremes?

Can you wholeheartedly soften yourself
And achieve the innocence of a baby?
Can you cleanse your mind
Of all prejudice and flaws?
Can you love your life and govern your bodily kingdom
And play no games or fantasies?
Can you inhale and exhale natural energy
As delicately as a gentle maiden
Without roughly breathing?
Can you achieve clarity of mind
And not engage the intellect?

From these questions, one should examine
You have received the gift of birth
But you do not own your life
You are raised by the Universe

---

[38] Esoteric Tao The Ching – Master Hua Ching Ni
[39] Two types of spiritual essence

But do not dominate the Universe

You grow together with all people

But you are not a lord over them

Therefore, in giving birth to other lives

One does not own them

In raising all lives

One is not lord over them

This is the spiritual virtue of the Universe

Other famous and pertinent Daoist works exist, and these are also relevant to our understanding of the foundations of 太极拳 tàijíquán. Some of these are included below. Many, many others could also have been included. Please note that the works mentioned in this book are chosen as good examples. There is not enough space in this volume for us to cover more examples. We do hope, however, that we have stimulated your appetite for further research.

淮南子 Huáinánzǐ – the Masters of Huainan

The 淮南子 Huáinánzǐ is another famous Daoist classic that is worthy of investigation. It comprises an eclectic collection of treatises, including one of the earliest mentions in a philosophical context of the principles of 阴 yīn and 阳 yáng and the 五行 wǔ xíng Five Phases theories.

The 淮南子 Huáinánzǐ describes three essential essences of man. These are, firstly, an ethereal component that is known as 精 jīng. Secondly, there is a condensed, coarser substance that gives instinct to Man and is known as 气 qì and, thirdly, there is that which is known as mind, spirit or conscience, 神 shén. These three essences are directed by a human's free will to choose their own path in life. Harmonising these three essences should be undertaken to best harmonise with the 道 dào. We will cover these themes in more detail later. These are also known as the Three Treasures.

The Universe also has three most important components, those of Heaven, Earth and Man that were covered earlier. This is known in Chinese as 天地人 tiān dì rén. 天 Tiān 地 dì and 人 rén emanate from the three essences and can only be sustained by a person through the process of identification and harmony with the Dao. It states that the human body itself is a microcosm. That is, it is a miniature of the Universe. It guides the way for a person to remain the same through the management of one's own mind. This harmonisation and self-cultivation is the most important task for a person.

If you maintain harmony with the Dao, you can become a 真人 zhēn rén True Man or a 至人 zhì rén Perfect Man. This state of being the Perfect Man and meant generically to mean person – and so includes women before anyone shouts! – is also mentioned by 庄子 Zhuāngzǐ, whose works we will cover later in the chapter and who is an abiding image in Chinese metaphysics. The True Man behaves as though others do not exist for his or her own sake. This gives him a freedom in life but does not mean that forgetting others is shorthand for neglecting them or not caring for others. If you are in harmony with the nature of the Dao, then you will obtain purity and tranquillity, and come to a state of not acting, 无 wú wéi. What this really means is that one should not view oneself as the subject and others as individual objects, referencing them within the prism of one's own outlook and prejudices. Rather, examine how the connection and mutual relationship affects one as the other.

Huai Nan Zi [40] [41]

When the spirit controls the body, the body obeys,
When the body overrules the spirit, the spirit is exhausted.
Although intelligence is useful, it needs to be returned to the spirit
This is called great harmony

---

40 Thomas Cleary, The Taoist Classics
41 Thomas Cleary, The Taoist Classics

The mind is the ruler of the body while the spirit is the treasure of the mind.

When the body is worked without rest it collapses

When the spirit is used without cease, it becomes exhausted.

Sages value and respect them, they do not dare to be excessive

呂洞賓 Lü Dongbin

Ancestor Lu, who is also known as 呂洞宾 Lü Dongbin, is an individual from the collective of what is known as the 八仙 Bāxiān or Eight Immortals of legend. By this stage, the reader will perhaps not be surprised to learn that each Immortal also represents one of the trigrams of the 易经 Yì jīng.

吴元泰 Wú Yuán Taì of the 明朝 Míng cháo Ming dynasty wrote *The Emergence of the Eight Immortals and their Travels to the East.* From that time on, their relationship with the Chinese people became more firmly established. An unknown author wrote *Eight Immortals Cross the Sea* 八仙過海 bā xiān guò hǎi, which tells the story of the Eight Immortals attending the 'Conference of the Magical Peach' 蟠桃會 pán taó huì. On this journey, they encounter an ocean, whereupon their leader, Lü Dongbin, decided that they should each should use their magical powers to cross over. Hence the Chinese proverb *The Eight Immortals cross the Sea, each reveals its divine power.* This means that each individual should apply their own powers to a common goal.

Most of the said eight are said to have been born in the era of the Tang or Song Dynasties.

The Eight Immortals are:

- 何仙姑 Hé xiān gū – Immortal Woman He
- 曹国舅 Cáo Guó jiù – Royal Uncle Cao
- 李铁拐 Lǐ Tiě guǎ – Iron Crutch Li

- 藍采和 Lán Cǎihé – Blue Eccentric Wanderer
- 呂洞賓 Lü Dongbin – Leader
- 韩湘子 Hán Xiāng zǐ – Philosopher
- 张果 Lǎo Zhāng guǒ lǎo – Elder Zhang Guo
- 钟离权 Zhōnglí Quán - True-Yang Ancestor-Master

Figure 1-10: The Eight Immortals as they cross the sea

Portrayed in a clockwise direction in the boat from the stern onwards are 何仙姑 Hé xiān gū, 韩湘子 Hán Xiāng zǐ, 藍采和 Lán Cǎihé, 李铁拐 Lǐ Tiě guǎ, 呂洞賓 Lü Dongbin, 钟离权 Zhōnglí Quán, and 曹国舅 Cáo Guó jiù, and outside the boat is 张果 Lǎo Zhāng guǒ lǎo.

呂洞賓 Lü Dongbin is considered the founder of the lóngménpài 龙门派 Dragon Sect of 全真派 quán zhēn pài Complete Reality Daoism. This school remains in existence today and has a history of practising internal alchemy and has fused Confucian and Buddhist beliefs if not their practices.

An example of this aspect of looking to harmonise with the Dao through the practising of internal alchemy, that is, through the attainment of Dao working through the body, the microcosm to reach the macrocosm, is provided below:

The human body is only vitality, energy and spirit. Vitality, energy and spirit are called the three treasures. Ultimate sagaciousness and non-contrivance are both attained from these. Few people know these three treasures, even by way of their temporal manifestations. What is inconceivable is their primordial state – is it not lost? If you lose these three treasures, you are incapable of non-contrivance, and so are unaware of the primordial.

Ancestor Lu – The Three Treasures[42]

庄子 Zhuāngzǐ

Another famous Daoist was 庄子 Zhuāngzǐ, a contemporary of 老子 Lǎozǐ. The word 子 zǐ means Master, hence the commonality between the two names. His work, the *Book of 庄子 Zhuāngzǐ*, provides us with yet more deep levels of philosophical Daoist wisdom. This classic work, which is named after him, is also known as the *Zhuangzi* 庄子. His philosophy is similar to that of 老子 Lǎozǐ but has a somewhat different emphasis.

The *Book of 庄子 Zhuāngzǐ* describes a sage who rejects the world of politics and the state and refuses all entreaties to join the government.

For 庄子 Zhuāngzǐ, who was writing during very turbulent times in China, the preservation of spiritual integrity was the most important aim in life. To achieve this spiritual integrity requires the sage or the true person to cultivate and retreat from the world of corruption and greed that are the hallmarks of an unnatural way of living.

---

[42] Thomas Cleary, *The Taoist Classics*

庄子 Zhuāngzǐ was born during the period of the Warring States 403–221 BCE dynasty. It is thought that he lived from around 370 to 301 BCE. He came from the town now known as 商丘, Shāng qiū, in the same province as 陈家沟 Chénjiāgōu, 河南 Hénán province.

During this time, the 周朝: Zhōu Cháo, Zhou dynasty had lost its power. It was a time of chaos as a result of a political power vacuum. This produced an increase in violence and ensuing warfare as the competing factions vied for control. It was within this turbulent period that the 百家 bǎi jiā, or Hundred Schools of Daoism, emerged. These schools advocated and explained the measures by which one could return to a state of harmony in such arduous times.

庄子 Zhuāngzǐ advocated the adoption of a flexible and non-dogmatic approach to life, eschewing convention and society's rules and dogma in order to live a simple and natural life. The key was spontaneity 自然; zìrán, self so. This means to act spontaneously, naturally, and in accordance with the situation. It is to be in harmony with the Universe.

This book, 庄子 *Zhuāngzǐ*, which is named after the author, is a witty, deep and philosophical text, exploring the attributes that are the constituent characteristics of the 真人 zhēn rén or true person. For the 真人 zhēn rén true person, much emphasis is placed upon the act of spontaneity and adherence to the natural order of life. What brings us to discuss the 庄子 *Zhuāngzǐ* with regard to our main topic is the reference that is made to meditation and 气功 qì gōng, for example, the inclusion of the phrases 'sitting in oblivion', 'fasting the heart-mind' and 'breathing through the heels'. The latter phrase touches on the significance of the most advanced breathing practices, which is the most natural way of all and consists of the absorption of the 阴 yīn energy from the earth. This is breathing with the whole body and without effort.

The current extant text of 庄子 Zhuāngzǐ is the result of the editing and arrangement of the 晋朝 Jìn Cháo Jin dynasty thinker and commentator, 郭象 Guō Xiàng, who lived around the 3rd century. He reduced what was then a work that had consisted of 52 chapters down to the current edition, which comprises just 33 chapters. He also included a philosophical commentary. This updated version is the work that is the most famous one today. His commentary on the text provides an interpretation that became highly influential over the subsequent centuries.

The book is divided into three sections: The Inner Chapters 内篇 nèi piān are numbered from one to seven, and the Outer Chapters, known as 外篇 wài piān, are numbered from eight to twenty-two. The Miscellaneous and remaining 杂篇 zá piān Chapters are numbered from twenty-three to thirty-three.

Chapter 2 齐物论 qí wù lùn of the book, translated below by Burton Watson – 'Discussion on Making All Things Equal' – gives a good illustration of its philosophy.

Men claim that Mao Qiang and Lady Li were beautiful, but if fish saw them they would dive to the bottom of the stream; if birds saw them they would fly away, and if deer saw them they would break into a run. Of these four, who knows how to fix the standard of beauty in the world? (2, tr. Watson 1968, p. 46)

This is an excellent description of the relativity of all things and on the perspectives, that are applied with regard to objective reality by both individuals and society in general. It also highlights how people see the world through their own prism of subjectivity.

Another famous section is that of the Butterfly Dream, which posits a question as to the state of reality and of the mind's perception. It is found at the end of Chapter Two and is one of the most famous stanzas in Daoism.

Once Chuang Chou[43] dreamt he was a butterfly, a butterfly flitting and fluttering around, happy with himself and doing as he pleased. He didn't know he was Chuang Chou. Suddenly he woke up and

---

[43] The Wade-Giles transliteration

there he was, solid and unmistakable Chuang Chou. But he didn't know if he was Chuang Chou who had dreamt he was a butterfly, or a butterfly dreaming he was Chuang Chou. Between Chuang Chou and a butterfly there must be some distinction! This is called the Transformation of Things. (Basic Writings, p. 45)

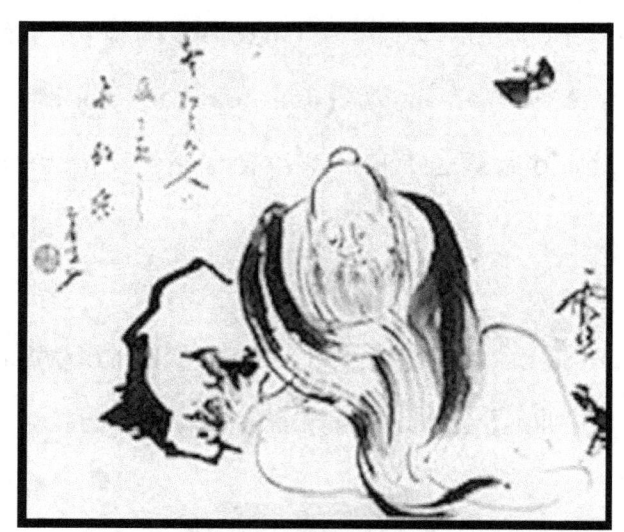

Figure 1-10 庄子 *Zhuāngzǐ*

The *Zhuangzi* 庄子, introduces the concept of 太极 tàijí in one of the Inner Chapters of the book.

The Way has attributes and evidence, but it has no action and no form. It may be transmitted but cannot be received. It may be apprehended but cannot be seen. From the root, from the stock, before there was heaven or earth, for all eternity truly has it existed. It inspirits demons and gods, gives birth to heaven and earth. It lies above the zenith but is not high; it lies beneath the nadir but is not deep. It is prior to heaven and earth, but is not ancient; it is senior to high antiquity, but it is not old[44]

In the chapter 'The Great Patriarch', it states that

Dao is a reality that can be trusted even though it has neither behaviour nor form; it can be transmitted from heart to heart, but not by words; it can be obtained but not seen. It is its own root, and existed prior to Heaven and Earth. It created the Spirits and Divinities, and gave birth to Heaven and Earth. It is higher than the Supreme Ultimate yet is not high; it is lower than the Six Directions

---
[44] Translated by Mair, 1994, p. 55

六极 Liuji *yet is not deep; it precedes all creation yet is not old; it is farther than the remotest antiquity yet is not distant.*

These profound words should both guide and caution us. We can understand that the Dao is difficult to describe. Were it not so then it would not be anything as sublime as it is. Just as the Dao needs to be experienced so must the practice of 太极拳 tàijíquán. The words in this book are, in the famous Zen saying, merely fingers pointing at the moon.

老子 Lǎozǐ and 庄子 Zhuāngzǐ are not part of any organised school of religion, even of the later developed Daoist ones, but are part of what is often referred to as the 黄老 Huáng Lǎo traditions or those of philosophical Daoism. They are both from a tradition of the outsider, the rebel and the poet mystic, and both advocate looking to nature as the best teacher.

It is of great advantage for the 太极拳 tàijíquán student to become familiar with these great works. It will enable you to use 太极拳 tàijíquán principles all of the time in every aspect of your life, regardless of whether you are actually practising the 太极拳 tàijíquán form.

The Jade Emperor's Mind Seal Classic

葛洪 Gě Hóng was one of the more famous Daoist practitioners, who lived during the fourth century CE. His most famous work was the 高上玉皇心印经. Gāo Shàng Yù Huáng Xīn Yìn Jīng translated it as *The Jade Emperor's Mind Seal Classic*, or *The Mind-Seal Scripture of the Exalted Jade Sovereign*. The Mind-Seal scripture is considered one of five most important scriptures and, accordingly, is recited daily by the Dragon Gate sect of the Complete Reality School of Daoism, founded by Master 吕洞宾 Lü Dongbin, as introduced earlier. This classic is recognised by both Confucians and Buddhists.

Daoist principles state that each human being inherently possesses the medicine necessary to achieve optimum health, to live a long life and, ultimately, to achieve immortality. Illness and death are viewed as arising from the dissipation of the Three Treasures. These are the three that were mentioned earlier and are a constant theme in both Daoist scripture and in health and martial arts applications.

To refresh our memories these three treasures are 精 jìng, including sexual and physical energy - the most elementary constituents that make up the human body; or the subatomic particles, or the substance. The 气 qì - breath and vital energy that maintains the body and keeps the cells healthy and dividing, and finally the 神 Shén, which translates as spirit, mind, consciousness and mental health.

The book states that there is a practice available to restore, accumulate and transform these three treasures and thus to create an internal elixir of immortality that will reverse the aging process. This method of working with 精 jìng, 气 qì and 神Shén will result in a spiritual awakening.

The Jade Emperor's Mind Seal Classic provides us with a straightforward and precise summary of the Daoist attitude to spiritually. As the 易经 Yì Jīng states, 'All things on Earth are first archetypes in Heaven'.

An example from this short classic as detailed below provide us with a good insight into the principles that we will need to embrace in our practice of 太极拳 tàijíquán.

1. The Supreme Medicine has three distinctions: Ching[45] (essence), qi (vitality) and Shén (spirit), which are elusive and obscure.

2. Keep to nonbeing, yet hold on to being, and perfection is yours in an instant.

3. When distant winds blend together,

in one hundred days of spiritual work

---

[45] Wade-Giles method; *pinyin* is *jing*

And morning recitation to the Shang Ti,

Then in one year you will soar as an immortal.

4. The sages awaken through self-cultivation;

Deep, profound, their practices require great effort.

* Transliterated by Stuart Olson

抱朴子 Bàopǔzǐ

Another important discourse on our subject is the 抱朴子 *Bàopǔzǐ, The Book of the Master Who Embraces Simplicity*, written by 葛洪 Gě Hóng. It consists of two parts. There is the Inner Book 內篇 nèi piān and the Outer Book 外篇 wài piān

The Inner Book is a Daoist treatise of 20 volumes and covers the topics of immortality and the medicine for immortality 方藥 fāng yào, the transformations of demons and ghosts, nourishing and prolonging life, exorcism and the avoidance of disasters.

The Outer Book is a treatise of 50 volumes and covers the successes and failures in the human world and the good and evil of affairs of human life. This is considered a Confucian work and expounds on the theories of the Immortals, such as 神仙家 Shén Xiān Jiā, since the time following the Warring States period. It establishes the Daoist theoretical system of Immortality from that point onwards.

It takes on 魏伯阳 Wei Boyang's theories on Refining Elixir 炼丹 liàn dān and is the culmination of Daoist alchemy from the 魏朝 Wèi cháo and 晋朝 Jìn Cháo Jin dynasties. It is a very good source for our analysis of Daoism and is included in the Supreme Clarity Section 太清部 Tài Qīng Bù of the Daoist Canon of the Zhengtong Era 正統道藏脏 Zhèng Dào Zàng Zàng.

Figure 1-11: 葛洪 Gě Hóng 陶弘景 Táo Hóng Jǐng

陶弘景 Táo Hóng Jǐng of the Liang dynasty compiled the *Commentary on the Book of the Master Who Embraces Simplicity* 抱朴子注 *Bàopǔzǐ Zhu.* This comprises some 20 volumes, but, alas, has been lost to this generation of scholars. His other work, *The Collation and Interpretation of the Inner Book of the Master Who Embraces Simplicity* 抱朴子內篇校释 *Bàopǔzǐ Neipian Jiaoshi*[46] is fortunately available to us.

Figure 1-12 陶弘景 Táo Hóng Jǐng Journey to the West 西游记 Xī Yóu Jì

---

[46] China Press, 1985, written by Wang Ming

The famous novel, *Journey to the West* 西游记 Xī Yóu Jì. This is a very famous Chinese classic that was written in the 16th century.[47] It provides us with a concise instruction from an enlightened Daoist patriarch who instructs the central character Sun Wukong, who is known as the Monkey with a poem, that begins:

Know well this secret formula wondrous and true: Spare and nurse the vital forces, this and nothing else. All power resides in the semen [jìng], the breath [qì], and the spirit [Shén]; Guard these with care, securely, lest there be a leak. Lest there be a leak! Keep within the body![48]

The term natural breathing is much stated in Daoist teachings. In its original sense, it is not a method but a return to the simplicity of breathing like a baby. 老子 Lǎozǐ's question 'Can you attain the pliability of a child?' is a reference to this subject. A baby breathes naturally from the abdomen, and in an unaffected way. There is no concept of fixing the breath in the abdomen; the breath is there naturally. The entire stomach expands during breathing and then contracts slightly so that it operates like a bellows.

The naturalness of the respiration of a baby is down to the fact that the breath is natural and concentrated in the abdomen. This causes the cheeks to be red, the bones to be soft and the joints to remain flexible. As we age, the opposite occurs, including the negative effect of the breath becoming concentrated in the chest and leading to shallow breathing. The Daoists seek to restore us to a more natural state of health and vitality and reverse the trend to a breathing pattern that constantly rises upward, away from the lower abdomen, as it was during childhood.

老子 Lǎozǐ also said: 'The whole of cultivation is in subtraction, not addition.' Daoism thus aims to a process of reversal, restoration and rejuvenation to go back to that state when we were children, to youthfulness. This is developed in 太极拳 tàijíquán and is, hence, a major reason why it benefits a

---

[47] It is one of the four great classical novels of Chinese literature and was published anonymously in the 1590s during the rule of the Ming Dynasty.
[48] tr. Yu, 1977, p. 88

person's health and wellbeing. The work involved is to undo bad habits rather than add extra layers to unnatural habits and mental stasis.

Thus, if you focus the mind on the 丹田 dān tián, to the abdomen, the breath will follow the mind. The mind does not follow the breath. Breathing naturally allows the breath to become deep, slow and harmonious. When you stand in 站桩 zhàn zhuāng or practise 缠丝劲 chán sī jìn you will notice that this occurs naturally. This is something that cannot be forced. You may think that you are learning a technique in 太极拳 tàijíquán, but in fact you are unlearning the unnatural ways of your posture, breathing and movement. When the breath is left to sink into the 下丹田 xià dān tián the 气 qì will move through the upper parts of the body and will circulate freely throughout the limbs.

As you may know there are many Daoist techniques and 气功 qìgōng exercises that use particular breathing exercises. Some of the most popular of these are embryonic breathing, abdominal breathing and reverse breathing. All of these exercises are valuable and have a specific aim. Natural breathing is, however, the basic foundation of all these techniques.

Natural breathing as the term implies will occur naturally. The Daoist way is not that of force or affectation. The more you pay attention to the lower abdomen, the deeper the breath becomes. Adopt the practice of 无为 wú wéi.

So what have we learned from all of the above topics covered in this chapter? The theme that runs through these few chosen examples from the Daoist works is that of seeking to return to a state that is exemplified by the Dao. These are naturalness, simplicity and cultivation through the refinement of our energies in order to dissolve into the source from which all energy derives. This is described in different ways and levels as we have shown. In the next chapter, we shall move onto the subject of internal alchemy, and by so doing, we can form the bridge between the theory and practice of 太极拳 tàijíquán.

The practice of 太极拳 tàijíquán is that of naturalness and harmony with nature, as has been described in this chapter. Although great force and power may be achieved through this practice, it is through a natural and soft approach that does not harm the body. The practice is an expression of the body, in the same way that the Dao manifests itself.

Whilst the instructions appear to be, and indeed are simple, it is easy to misunderstand. Paradoxically the simpler the message, the more difficult it is for us to understand. We are wedged into the intellect and are constantly seeking something more difficult. It is like looking everywhere for your spectacles when they are already on the bridge of your nose. In this case the message is that the breath is the mind and the mind is the breath.

Chapter 2 內丹, Nèi Dān, Internal Alchemy

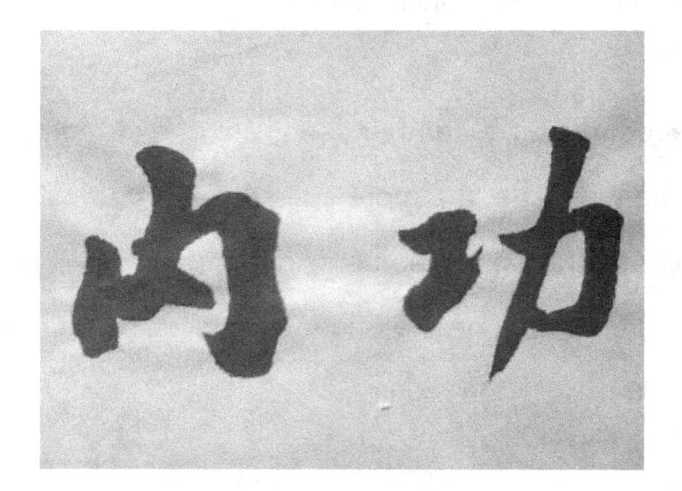

內功

巽 xùn

All of the universal things have a common origin.

We regard this as the mother of all things.

Recognising the mother, we can know its children,

Knowing how the children have come into being,

We may return to hold on to the mother.

Thus ensuring no danger to life.

Closing the eyes and shutting up the mouth,

One can keep from sickness forever.

Opening the eyes to chase after desires and passions,

One can be helpless forever.

Watching less and less we can be enlightened.

Holding on to the weak we can be strong.

Using the external seeing,

To turn to illuminating the internal,

Thus, there can be no hazards for the body.

This is called practising the enduring Dao.

Dao de Jing – Chapter 52

Background

This chapter outlines Taoist internal alchemy practices merely to provide the reader with an academic oversight of one of the important influences of Taijiquan. If the reader wishes to go straight to the actual practice and theory of Taijiquan then please continue to chapter 4.

We have introduced you to some of the fundamental Daoist principles and described the concept of the relationship that lies between the inter-connection of Man as the microcosm and the Universe as the macrocosm. As well as showing you how this connection is represented in both a philosophical way by the works of 老子 Lǎozǐ and 庄子 Zhuāngzǐ and by the more scientific form depicted in diagrammatical forms in the 易经 Yìjīng let us now examine how this directly links to our subject matter, 太极拳 tàijíquán.

This relationship between concept and form is provided through the examination and explanation of the theory and practice of 內丹 nèi dān or, as it is translated in English, Internal Alchemy. In this chapter, we therefore seek to explain in some detail this profound and sublime topic. In so doing, we will perform the service of highlighting the depth of our art.

內丹 nèi dān, Internal Alchemy is a subject we could write dozens of books about (if we had the time!) in order to be fully able to provide you with a comprehensive coverage of the subject matter. As well as the limitations of time, there is also perhaps the lack of propensity in you the reader to consume all of this in one go. Hence, we are inhibited from following this route. What we shall seek to provide will hopefully be a sufficient explanation to enable you to connect the theory to the practice, and thereby to experience a synthesis from both an intellectual and a physical perspective.

內丹 nèi dān, Internal Alchemy was developed by what is known as the 內丹术 nèidān shù schools of Daoism. These schools taught methods to practise and harmonise the mind-body-spirit aspects of the human being in exercises that collectively comprise what is known as cultivation, and which is based, of course, on Daoist metaphysics. This cultivation is a path that ultimately leads to the realisation of the Dao.

A word of caution is required. There are many Daoist schools. None of these schools abjure the body in the way some spiritual traditions do. But there are both different degrees of emphasis and levels of practice as the body and mind are used to refine energy from the coarser to the more subtle levels.

太极拳 Tàijíquán is both a physical exercise and a process of internal energy refinement, and, eventually, practised at the highest level, it is meditation in motion. Whilst some traditions purely emphasise the stationary or seated meditation practices, 太极拳 tàijíquán includes meditation in movement. If you are able to synthesise your breath with your mind and use your intent to guide the movement, then you have reached a very high level of mental and physical awareness.

As you can probably deduce, the language that we encounter is difficult to understand without referring to less esoterically worded explanations. An acceptance of the tendencies of these ancient authors to use archaic language and imagery underlies the intention they had to place a veil over that which was intended to be kept secret Nevertheless, the underlying objective is to refine energy and awareness – to achieve union with the Dao It is the process of transcending the self and ascending to a state of experiencing and ultimately operating from the Self by harmonising the mental and physical energies.

Unlike some methods adopted by some schools of spiritual cultivation, the Daoists did not abjure or exclude the use of the body. Whilst ultimately the body, like everything else in the Manifest world, is ephemeral and bound to die, nevertheless, it can be used a vehicle to return to the Dao. The body is both a hindrance and a vehicle for cultivation to realisation, depending on how you handle it.

Like many of the Chinese words that we have encountered thus far, the word(s) nèi dān is a composite, consisting of two parts – 內 Nèi meaning inner and dān 丹, translated as alchemy. The 內丹 Nèidān schools developed their practices from the Daoist understanding of the human body and its energy pathways. This knowledge is also deeply applied in traditional Chinese medicine theory, emanating as it does from the same Daoist classics.

As we have previously mentioned, the 易经 Yìjīng describes the Manifest state and its relationship to the Unmanifest. 內丹 nèi dān, Internal Alchemy seeks to achieve the state of journeying to and arriving back to the Unmanifest or Dao through the cultivation of the human body.

In actual fact, whilst it may be described as a journey, it is a shift in consciousness and realisation of the True Self, the Real Person, known as 真人 zhēnrén. It is a process of using the human body as a starting point to refine the energies and hence the consciousness so as to arrive back at the pristine state and merge with the Dao. As with most of the higher spiritual traditions, Daoism posits the belief that the human being is not separate from the Universe. The person is connected, not as separate parts of a whole that it just the sum of those parts. It is best described in metaphor. He or she is like a wave in the ocean, only seemingly separate and with each rise and fall of that wave is both simultaneously affected by and also affecting the whole expanse of water. It is mere consciousness at play, seemingly subject and object it is only the Dao expressing itself.

Being an eminently pragmatic people, the Chinese realised that even if you were to accrue just the physical benefits of the practices this in itself would be of enormous benefit. Hence, the cultivation was of a practical as well as spiritual benefit. Central to the method of the cultivation of 內丹, nèi dān, Internal Alchemy is the importance of 气 qì on the function of the mind, body and spirit. Hence, the logic is simple. If the human body mirrors that of the Universe and the energies that move each are one and the same, it should be possible to cultivate these energies in order to achieve that supreme state of being.

Centuries of practical experience, experimentation and cultivation form the basis for the development and refinement of the practices. The ones that were developed in order to achieve the purification of the mind, body and spirit are not dissimilar to those found in other parallel metaphysical developments within other cultures, for example, the Indian yoga and the Tibetan Esoteric Buddhist schools. Indeed, there is evidence of some cross fertilisation of ideas between these different sources.

References to 內丹, nèi dān, Internal Alchemy go back many centuries. There is reference to this in the 道德经 Dào dé jīng by 老子 Lǎozǐ, who we introduced you to in the previous chapter. 老子 Lǎozǐ introduced the concept of the bellows or 橐籥 tuó yuè, using the metaphor for an explanation of the physical practice of external alchemy. This involves the search for the means of making artificial gold and silver, with a bellows being used as the forging tool. This was called a 橐籥 tuó yuè. He describes how Heaven and Earth function like a bellows. Man also uses a bellows to manifest vital energy or 真气 zhēn qì. This bellows operates through breathing. The nostrils are 籥 yuè and the rest of the body is 橐 tuó.

In Chapter 5 of the 道德经 Dào dé jīng, it states in the second stanza:

其猶橐籥乎？虛而不屈，動而愈
出。多言數窮，不如守中

Can not the space between heaven and earth be compared to a bellows? It is empty, yet it does not lose its power;
It moves, and thus more and more comes forth. How a gossip's fulsome talk is soon exhausted! And would we not prefer to walk on the middle path?

The purpose of practising Inner Alchemy is to extract the intangible 真气 zhēn qì or vital energy by the process of breathing. 元气 yuán qì or original or primal energy forms the structure of a person. If the 元气 yuán qì is exhausted, life ends. 內丹, nèi dān, Internal Alchemy is thus a method to preserve life through fully utilising the yang spirit, and ultimately to merge spiritually with the Dao.

Nèidān became a separate school in its own right during the Tang dynasty, and therefore at that time became distinct from the external or 外丹 wài dān schools. The famous classic on 內丹, nèi dān, Internal Alchemy, 参同契 Cāntóng qì or *The Kinship of the Three* by Wei Boyang is discussed later in this chapter. It is the earliest known book on theoretical 內丹, nèi dān, Internal Alchemy ever discovered in China. Hence, it forms a strong link in our journey between the signposts of Daoist philosophy, the Yijing and 內丹, nèi dān, Internal Alchemy, and thereby allows us to absorb this into our 太极拳 tàijíquán theory and principles and ultimately into the practice itself.

Underpinning all of the esoteric language is the premise that, 人 rén Man is centred between Heaven and Earth and has the ability to cultivate the three 丹田 dān tiáns or energy centres in the body in order to return to the pristine Dao. This can be achieved by the cultivation of the yin and yang energies of our body, from the lower energy frequency of the 下丹田 xià dān tián or lower energy centre through to the correspondingly higher levels at the 中丹田 zhōng dān tián middle energy centre and the Upper 上丹田 shàng dān tián – upper energy centre. Hence, the process is one of refinement, that of returning from the Manifest to the Unmanifest through the transformation of the energies at our disposal through the direction of the mind.

內丹 Nèidān uses a variety of techniques. These include meditation, visualisation, breathing, body postures and movement. There are three cultivations used in the practice. These are known as the 三宝 sān bǎo, the three treasures that are central to 內丹, nèi dān, Internal Alchemy. These are 精 Jīng, 气 qì and 神 shén. We can develop these three treasures as the basis for 太极拳 tàijíquán internal practices.

The three treasures of 精 Jīng, 气 qì and 神 shén can be loosely translated into English as Essence, Vitality and Spirit. These treasures are the essential energies that sustain all human life and have the following attributes:

- 精 Jīng – nutritive essence, refined, perfected; extract; spirit, demon; sperm, seed
- 气 qì – vitality, energy, force; air, vapour; breath; spirit, vigour; attitude
- 神 shén – spirit; soul, mind; god, deity; supernatural being

A famous depiction of 內丹 Nèidān, Internal Alchemy was produced in the form of a diagram.

This is known as the 內经图 Nèijīng tú or Internal Map. The one shown below is taken from the White Cloud Daoist monastery in Beijing, which is believed to originate from the late nineteenth century, but is based on one that was created around a thousand years ago. The White Cloud Temple is a centre for the 龙门派 lóngménpài Dragon Gate Sect of Daoism.

The 內丹 Nèidān tradition of internal alchemy is practised by working with the energies that are already present in the body. This is wholly different to that of using natural substances, medicines or elixirs emanating from outside of the body and used by the 外丹 wài dān or external alchemy schools. The 上清 Shàngqīng, Supreme Clarity tradition of Daoism played an important role in the emergence of 內丹 Nèidān alchemy, having used 外丹 wài dān mainly as a meditative practice and, therefore, turning it from an external into an internal art.

The Complete Reality School of Daoism merged Confucian ethics with Buddhist values to create a composition of teachings in a school that survives to this day. Nowadays, it is difficult to identify the individual strands of each religion such is the success of the synthesis. Over the ensuing centuries, many sects emerged and some have disappeared altogether.

The 內经图 Nèijīng tú diagram is based on a meditation process and shows in visual form both the theory and the application of the energy circulating through the Microcosmic Orbit, which is made up of the 任脉 rèn mài and 督脉 dū mài channels. These are two of the most important of the eight extraordinary channels through which 气 qì flows through the body. This energy circulation moves 气 qì up the spine and passes through three points upwards through to the cranial cavity. The three 丹田 dān tiáns are represented in the diagram as the ox, the cowherd and the old man.

The 內经图 Nèi jìng tú is translated into English as the Chart of Inner Luminosity and was carved in stone at the White Cloud Temple in Beijing. It was created by Liu Cheng Yin, a Daoist priest. It is a chart that depicts the human body as consisting of a number of spirits. All of the body is living and energetic and the spirits are akin to members of a kingdom, ruled by a benevolent ruler.

There are nine sacred mountain peaks in the picture, each with a cave to reflect how the heavenly energy can be captured, transformed and stored in the caves. These represent the nine storage points in the human body. These nine points are as follows:

The Immortal Realm

This is located in the head, in a position just in front of the crown and can be used to ingest the universal, heavenly energy through sublime alchemical inhalation.

The Top of the Giant Peak

This is also located in the head, but at the back and can ingest the energies from the North Star and is related to the thymus gland.

The Mud Pill

This is located in the 百会 bǎi huì point that we have encountered earlier in the book. It is connected to the Big Dipper and the hypothalamus gland.

The House of the Rising Sun

This is the third eye, in the centre of the forehead, between the two physical eyes (It is also a song made famous by the Animals). This third eye is known to many esoteric traditions and is known in Hindu yoga as the ajna or brow chakra.

The Nine Peaks Mountain

This is connected to the mid eyebrow and is related to the pituitary gland.

The Obscure Spiral Alter

This is a position located between the Mud Pill and in front of the Giant Peak

The Cave of the Spirit Peak

This is known as The Jade Pillow and is the first cervical vertebra at the base of the skull.

The True Jade Upper Gate

This is located near to the throat and connects to the brain.

The Source of Rising Law

This is located behind the soft palate and is connected to the pituitary gland.

The Nèi jìng Tú representing 內丹, nèi dān, Internal Alchemy and Man as a microcosm of the Universe is shown in Figure 2-1.

At the bottom, there is a boy and girl working a water treadmill, representing the left and right kidneys, the seat of the water element and sexual energy. This sexual energy can be reversed from its normal trajectory outwards and usually lost through sexual excess, and can then be transmuted internally and upwards through and up the spine through the practice of 內丹, nèi dān, Internal Alchemy.

Further up, a man toils with a plough and an ox. This signifies that stamina is required in order to prevail with 內丹, nèi dān, Internal Alchemy. Planting the seeds represents the acquiring of energy in the spleen through a balance of diet and living healthily.

The four 太极 tàijí symbols above the cauldron with a fire underneath represent the balance of kǎn and li, water and fire in the 下丹田 xià dān tián, the lower dantian. The processing of the energies to transmute jìng and 气 qì is central to the practice of 內丹 nèi dān, Internal Alchemy.

The weaving maid is depicted as yin and the cowherd boy as yang and symbolises the balance of these two energies rising up the spine. The boy is yang energy in the middle 丹田 dān tián. The girl and the boy are ex-lovers who can yet be re-joined to achieve energetic balance. This depicts that at the level of experience at the Manifest, the two energies are separate. Uniting these two energies is to return to the Unmanifest Dao.

The stars of the Big Dipper protrude from the crown of the cowherd. Absorption of the energies of the stars seeks to extract powerful energies. As a microcosm of the Universe, a person consists of those latter energies.

The forest represents the wood element that resides in the liver and which controls the flow of 气 qì, and it is central to ensuring that the Five Phases of energy are balanced.

The twelve-tiered pagoda is the throat and the back of the neck, where the energy must flow freely in order not to get stuck and create a blockage, as the energy flows around the small heavenly circle through the 督脈 dū mài and 任脈 rèn channels or meridians.

To the left of the pagoda is a pool of water that represents saliva created when the tongue touches the roof of the mouth, and it acts as a bridge between the 督脈 dū mài and 任脈 rèn mài meridians. The saliva is swallowed down to the 丹田 dān tián.

The two circles above the pagoda are the eyes representing the sun and moon and also depict Bodhidharma and 老子 Lao Zi, highlighting the importance of meditation and turning inwards. It also shows the incorporation of Buddhism within the Dragon Gate school of Daoism that also incorporated Confucian principles.

The series of peaks represent the head and the absorption of the yang heavenly energies into the body. Above the head, it states that there is the realm of enlightenment and longevity.

太乙金华宗旨 Tài Yǐ Jīn Huá Zōng Zhǐ – The Secret of the Golden Flower

The Secret of the Golden Flower is a manual for meditation. It is a treatise that represents how the light of the Mind may be refined and so blossom. This is achieved through the process of refining the spirit and enables you to attain a state that is that of pure awareness. It is from the 全真 Quán zhēn Complete Reality School of Daoism. The school was founded by Wang Chongyang in 1167 CE, which was during the 晉朝 Jìn Cháo, Jin Dynasty, in present-day 寧海縣 Nínghǎi Xiàn, Ninghai County, Shandong Province.

This form of 內丹, nèi dān, Internal Alchemy is based upon attaining the Dao through escaping from the state of being a prisoner of contrived habits, where one is doomed to a self-perpetuating state of mechanical actions, controlled by untrammelled thoughts fuelling the monkey mind. This is alas a habitual condition, and is akin to that of a self-hypnotic state. It is the fate of most human beings, who are unable to attain the state of 无为 wú wéi. They therefore remain imprisoned by repetitive and conditioned actions. All serious spiritual traditions caution against this state. The cure is to practise reflection. 孔子 Kǒng Zi describes how it is the act of perceiving that brings one to the goal. The Buddha describes it as the vision of the heart and 老子 Lǎo Zǐ as the Inner vision.

The absolute unity refers to that which cannot be surpassed. There are very many alchemical teachings, but all of them make temporary use of effort to arrive at effortlessness, they are not teachings of total transcendence and direct penetration. The doctrine I transmit directly brings up working with essence and does not fall into a secondary method. This is the best thing about it.[49]

Hence the Secret of the Golden Flower is a practice that goes to the very direct method, that of working with the source rather than anything superficial.

The Immortal Fetus
From *The Secret of the Golden Flower*
Harvest/HBJ, New York and London, 1962

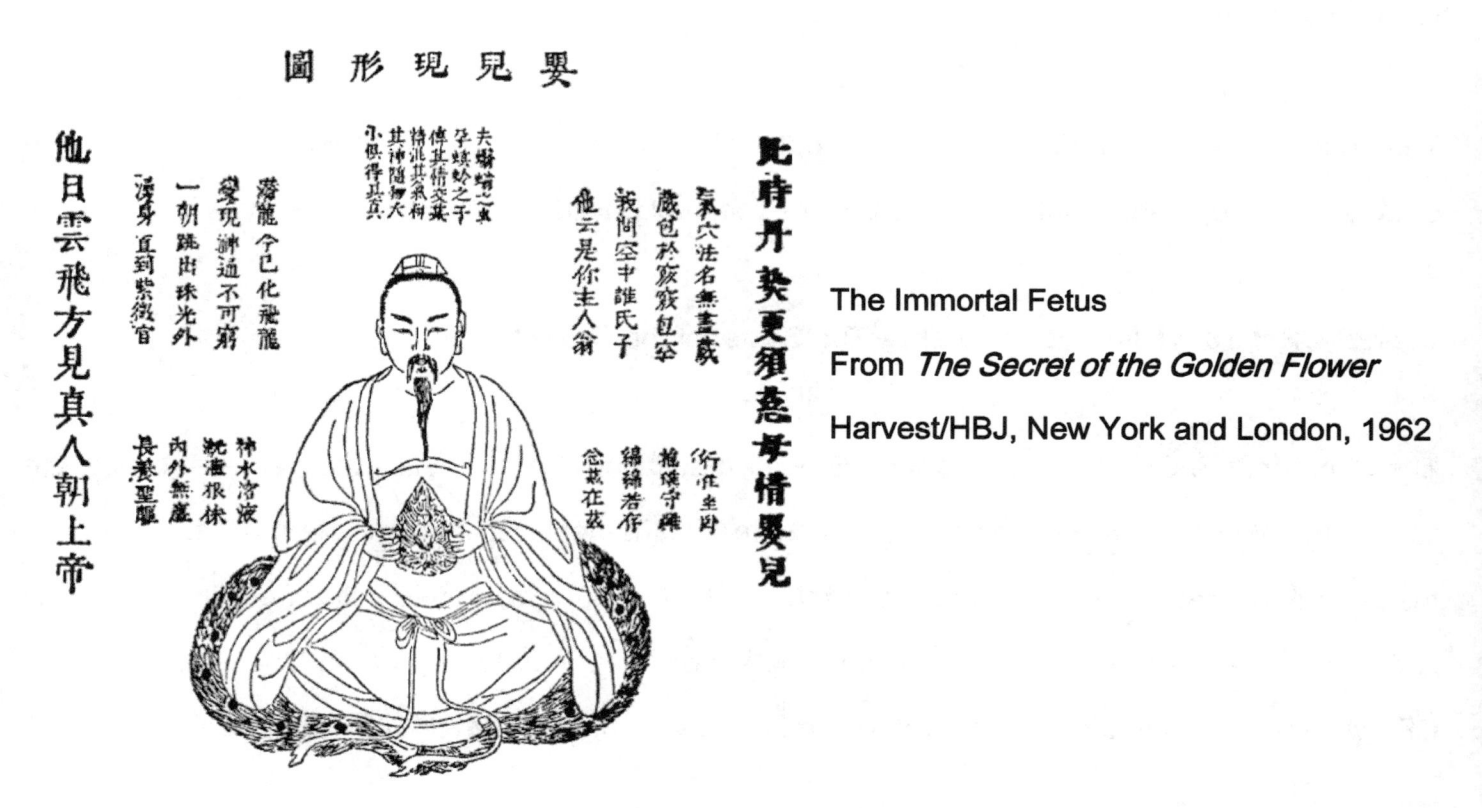

Figure 2-2: Gathering the Light, a Daoist meditation from The Secret of the Golden Flower

The practice requires an inner contemplation so that the circulation of the light enables your

---

[49] The Taoist Classics – The Secret of the Golden Flower

thoughts, the place of heavenly consciousness, and the heavenly heart to be amalgamated as one. The position of the heavenly heart lies between the two eyes at the 上丹田 shàng dān tián Upper Dan Tian. This is the real Emperor, rather than the Commander who often takes charge, and is located within the middle dantian. This differentiation is that of the 'self' and the 'Real Self' and hence reflects the fact that few people are able to govern their lives through the Real Self, which is ignored and by-passed.

The whole process whilst seeming simple in this admittedly abridged explanation does in fact require much effort – an effortless way, by being relaxed and allowing the energy to circulate throughout the body whilst concentrating on the primal spirit located between the eyes at the 上丹田 shàng dān tián, Upper Dan Tian.

As we have explained, according to Daoist doctrine, the Three Treasures can be described as three types of energy available to humans. In the 道德经 Dào dé jīng 老子, Lǎozǐ states in Chapter Forty-Two that:

The 道 Dao gives birth to the One, the One gives birth to the Two (太极 tàijí or Yin and Yang 阴阳) and the Two gives birth to the Three; (which some interpret to mean 精 jīng, 气 qì and Shen 神, or sometimes Heaven: Tian 天, Earth: Di 地 and Man: Ren 人).

The transmutation of the three treasures is through the three phrases:
炼精化气 liàn jīng huà qì, refining essence into breath
炼气化神 liàn qì huà shén, refining breath into spirit
炼神还虚 liàn shén huán xū, refining spirit and reverting to emptiness.

The three treasures can also be described as belonging to 先天, xiān tiān, Pre-Heaven, or the Unmanifest and the 后天 hòu tiān Post-Heaven, terms we explained in the previous chapter. This is taken from 易经 the Yì Jīng and correlates with 元精 Yuán jīng 'Original Essence', 元气 yuán qì, the Original Breath, and 元神 yuán shén, the Original Spirit.

So in explaining the workings of the Universe from the Unmanifest to the Manifest, these principles also relate to Man as the microcosm of the Universe. 內丹, nèi dān, Internal Alchemy is therefore a method that aims to return to the Unmanifest state, the return to the pristine Dao.

Other practices such as Tantric Buddhism and Yoga also aim to unite with God or the Void, or whatever term is used. These are all practices that seek to harmonise the individual with the Universe, to unite all of the constituent elements of which a person is on one level individual and yet equally part of the whole. In actual fact, all are connected in true terms as there is no division or separation, merely the appearance or feeling that there is.

This cultivation, which is that of returning to the interconnectedness of all things, is of course easier said than done. It can only be achieved by starting at the level of the physical, the Manifest, where things most obviously appear disconnected. There is then the process of refining the energies to the state where things are most obviously connected. This is undertaken whilst at the same time understanding that all of the states are paradoxical, simultaneously existing and perceivable and yet transcendable through shifts in consciousness. This notion is at the centre of these practices.

精 Jīng

精 Jīng can be translated as life essence. Each person is born with 精 Jing and is responsible for the growth processes in the body from birth to maturity. It is akin to having a bank account with some money deposited in it when you are born. A person can protect the ongoing loss of their 精 Jing through following an appropriate diet and exercise. This is like investing their initial deposit and accumulating extra funds. Of course, we know that it is entirely feasible that some people will spend more than they save and, hence, deplete these resources.

精 Jing is stored in the kidneys and transformed through the 下丹田 xià dān tián or lower dantian or energy field, which is located roughly three fingers in length just below the navel. The 下丹田 xià dān tián is in fact an energy field that resides within the body, rather than existing on the surface of the skin. There are in fact three 丹田 dantians in the body. In addition to the lower, there is the middle one, situated roughly between the two nipples, and the upper one, which is situated approximately between the two eyes. There are meditation techniques that focus on each of the three 丹田 dantians. We will cover the role of the 下丹田 xià dān tián in the practise of 太极拳 tàijíquán later on.

精 Jīng is the energy that creates both bone marrow and the grey matter of the brain, and it derives from three main external sources. The first is acquired as an inheritance from the parents. The second is the nutrition derived from eating pure and refined food where the nutritional value is not degraded by modern processing methods. The third is from a method used by some schools of Daoism, often referred to as the Left Hand Path that uses sexual techniques, similar to tantric sex, and is also known as dual cultivation.

This aim is to develop and enhance the sexual essence. This is regarded as a major source of cultivation. Hence, this is the reason why most spiritual traditions place great emphasis on it. This ranges from adoption of pure celibacy on the one hand to embracing the practises of the Left Hand path. The purpose of the practice is to exchange the yin energy of the female with that of the yang of the male. This is similar to the practices of Tantra and tantric yoga, using the body to refine the yin and yang energies.

Within the practice of 內丹, nèi dān, Internal Alchemy, there are specific exercises that work with this sexual connection. A famous one is that known as 还精补脑 huán jīng bǔ nǎo. The primary purpose of this exercise is to reverse the normal flow of 精 Jīng from the genitals which is usually downwards that usually occurs through sex or natural loss. The exercise works to internally re-cycle the 精 Jing in order to repair the brain by the prevention of the degeneration of the brain cells as one ages.

This notion of reversing the sexual energy to counter the normal cycle of this leaving the body and instead being circulated to the brain is mentioned in the 道德经 Dào dé jīng. These practices are really intended to be undertaken at the energetic level for cultivation, but are also effective for the more physical advantages that accrue.

Another related practice is that which is known as dual cultivation. In this sexual practice, the energies are blended: yin from the female, and yang from the male. These can be mutually absorbed from each other's sexual partner. 精 Jing is reduced by the avoidance of ejaculation in men, and menstruation in women.

These are very difficult practices to learn and subsequently perform, and can be dangerous for the uninitiated. The techniques are similar to Tantric yoga. For the male, the process involves breathing and concentration and a lot of self-control. He becomes aroused almost to the point of ejaculation, then by mental concentration and physically contraction, the energy is reversed flow upwards along the spine and then into the head area. The practice is really one of control of energies and, with the sexual one being the most powerful, this is a discipline that is difficult to master.

Within the body, 精 jing is stored in the 腎 shèn, which is a traditional Chinese medicine term for that part of the body and system that includes the kidneys, the adrenals and the urogenital system. This word 腎 shèn should not be confused with the other word 神 Shén, meaning spirit, and which is covered below. As noted in the introduction, only the use of the original Chinese characters can precisely distinguish between the two words.

The amount of 精 jing stored in our bodies will determine the length of our life. When the level of jing is seriously depleted, people become sick and eventually die. This is seen with older men, who lose strength in their legs as their 精 jing depletes. 太极拳 tàijíquán is excellent for building up and retaining strength in the legs and opening up the 胯 kuà area, which is the area of the inguinal crease between the thigh and the pelvic area and internally comprises the ball and socket joint of the femur and the pelvis. This is a key element of our 太极拳 tàijíquán practice and hence helps with our health and longevity.

气 Qì

Taking the traditional Chinese characters of qì – 氣, we can analyse the components that comprise this: 气 is steam rising from rice 米 as it is cooked. This provides us with a clue to the alchemical process. It is that of applying heat or fire to water to produce steam. This underpins the 內丹, nèi dān, Internal Alchemy practices and 气功 qì gōng and 內功 nèi gōng. Underpinning all of the various and sometimes seemingly odd ways of describing 內丹, nèi dān.

Forming the basis of all of the various and sometimes seemingly odd ways of describing 內丹, nèi dān, Internal Alchemy is the theme of cultivation that consists of transformation through the refinement of energy within the body. These energies within the body are universal and hence form the connection between the person and the Universe. In fact, the person is the Universe.

Sources and types of 气 qì

There are different types of 气 qì, or, more accurately, many levels of categorisation because as we know, there is only actually one all-encompassing energy. Nevertheless, this level of analysis was used by the Daoists to explain the energy expressions and how they relate to the human being. The three main sources of 气 qì come from the breath, food and the body itself. Qì travels through the body and acupuncture meridians and from there, internally throughout the whole body.

Traditionally, Daoists have used astrology and 风水 fēng shuǐ to explore the most optimal times to reproduce, based upon the environment and the state of energy at the time. Whilst being outside the scope of this book, it should be acknowledged that both internal and external energies (although ultimately being just one) have an impact on the individual.

This universal 元气 yuán qì can be absorbed from the universe by practising 太极拳 tàijíquán and other complementary 气功 qì gōng exercises. This acquired 气 qì is known as 后天 hòu tiān or post-natal 气 qì from the 易經 Yìjīng classification of post-heaven and needs to be built up slowly and consistently. The body needs to be strong to absorb the energies, like a cable needs to be strong enough to take the higher currents of electricity in order to avoid damage and burnout.

As an analogy, some people are lucky enough to inherit lots of money but dissipate it through unwise choices or bad luck. Others are born with little but through effort and consistency build up a fortune. Note that we are describing this from a one-dimensional perspective in order to provide a framework. The subject of fate, predestination, karma, luck, chance, etc. has to remain the topic for another book. What we can say is that fate is something that can be changed and altered by cultivation. Hence, there is such a thing as fate and, equally, we are not robots totally without influence on our outcomes. This is related to the subject of karma.

内气 nèi qì or internal qi that moves within the body can thus be increased through the practice of 太极拳 tàijíquán and 气功 qì gong, and the circulation thereof is greatly improved. External 气 qì or 外气 wài qì as a noun means that which surrounds the body, known as the aura, and as a verb, it means the externalisation of the 气 qì or projecting the 气 qì for healing or other purposes.

The reader should be aware that 气 qì can be broken down into endless types, just as the human body can be in anatomy studies, especially when it is mentioned within the context of Chinese medicine and the martial arts. It is however a categorisation based upon activity and flow rather than discrete concrete and separate parts.

As mentioned, the separation of the whole into parts is more of an abstract than a concrete separation of individual parts. Like waves that are seemingly separate but yet part of the ocean, the 'types' of 气 qì are individuated merely for the purposes of explanation and understanding.

元气 yuán qì – Prenatal Qi

This has as its location the 命门 mìng mén, at the acupuncture point GV-4 between the two kidneys. It is translated as 'Door of Life'. It forms the root for all yin and yang energies within the body. It is hereditary, being derived from one's parents. It is therefore a very important point in the body for our practices and in the movements of the form it is important that this part of the body is open.

后天之气 hòu tiān zhī qì – Postnatal Qi

This qi is derived during one's lifetime and, following birth, is cultivated from food, drink and air.

脏腑之气 zàng fǔzhī qì – Zang and Fu Organs Qi

The Yang Fu and Yin Zang organs of the body transform the vital essences taken from food and drink.

谷气 gǔ qì – Food Qi and 中气 Zhōng Qì – Central Qi, 清气 qīng qì – Clear Qi and 浊气 zhuó qì – Turbid Qi

Food enters through the stomach and is transformed by the spleen, where the Yang Qi that is pure and the Yin Qi that is impure are separated and transported through the body through the Middle and Lower Burners. This qi from the stomach and spleen is transported to the chest, where it is combined with the 气 qì of the heart and lungs. The 清气 qīng qì clear qi is the pure energy extracted from the 谷气 gǔ qì and the impure or turbid 浊气 zhuó qì – Turbid qi.

宗气 zōng qì – Gathering Qi

宗气 zōng qì – Gathering Qi宗气 zōng qì is formed from the 谷气 gǔ qì in the lungs in combination with the 元气 yuán qì. It nourishes the lungs and heart and drives the pulse and respiratory activity. Zōng qì collects in the chest and is called the 'Sea of Qi'. It is controlled by the acupuncture point 膻中 shān zhōng Acupuncture Point Ren-17.

真气 zhēn qì – True Qi

The last stage in the transformation and refinement of qi is the formation of 真气 zhēn qì that circulates in the meridians inside the body as well as outside the body.

营气 yíng qì – Nutritive Qi

Yíng Qì nourishes the whole body and energises the internal organs in a cycle of two hours, travelling through each of the 12 meridians. Hence the cycle is one day and is known as the Horary Cycle.

卫气 wèi qì – Protective Qi

This flows between the muscles, just under the skin. It's main purpose is to protect the body from pathogens by warming the body and enabling the pores of the skin to open. Practising the form will help to strengthen this function.

正气 zhèng qì – Upright Qi and 邪气 xié qì – pathogen

This is not a type of 气 qi as such but more of a function or an ability to provide resistance against disease or what is known as 邪气 xié qì.

The aim of 內丹, nèi dān, Internal Alchemy is to prolong the life of the body, which acts as the host for an immortal spirit body that can survive after death. The human body therefore acts as a cauldron for blending and refining the energies. The Three Treasures of 精 jīng, 气 Qì and 神 Shén are cultivated for the purpose of improving physical, emotional and mental health, and, ultimately – if you are conscientious enough in merging with the Dao, that is – of going to the level of becoming an Immortal.

This process is often erroneously considered to be the act of prolonging the human body but is in fact the synthesis of the more refined energies so that the person can travel at the time of death. There have been numerous examples in China and Tibet of enlightened masters leaving behind just finger and toe nails at death s to the delight of chiropodists no doubt. 內丹, nèi dān, Internal Alchemy is an important form of practice for most schools of Daoism.

神 Shén

Shén is translated as either spirit or mind. This is mind in its larger sense rather than the small mind of involuntary thinking. It is the energy that is used in mental, spiritual and creative expression, and it resides in the Upper 丹田 dān tián, which as we have stated, is located between the eyes at the mid-brow point. The Chinese character for 神 Shén assimilates the meaning of both the noun 'spirit' and the verb 'to stretch'.

神 Shén can be cultivated internally from within the body or externally through absorption of the energies of the universe. It is associated with the heart and liver. In the -太乙金华宗旨 Tài Yǐ Jīn Huá Zōng Zhǐ – Secret of the Golden Flower that we have just covered, it states that the 神 Shén is located in the heart and resides in the liver at night when it dreams.

It can also be defined as physic energy. 神 Shén is lost by too much self-reflection through the mental chatter of endless external thoughts from a restless and chattering monkey mind.
It circulates through the 奇精八脉, qí jīng bā mài that are explained in further detail below.

The activities of the 神 Shén include all of a person's mental activities, including those of consciousness and of the nervous system. The nervous system includes the 'original spirit' and those activities that are essential for the body's survival, such as respiration and the heartbeat. These would be described as the subconscious in Western approaches to such an analysis. A person's consciousness is comprised of the spirit of awareness, the conscious activities, and the self-reflection of the thinking process. Internal alchemists focus on the original spirit of 神 Shén.[50]

神 Shén is known to reside mainly in the heart or, more specifically, the blood which relies on the heart. It is believed that 神 shén sleeps at night and if it is disturbed, the result can be insomnia. Healthy 神 shén can be seen in a person's physical appearance through the brightness of the eyes. If the eyes are bright and shining, a healthy 神 shén is indicated. Conversely, If one's 神 shén is unhealthy, the eyes will appear dull. 神 Shén is dependent on both the 精 jing and the 气 qì. If both

---

[50] Neidan: The Traditional Meditative Practice 14.

are harmonious, the 神 shén will be in a good state and balanced, hence, one can see the inter-relationships of the whole constitution of the human being, from the coarser physical to the more refined spiritual. This relationship of mind-body-spirit is now recognised by almost everybody.

神 Shén may be thought of as either a singular concept or as a pluralism. When viewed singularly, 神 Shén is defined as being located in the heart and is known as the heart 神 Shén. When it is considered in the plural, it is described as residing within five of the yin organs: the liver, the heart, the spleen, the lungs and the kidneys. The singular 神 shén relies upon the others as equally as the others depend on it, hence the prevailing and constant paradigm of mutual dependency endures. If the heart 神 Shén malfunctions, this can damage the other 神 shén and lead to problems such as mental illness.

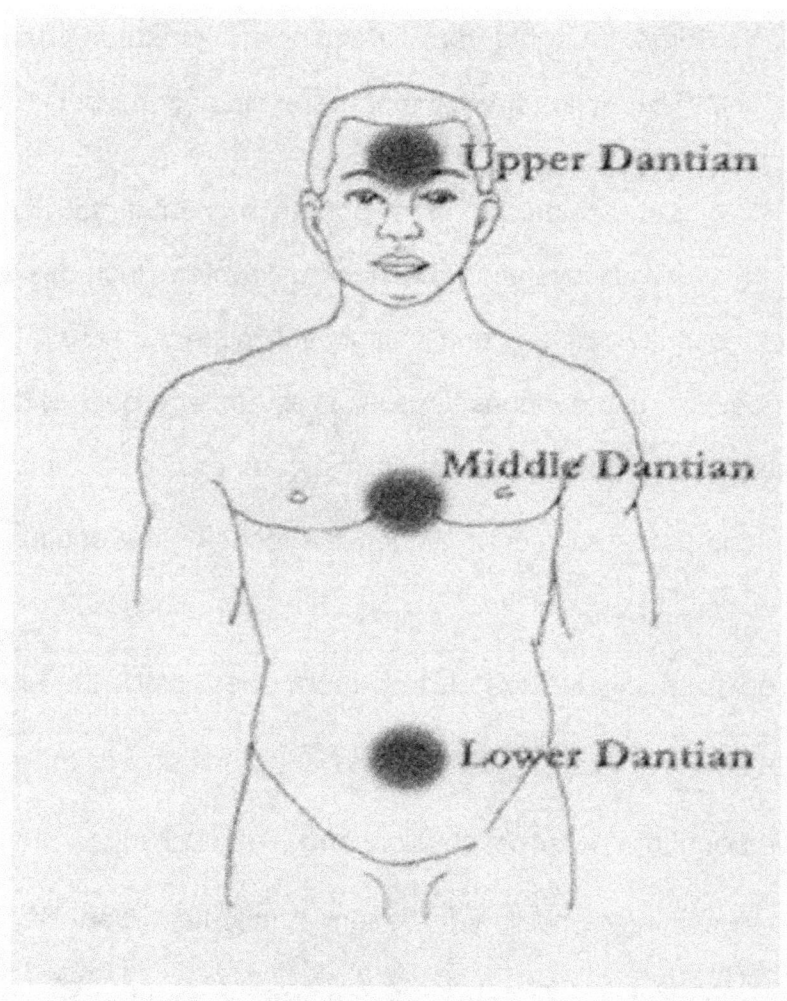

Figure 2-3: The three 丹田 dān tiáns

Meridians in Traditional Chinese Medicine

The origins of traditional Chinese medicine can be traced back more than 5,000 years. There are three legendary figures associated with it, Fu Xi, who we mentioned during our discourse on the Yì jīng, 神农 Shen Nong and 黄帝 Huáng Dì. 神农 Shen Nong was an emperor who is recorded as being the founder of herbal medicine.

He is said to have experimented on himself by ingesting many types of plants, passing his results on to his contemporaries, and subsequently this knowledge was passed on verbally through successive generations.

The earliest recorded text on traditional Chinese medicine is the 黄帝内经 Huáng Dì Nèi Jīng, *The Yellow Emperor's Canon of Internal Medicine*, written between 800 BC and 200 BC. It covers a wide range of topics on medicine, such as the meridian theory, diagnosis, treatment, prevention physiology, pathology, acupuncture, moxibustion and 推拿 tuī ná.

Another famous classic, written by 神农 Shen Nong during the Spring and Autumn Warring States Period was his 神农本草经 Shén nóng běn cǎo jīng or Classic of Material Medica. This was the first the pharmacopoeia of traditional Chinese herbal medicine, describing the uses of 365 herbs. Other classics within the history of traditional Chinese medicine are described briefly below.

All of this is based upon the same paradigms that we have encountered hitherto – the observance of energy and flows of 气 qì through the human body and the effect that a blockage in the circulation of 气 qì can have on the individual's health.

Shāng Hán Lù

The Shāng Hán Lù was written by 張仲景 Zhang Zhong during the Han dynasty period. It relates exclusively to the preparation and use of herbal formulae as a treatment method and uses the six-channel theory differentiation. It is best or least inaccurately translated as *cold damage.* It covers the subject of externally contracted diseases. It originally formed part of the treatise known as the 傷寒雜病論 Shāng Hán Zá Bìng Lùn, *On Cold Damage and Miscellaneous Diseases*. This was written by Zhāng Jī, a physician who originated from East Hàn around 1,800 years ago.

金匮要略 Jīnguì Yàolüè, Synopsis of the Golden Chamber

The 金匮要略 Jīn Guì Yào Lüè, translated as *Essential Prescriptions of the Golden Coffer*, is the accompanying volume to the *Shāng Hán Lùn*. The *Jīn Guì Yào Lüè* reconstitutes parts of the text of the *Shāng Hán Zá Bìng Lùn, On Cold Damage and Miscellaneous Diseases*, which has unfortunately been lost and is nowadays sadly only referenced in other works. This was written by the Hàn Dynasty physician 张机 Zhāng Jī, who is widely acknowledged to be the most brilliant medical mind that China ever produced,

张机 Zhāng Jī's influence on the development of Chinese medicine is difficult to exaggerate. Using his vast knowledge and practical experience of medicine, Zhāng integrated what at the time were the relatively new theories of the systematic correspondence of the *Nèi* and *Nànjīn*. Only sometime later during the Song dynasty did the importance of his works become fully appreciated and they have subsequently stood the test of time as being of practical value when put to the test by later Chinese doctors.

A further two classics worth mentioning here briefly are the 温病学 Wēn Bìng Zué and the 难经 Nán Jīng. The 温病学 Wēn Bìng Zué classic discusses the theory and treatment of infectious diseases that are caused due to the presence of an excess of heat in the body. Heat is deemed to be a primary cause of externally acquired diseases. The Nan Jing is known as the *Classic of Difficulties*, from the word nán Jīng, meaning hard or difficult. Written during the Han Dynasty, it seeks to amplify and provide clarification to the 皇帝内经, Huáng Dì Nèi Jīng. In so doing, it further developed the theory of the 奇經八脈; qí jīng bā mài, the Eight Extraordinary Meridians, and Five Shu points' theory which provides the foundation of Five Elements Acupuncture.

Meridian System

The meridian system, through which 气 qì flows within and through the body, is known as 经络 jīng luò and is the theory that is applied within traditional Chinese Medicine, 气功 qì gōng and 太极拳 tàijíquán. The meridian system is composed of channels through the body and these are depicted in the acupuncture charts that one usually sees on display in traditional Chinese medicine clinics.

There are 20 meridians in total, and these are classified according to two main types: the Twelve Regular Meridians and the Eight Extraordinary Channels, 经络 Jīng luó or meridians, two of which have their own sets of points, the remaining ones connecting points on the other channels.

The theory of the channels or meridians – and these terms tend to be used inter-changeably – is associated with the internal organs. These internal organs, although named after the physical organs of the body as we know them and used in Western medicine, and, thus, familiar to us all (at least by name, if not sight), are not regarded as independent anatomical entities. In traditional Chinese medicine theory, the organs are described functionally with regard to how they interact with each other. This description and categorisation is within a holistic system and is in relation to the five essences of the body: the Three Treasures, 精气神 jìng, qì, and Shén, 血液 xuè yè, blood and 津液 jīn yè, a collective term for bodily fluids. We have previously discussed the Three Treasures, and so now need to describe xuè yè, blood and 津液 jīn yè, bodily fluids.

血液 xuè yè is translated as blood, but it is not strictly blood as derived from the English medical definition. The heart is the ruler of the blood, the liver is the storage and the spleen is the governor. The 津液 jīn yè is a collection of all of the remaining bodily fluids, such as saliva, sweat and urine. 津 jīn means light and 液 yè means heavy, hence the collective term 津液 jīn yè denotes the light and heavy fluids. The function of 津液 jīn yè is to lubricate, and it involves and affects the skin, hair, follicles, membranes, etc.

The 经络 Jīng luò system accounts for the main channels of communication and energy distribution within and throughout the whole body. They also inter-connect with the entire musculoskeletal system. Manipulation of the energy system through key points in the meridian system is undertaken by the use of needles. This is known as acupuncture. In modern acupuncture, the twelve primary meridians and two of the 奇經八脈 qí jīng bā mài, the 督脈 dū mài and 任脈 rèn mài are accessed. These latter two are used due to their importance as circulators of energy and 內丹, nèi dān, Internal Alchemy places great importance on the circulation between these two, 督脈 dū mài and 任脈 rèn, which is known as the Small Heavenly circulation.

The Small Heavenly circulation through the 督脈 dū mài and 任脈 rèn mài is also known as the microcosmic circulation or Small Orbit circulation. Placing the tongue at the top of the upper palate next to the top teeth acts a bridge between the yin and yang energies of the two channels. This is known as the Magpie Bridge and is something that we adopt in our 太极拳 tàijíquán practice. There are in fact five different positions where the tongue may be placed within the upper palate. This action and the extent to which the energies are flowing freely can be evidenced by the amount of saliva that is produced. The saliva is known by many names, including; Divine Juice 灵 汁 líng zhī and Heavenly Dew 天 露 tiān lù

This saliva produced is one of two kinds, that of the kidneys, of the 精 jīng, which is pure and refined, and that of the digestion, which is less so. Saliva is known to contain enzymes. In some Indian yogic traditions, water is only consumed after the food so that the saliva is not washed away during eating, thus optimising the digestive process.

There are five yin organs of the body: 肝 gān – liver, 心 xīn – heart, 脾 pí – spleen, 肺 fèi – lungs, and 肾 shèn – kidneys. Collectively, these are known as the 五脏 wǔ zàng. In addition there is a sixth. This is known as the 心包 xīn bāo or pericardium. Again we must emphasise that these descriptions relate to function. That is, it is movement, described by verbs. This differs somewhat with the Western medical system use of a static model, using nouns. The yin organs produce, transform and store. Their relevance and understanding relates directly to the Five Phases (Elements) system.

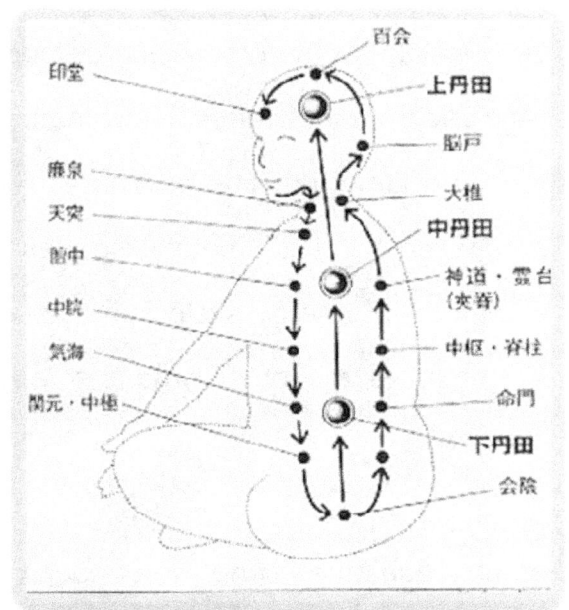

Figure 2-4: The Microcosmic Orbit

The liver is responsible for the movement of body fluids within the body, adjusting the 气 qì and 血液 xuè yè, blood and to ensure that it circulates smoothly and does not stagnate. The heart governs the vessels that move blood around the body, and it stores the 神 Shén, 血液 xuě yè, blood and 津液 jīn yè, bodily fluids. The spleen is the primary organ of digestion and moves extracted 气 qì from the food to the lungs. The lungs account for movement and circulation and that of the descending and liquefying function. The kidneys store jìng and are the source of life, reproduction and longevity. The pericardium is a defence against external diseases.

As you will by now no doubt have grasped, there are also six yang organs to complement the 五脏 wùzàng. These are known as the 六腑 liù fǔ Liu Fu. The 六腑 liù fǔ consists of 胆囊 dǎn náng – the gall bladder, 小肠 xiǎo cháng – the small intestine, 胃 wèi – the stomach, 大肠 dà cháng – the large intestine, 膀胱 páng guāng – bladder and 三焦 sān jiāo – the Triple Burner. These yang organs perform a draining function, receiving and breaking down substances, including food, and transporting and excreting waste from the body.

Individually, the yang organs function in an interrelated fashion, as follows. The stomach governs digestion in that the energy moves in a descending fashion. The gall bladder both stores and releases bile but is not of itself responsible for the processing of food or liquids. The small intestine acts as a purifier of food that is received from the stomach, eliminating what is not required and then sending nutrients onto the spleen. The large intestine eliminates waste whilst reabsorbing water. The urinary bladder acts as a regulator of fluid metabolism, as it processes urine before its elimination. The Triple Burner is akin to an energy system in itself, comprising three burners: the upper, middle and lower, and it regulates the other organs through intake, transformation and elimination.

The yin and yang organs complement each other – one yin for each yang as follows:
- Liver – gall bladder
- Heart – small intestine
- Spleen – stomach
- Lungs – large intestine
- Kidneys – bladder
- Pericardium – triple burner

To maintain health, 气 qì must flow uninhibited in four ways: ascending, descending, entering and leaving. Energy circulates through the body in a distinct pattern, over a cycle every day. This is shown in Table 2-1.

Table 2-1: The circulation of energy through the primary meridians

Organ	Time
Lungs	3am to 5am
Large intestine	5am to 7am
Stomach	7am to 9am
Spleen	9am to 11am
Heart	11am to 1pm
Small Intestine	1pm to 3pm
Bladder	3pm to 5pm
Kidney	5pm to 7pm
Pericardium	7pm to 9pm
San Jiao	9pm to 11pm
Gall bladder	11pm to 1am
Liver	1am to 3am

Table 2-2: The twelve standard meridians

No.	Meridian name (Chinese)	Yin / Yang	Hand / Foot	5 elements	Organ
1.	Tài yīn Lung Channel of Hand 手太阴肺经 or Tài yīn Lung Meridian of Hand	Tài yīn –Greater yin	Hand 手	Metal 金	Lung 肺
2.	Shǎoyin Heart Channel of Hand 手少阴心经 or Shǎoyin Heart Meridian of Hand	Shǎoyin –Lesser yin	Hand 手	Fire 火	Heart 心
3.	Jueyin Pericardium Channel of Hand 手厥阴心包经 or Jueyin Pericardium Meridian of Hand	Jueyin – Absolute yin	Hand 手	Fire 火	Pericardium 心包
4.	Shǎoyang Sanjiao Channel of Hand 手少阳三焦经 or Shǎoyang Sanjiao Meridian of Hand	Shǎoyang – Lesser yang	Hand 手	Fire 火	Triple Heater 三焦
5.	Taiyang Small Intestine Channel of Hand	Taiyang –	Hand 手	Fire 火	Small

	手太阳小肠经 or Taiyang Small Intestine Meridian of Hand	Greater yang			intestine 小肠
6.	Yangming Large Intestine Channel of Hand 手阳明大肠经 or Yangming Large Intestine Meridian of Hand	Yangming – Yang brightness	Hand 手	Metal 金	Large intestine 大腸
7.	Taiyin Spleen Channel of Foot 足太阴脾经 or Taiyin Spleen Meridian of Foot	Taiyin –Greater yin	Foot 足	Earth 土	Spleen 脾
8.	Shǎoyin Kidney Channel of Foot 足少阴肾经 or Shǎoyin Kidney Meridian of Foot	Shǎoyin –Lesser yin	Foot 足	Water 水	Kidney 腎
9.	Jueyin Liver Channel of Foot 足厥阴肝经 or Jueyin Liver Meridian of Foot	Jueyin – Absolute yin	Foot 足	Wood 木	Liver 肝
10.	Shǎoyang Gall Bladder Channel of Foot 足少阳胆经 or Shǎoyang Gall Bladder Meridian of Foot	Shǎoyang – Lesser yang	Foot 足	Wood 木	Gall bladder 胆
11.	Taiyang Bladder Channel of Foot 足太阳膀胱经 or Taiyang Bladder Meridian of Foot	Taiyang – Greater yang	Foot 足	Water 水	Urinary bladder 膀胱
12.	Yangming Stomach Channel of Foot 足阳明胃经 or Yangming Stomach Meridian of Foot	Yangming – Yang brightness	Foot 足	Earth 土	Stomach 胃

The table shows the relationships between the Five Phases, the Yin and Yang expressions corresponding to the meridians and the associated organs. We shall investigate each of the individual meridians and the later move onto the topic of the 奇精八脉, Qì Jīng Bā mài that is fundamental to our work in 內丹, nèi dān, Internal Alchemy. The following table lists the associated emotions that are stored as energy types in the internal organs and which if negative cause blockages and stagnation, leading to illnesses.

Emotions are stored as energy, and therefore negative ones influence the meridian system and the organs themselves. Practices to redress these adverse effects are undertaken in the form of transforming negative to positive emotions and use of sound to reduce excess.

Each organ has a yin and a yang aspect. The 脏 zàng organs are yin, and the 腑 fǔ organs are yang. As always, we must further emphasise the fact that these are relative descriptions, that all organs are interrelated rather than isolated individual components and that these energy systems are dynamic, not static.

Table 2-3: The organs and associated emotions

Element:	Wood/Liver	Fire/Heart	Earth/Spleen	Metal/Lungs	Water/Kidney
Zang Organ:	Wood （木）– Liver 肝	Fire 火 – Heart 心	Earth 土 – Spleen 脾	Metal 金 – Lung 肺	Water 水 – Kidney 肾
Fu Organ	Gall bladder 胆	Small intestine 小肠 *secondarily*, Sānjiāo 三焦 – Triple Burner and pericardium 心包	Stomach – 胃	Large intestine – 大肠	Bladder – 膀胱
Harmful Emotions:	Anger	Hate, excess joy	Worry, anxiety	Anxiety, sorrow	Fear
气 qì Effect:	Rising	Scattering	Knotting	Constriction	Descending
Positive Emotions:	Acceptance	Kindness and compassion	Fairness, Openness	Integrity	Wisdom

The organs and their associated emotions and healing sounds

Each of the five organs has a negative emotion associated with it. There is also a positive one. That is to say, the negative emotions that we develop and retain have an association with these organs (systems). These negative emotions can be transformed by the vibration of a particular sound that is uttered internally. This has the effect of transforming the negative energy into a positive one and any corresponding blockage stored within it. Thus the negative emotion is turned into a positive one and any ill health that has arisen is alleviated. There are also colours associated with the organ systems and even animals (e.g. Green Dragon for the liver).

The liver is associated with anger, and this can be transformed into the positive emotion of acceptance and harmony. Physically, the liver function of processing is important for the storage and release of glucose into the blood.

The heart is associated with the negative emotions of both excess joy (yes, this does sound counter intuitive, but it is an excess of energy nevertheless) and that of shock. While it may seem perverse to mention an excess of joy, a harmonious balance needs to be struck. And no doubt you have heard of the phrase, 'it will all end in tears...' These negative emotions can be transformed into kindness and compassion.

The stomach/spleen, which includes the pancreas, may hold the negative emotions of worry and anxiety which can be transformed into those of the positive: fairness and openness. These feelings are common to us all as butterflies or knots in the stomach.

The lungs store the negative emotions of sadness and depression, which may be transformed into courage. Emotional depression often leads to a physical depression and adverse effects on the chest and lungs.

The kidneys are negatively associated with fear and may be transformed into the positives of gentleness and kindness. Fear often adversely affects the adrenal glands, which are located slightly above the kidneys. The adrenal glands secrete adrenalin and noradrenalin if the negative emotion of fear causes the fight-or-flight response.

As all manifestations, including those of human beings, are formed from energy, which is in turn based on different vibratory levels, a series of healing sounds were devised for each of the organs. As we know, the relationship of the organs is based on the Five Phases theory, which has two cycles: a positive and a negative. Hence, blockages or excess energy in one organ will affect the others in a vicious cycle.

The six healing sounds comprise the five mentioned above: the liver, heart, stomach/spleen, lungs and kidneys, plus the Triple Burner. The Triple Burner is a functional energy system in itself, along with the others. As it is not so well known as the recognised Western organs, it is worth providing a brief description. The three burners relate to the body's three main cavities, the thorax, the abdomen and the pelvis. The Upper Burner controls intake, the Middle Burner controls transformation and the Lower Burner controls elimination.

The Upper Burner commences at the base of the tongue and descends to the stomach. It controls the intake of air, fluids and food. It balances the energy of the heart and lungs, as well as governing respiration, and regulates the distribution of guardian 气 qì, which protects the body. The Middle Burner commences at the entrance to the stomach and descends through to the pyloric valve. It controls digestion by balancing the energies of the stomach, spleen and pancreas. It extracts nourishing energy from what we take in, food and fluids, and circulates them through the meridians.

The Lower Burner commences at the pyloric valve and descends to the anus and urinary tract. It separates the pure from the impure products of digestion; it absorbs nutrients, and eliminates solid and liquid wastes. It balances the energies of the liver, kidney, bladder, and large and small intestines. It also harmonises the sexual and reproductive energies.

The sounds for each of the six organ energy systems are listed in Table 2-4. The sounds are made internally whilst focussing internally on that part of the body where the organ resides. An energetic connection is made and the sounds are uttered sub-vocally in a state of relaxed awareness and intention.

Table 2-4: The healing sounds

Phase	Wood	Fire	Earth	Metal	Water	N/A
Organs	Liver, gall bladder	Heart, small intestine	Spleen, stomach	Lungs, large intestine	Kidneys, bladder	Triple Burner[1]
Emotional Excess	Anger	Joy (excitement)	Brooding	Sorrow	Fear	N/A
Sound	Shhhhhh	Hawwwwwww,	Hoooooo	Sssssss	Foo	Heeeeeee,

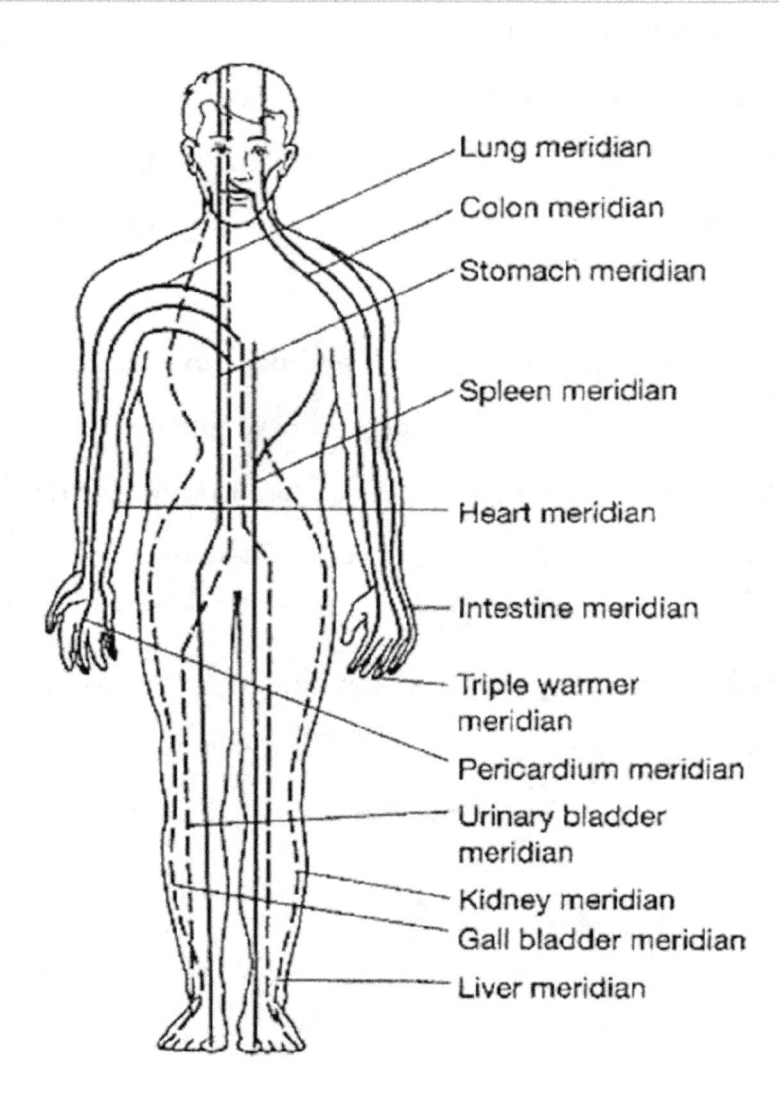

Figure 2-5: The twelve standard meridians

The meridians allow energy to travel between one part of the system to another. They are akin to a motorway or highway built to connect places together and allow efficient travel. We take each meridian in turn and explain in some detail how the energy flows through and the main course of travel.

Yin and Yang organs are paired according to the Five Phases. The Yang meridians move through the external side of the limbs whilst those of the Yin are on the inside. The Yang organs are also categorised as follows: Greater Yang 太阳 tài yáng, Lesser Yang 少阳 shǎo yáng, and Yang Brightness 阳明 Yáng míng. The Yin organs are categorised as Greater Yin 太阴 tài yīn, Lesser Yin 少阴 shǎo yīn, and Absolute Yin 厥阴 jué yīn.

1. The Lung Meridian of Hand 手太阴肺经

The lungs are one of the six Yin organs. In terms of the Five Phases (Elements), they relate to the metal element and are linked with the corresponding yang organ of the large intestine. As the metal characteristic within the Five Phases, they relate to the West as a direction and to autumn as the season. For 內丹, nèi dān, internal alchemy signifies that the White Tiger is invoked through the imagination as the ruler of the lungs organs. The opening is the nose and proper breathing will strengthen the guardian 卫气 wèi qì, and your overall body will be protected. Signs of healthy lungs are the state of the hair and skin.

The lung meridian commences in the Middle Burner region. This is in the solar plexus region around the stomach and spleen. It ascends through the large intestine and the opening of the stomach, moving through the diaphragm and then entering the lungs.

Moving transversely from the 中府 zhōng fǔ and 雲門 yún mén points, the meridian ascends through the medial side of the upper arm to the pericardium channel of the Hand-Jueyin. It thereupon continues through the 天府 tiān fǔ point to the 尺澤 chǐ zé point at the elbow and runs along the anterior portion of the forearm by way of the 孔最 kǒng zuì, and 列缺 liè quē points to the 經渠 jīng qú and 太淵 tài yuān points above the main artery of the wrist. Finally it runs through to the 魚際 yú jì and 少商 shào shāng points along the radial border and ends at the tip of the thumb.

As we have indicated, each of the organs has a corresponding positive and negative emotion. For the lungs, these negative emotions are sadness and anxiety. Interestingly, the English word 'anxiety' is derived from the German root angst, meaning 'narrow', which in that language relates to a description of the narrowing of the bronchial passages.

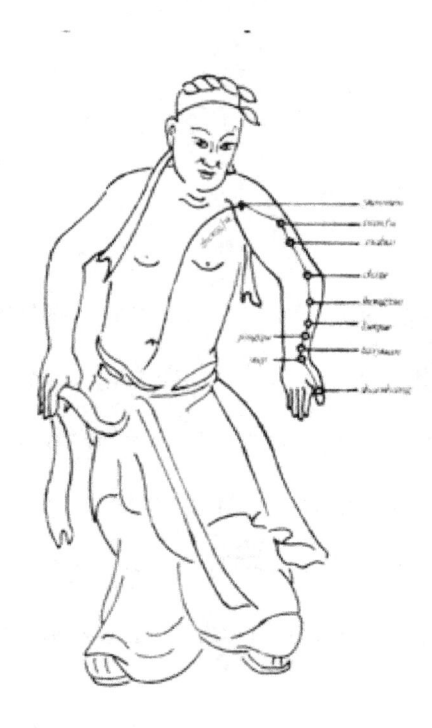

Figure 2-6: The Lung Meridian

2. The Large Intestine Meridian of Hand-Yangming 手阳明大肠经

The large intestine is a Yang organ and is paired with the lungs as we have just seen. The main function of the large intestine is to process the waste material from the small intestine, and send it through the urinary bladder, to be excreted as waste.

This meridian starts at the tip of the index finger. This point at the tip of the index finger is known as the 商陽 shāng yáng point. It ascends from the index finger on the radial side. It moves through to the points of the 二間 èr jiān, 三間 sān jiān and 合谷 hé gǔ and on through to the wrist. It then ascends laterally up the forearm through to the elbow. From there, it continues through the anterior border of the upper arm and on to the 肩髃 jiān yú point on the shoulder. It moves on to the supraclavicular fossa and to the lungs, diaphragm and the large intestine. It branches out along the neck, the cheek and the gum of the lower teeth, onto the 迎香 yíng xiāng point by the side of the nose.

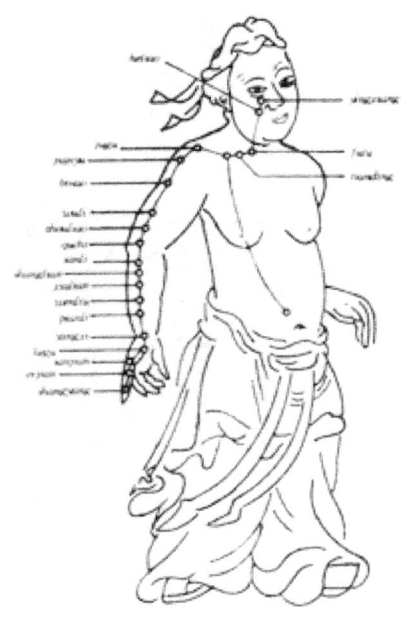

Figure 2-7: The Large Intestine Meridian

3. The Stomach Meridian of Foot-Yangming 足陽明胃經

This belongs to the element Earth, the central direction, and the season of long summer or the end of summer, that of dampness and the colour yellow. The opening is the mouth and it controls the flesh and the limbs.

The route of this meridian starts with an ascent to the base of the nose. It then descends to the upper gum and then moves around the lips to ascend through to the front of the ear and to the 頭維

tóu wéi point on the forehead, just under the hairline. It then branches out as one circulation at the 大迎 dà yíng point and descends to the 人迎 rén yíng point. It moves along the throat before circling through the diaphragm to meet the spleen. On the other route, it descends from the supraclavicular fossa to the 氣沖 qì chōng point that is located on the lateral side of the lower abdomen and continues through the nipple and along the navel. Still another branch emerges from the lower orifice of the stomach and runs down to the 氣沖 qì chōng point and the thigh and the knee and further down to the lower leg and the instep, ending at the 厲兌 lì duì point at the tip of the second toe.

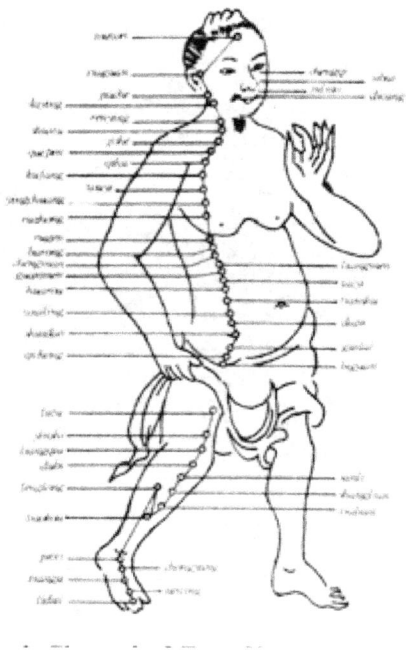

Figure 2-8: The Stomach Meridian of Foot-Yangming

4. The Spleen Meridian of Foot-Taiyin 足太阴脾经

In traditional Chinese medicine theory, the spleen as it is so named also includes the pancreas. The spleen is a Yin organ and is paired with the stomach which is Yang. It is the main organ of digestion. It transports nutrients from the food digested into nourishment for the body to regulate energy. It belongs to the element Earth, the central direction, and the season of long summer or the end of summer, that of dampness and the colour yellow. The opening is the mouth and it controls the flesh and the limbs.

The route commences and ascends from the tip of the big toe along the medial side of the foot and through the acupuncture points of the 隱白 yǐn bái, 太白 taì bái, 公孫 gōng sūn 商丘 and shāng qiū points. It then moves along the front of the inner ankle bone and the posterior surface of the lower leg. It continues ascending through the 三陰交 sān yīn jiāo, 漏谷 loù gǔ, 地機 dì jī, 陰陵泉 yīn líng qúan, 血海 xuè hǎi, 箕門 jī mén and 沖門 chōng mén points and up to the abdominal cavity. It runs through the spleen and the stomach and ascends through the diaphragm to the root of the tongue.

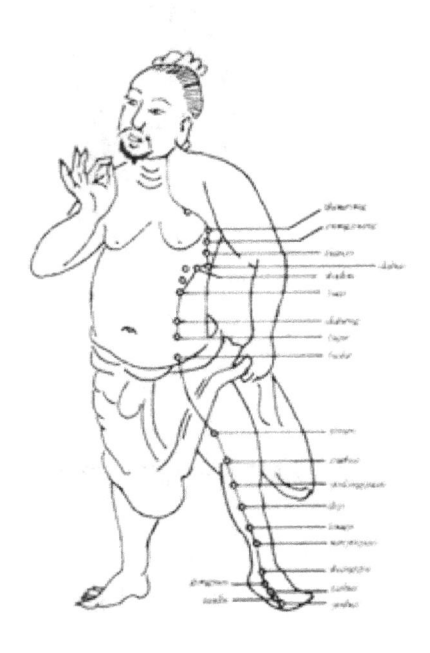

Figure 2-9: The Spleen Meridian of the Foot-Taiyin

5. The Heart Meridian of Hand-Shǎoyin 手少阴心经

The heart is paired with the small intestine, and it is Yin to that of the small intestine, which is Yang, It corresponds to the element Fire, the southerly direction, and the summer season. It has the condition of heat and the colour red. The point of entry is the tongue, and it has the function of controlling the blood vessels.

This meridian commences at the heart, whereupon it branches in three directions. The main branch ascends to the lungs and then descends to the armpit and through the forearm and onto the palm along the posterior border of the medial side of the upper arm. It continues to descend to the 少冲 shào chōng point along its medial side of the little finger, where it meets the Small Intestine Channel of Hand-Taiyang. Another then ascends to the side of the throat and meets with the eye.

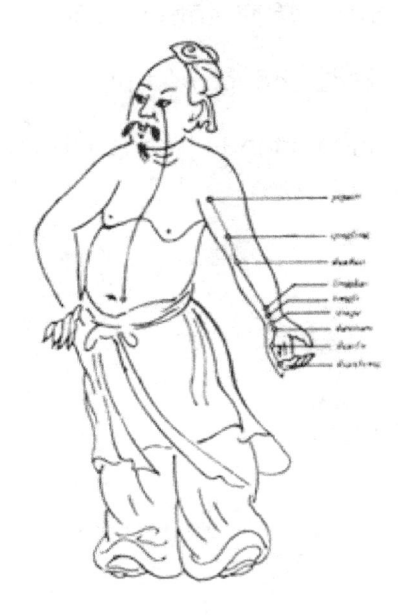

Figure 2-10: The Heart Meridian of Hand-Shǎoyin

6. The Small Intestine Meridian of Hand-Taiyang 手太阳小肠经

The small intestine is paired with the heart which is considered Yin, as opposed to the small intestine which is Yang. It is of the Fire element and is of the southerly direction, the summer season with the condition of heat and is of the colour red. The major function of the small intestine is to extract the waste substances from the nutrients in the food. The latter is then distributed throughout the body and the former is sent on to the large intestine.

This meridian commences from the 少澤 shào zé point. This point is located at the tip of the little finger and ascends the ulnar side of the dorsal hand through the acupuncture points of 前谷 qián gǔ, 後谿 hòu xi, 腕骨 wàn gǔ and 養老 yǎng lǎo. It thereupon continues along the posterior aspect of the forearm through the 支正 zhī zhèng point at the elbow. Then it ascends along the posterior border of the lateral side of the upper arm to the 肩貞 jiān zhēn and the 臑俞 nāo shū points behind the shoulder joint, and then along the scapular region through the 天宗 tiān zōng, 秉風 bǐng fēng, 曲垣 qū yuán, 肩外俞 jiān wài shū and 肩中俞 jiān zhōng shū points to meet with the 督脈 dū mài, Du meridian. From there, it further ascends into the supraclavicular fossa and on down to meet with the heart before running through the diaphragm along the oesophagus and to the small intestine by way of the stomach. Its branch moves up along the side of the neck to the cheek and the outer corner to the ear.

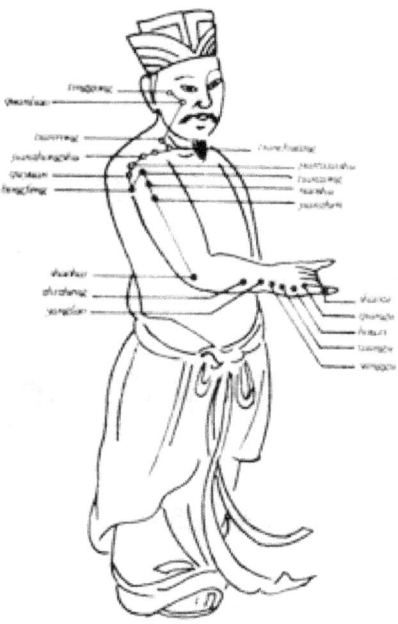

Figure 2-11: The Small Intestine Meridian of Hand-Taiyang

7. The Kidney Meridian of Foot-Shǎoyin 足少阴肾经

The kidneys are Yin and are linked to the Yang urinary bladder. The kidneys are of the Water element, that of the winter, the cold and, the southerly direction, with a colour of black.

The meridian commences at the interior side of the little toe and then ascends through the heel, along the medial malleolus and through the 涌泉 yǒng quán point. This is a very important point in our practice as it provides the source of yin energy and grounding. It then joins the Spleen channel of Foot-Taiyin and the Liver channel of Foot-Jueyin at the 三陰交 sān yīn jiāo region. It continues to ascend to the 陰都 yīn dū point at the inside knee and through the medial side of the lower leg. It then traverses the base of the spine through the inner thigh. There it joins the kidney and the bladder. It ascends to the kidney through the liver and diaphragm and on to the lungs, the throat and to the tongue.

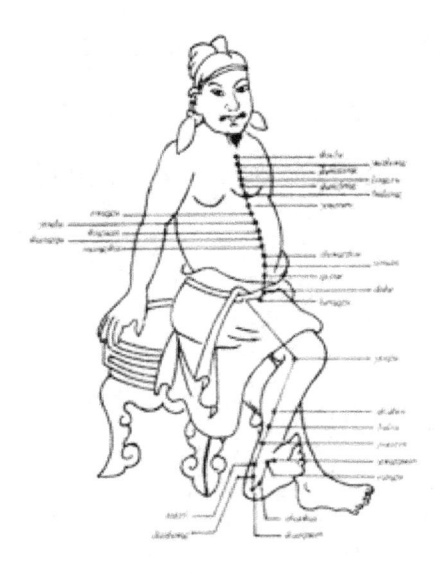

Figure 2-12: The Kidney Meridian of Foot-Shǎoyin

8. The Bladder Meridian of Foot-Taiyang 足太陽膀胱經

The urinary bladder is Yang and paired with the kidneys. It is of the Water element, that of the winter, the cold and the southerly direction, with a colour of black. It controls the bones, the marrow and the brain.

Commencing at the 睛明 jīng míng point on the inside of the eye and ascending through the 攢竹 cuán zhú, 曲差 qū chā, 五處 wǔ chǔ, 承光 chéng guāng and 通天 tōng tiān points, the meridian continues across the forehead and up to the 百會 bǎi huì point on the top of the head. The 百會 bǎi huì point is a very important point and is the source of yang energy and equates to the crown chakra in other esoteric systems. The main merdian crosses through the 百會 bǎi huì point to the brain. It then continues through the 絡卻 luò què, 玉枕 yù zhěn and 天柱 tiān zhù points to the top of head. It then descends the spine to the 腎俞 shèn shū point and the abdominal cavity. There it joins with the kidney and the bladder. One branch joins the 百會 bǎi huì and the temple. Another descends from the 腎俞 shèn shū to the 白環俞 bái huán shū point. It thereafter ascends to the 上髎 shàng liáo point before descending to the huiyang point close to the coccyx. It crosses the buttocks and the back of the thigh to the weizhong point. Another branch descends from the nape of the neck and through the gluteals close to the 至陰 zhì yīn point at the lateral side of the tip of the little toe.

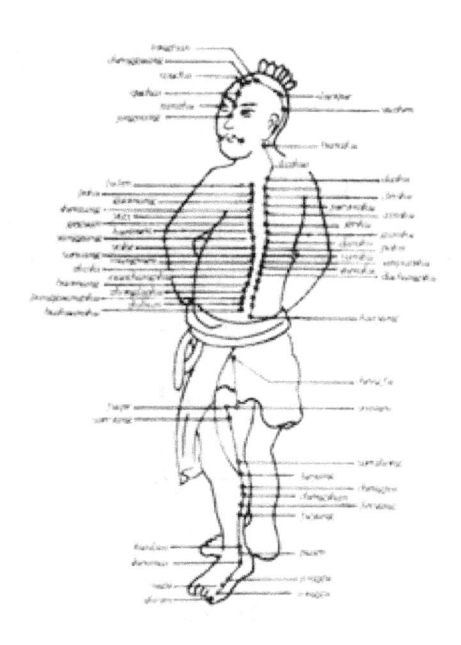

Figure 2-13: The Bladder Meridian of Foot-Taiyang

9. The Pericardium Meridian of Hand-Jueyin 手厥阴心包经

This is a Yin organ and is paired with the Triple Burner which is Yang. The meridian commences from within the chest and pericardium and descends through the diaphragm. It crosses the 天池 tiān chí point near the armpit and runs up along the chest and the medial side of the arm and down through the 天泉 tiān quán, 曲澤 qū zé, 郄門 xī mén, 間使, 內關 nèi guān and 大陵 dà líng points to the 勞宮 láo gōng point at the centre of the palm before ending at the 中衝 zhōng chōng point at the tip of the middle finger. The 勞宮 láo gong point is also very important in our practice as it is the centre for releasing energy from the hands.

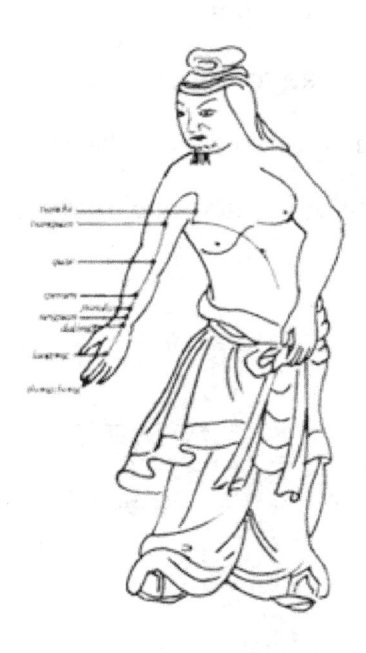

Figure 2-14: The Pericardium Meridian of Hand-Jueyīn

10. The San jiao Meridian of Hand-Shǎoyang 手少阳三焦经

This is an organ that can be described as having a name but no form. It has the purpose of co-ordinating all of the water metabolism. The Lower Burner covers the lower abdomen, and controls the functions of the kidneys and urinary bladder. The Middle Burner covers the area between the chest and the navel, and controls the functions of the stomach, liver and spleen. The Upper Burner includes the chest, neck and head, together with the functions of the heart and lungs.

The meridian commences at the 關衝 guān chōng point at the tip of the ring finger. It then moves across the arm to join the Du 14 point at the 大椎 - dà zhuī. It then ascends across the shoulder to the clavicle and moves through the chest. It thereupon joins with the pericardium, and moves through the diaphragm to the three jiao – the upper, middle and lower. One part branches and ascends through the neck to the posterior border of the ear before crossing through the face and ending below the eye. Another branch commences behind the ear and goes through the ear before joining the other branch on the face and through to the lateral angle of the eye.

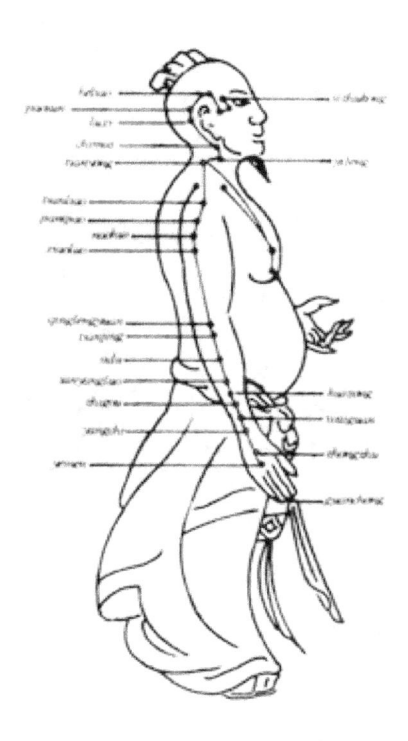

Figure 2.15 The San Jiao Meridian of Hand-Shǎoyang

11. The Liver Channel of Foot-Jueyin 足厥阴肝经

The liver is a Yin organ and paired with the gall bladder, which is Yang. The element is Wood, the direction is East and the season is Spring. The related colour is green. The main responsibility of the liver is to regulate 气 qì throughout the entire body.

Commencing from the 大敦 dà dūn point, which is situated at the tip of the big toe, the median ascends through the instep of the foot and then through the medial side of the leg to the pubic region. From there, it passes through the genitals and the lower abdomen and continues on through to the gall bladder. Continuing through the diaphragm it passes through the ribcage and then ascends to the throat and on to the eyes. It then ascends to the forehead, where it joins the 督脉 dū mài. One branch commences at the liver and ascends to the diaphragm and on to the lungs. Another branch commences at the eyes and descends through the cheeks to the inside of the lips.

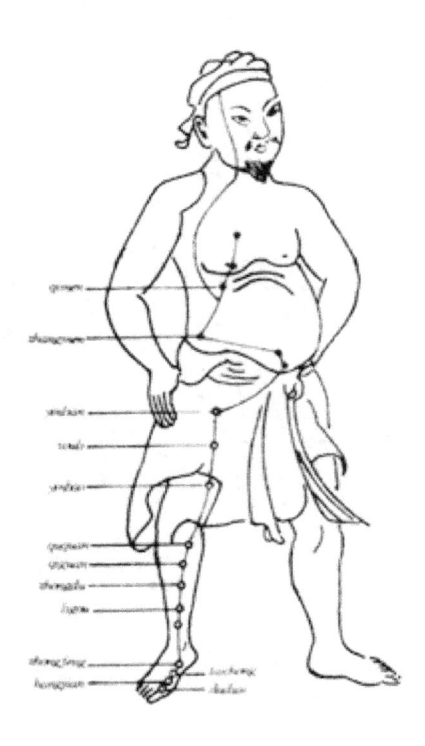

Figure 2.16: The Liver Meridian of Foot-Jueyin

12. The Gall Bladder Channel of Foot-Shǎoyang 足少阳胆经

The gall bladder is Yang and is an organ paired with the liver. The element is Wood, the direction is East, and the season is Spring. The related colour is green. The main function of the gall bladder is to both store and excrete the bile produced by the liver.

Commencing at the 瞳子髎 tóng zǐ liáo point of the eye, the meridian ascends through the temple to the back of the ear. It then ascends through the shoulder to meet with the 督脈 dū mài Du meridian at the 大椎 dà zhuī. One branch commences from the back of the ear and crosses over and into the eye. Another branch starts at the eye and then descends to the neck and the chest and through the diaphragm. There it joins the liver and gall bladder before moving through the side of the lower abdomen, passing through the pubic region and on to the hip region. Another branch starts at the clavicle and descends through the armpit and through the side of the chest and the 日月 rì yuè point. It then descends through the thigh and the knee to the 丘墟 qiū xū point near the ankle and ends at the 足竅陰 zú qiào yīn point on the tip of the toe.

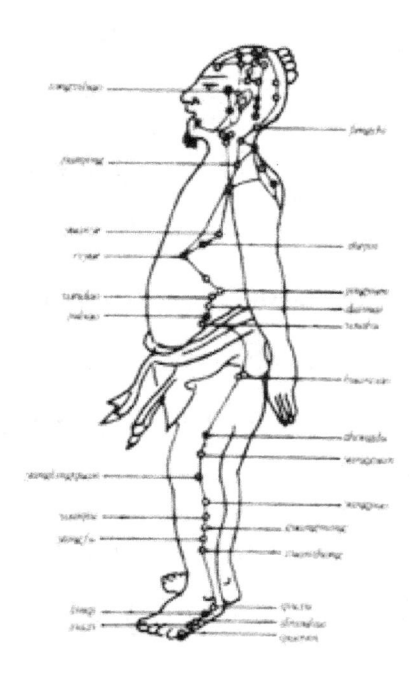

Figure 2-17: The Gall Bladder Channel of Foot-Shǎoyang

We appreciate that the above topic covering the 12 meridians may be somewhat of a dry subject for most of you. It has been included merely for the sake of completeness and for reference purposes. It is a very deep subject and one in which doctors qualified in traditional Chinese medicine need to undertake many years of study to be able to understand and effectively treat patients.

The Eight Extraordinary Meridians 奇经八脉 Qì Jing Ba mài

The eight extraordinary meridians, which are additional to the twelve that we have already discussed and known as 奇经八脉 qì Jing Ba mài, are of major importance to 气功 qìgōng, 太极拳 tàijíquán and Chinese 內丹, nèi dān, Internal Alchemy. The word 奇 qí in this context is a different word to the 气 qì we have already encountered – as you will see in the difference in the characters. In this context, it means odd, strange or mysterious; 經 jīng means meridian or channels; 八 bā means eight, (as the linguists amongst you will have no doubt grasped); and 脈 mài means vessels.

奇经八脉 qì Jing Ba mài is thus translated as *Odd Meridians and Eight Vessels* or as *Eight*

Extraordinary Meridians. Other names for these 奇精八脉, qì Jīng Bā mài include the Eight Curious Vessels, the Eight Marvellous Meridians, and the Eight Irregular Vessels. As you may have guessed, the number is derived from the eight trigrams of the Yijing.

The Eight Extraordinary Vessels are the conduits of 元气 yuán qì or original 气 qì and Jīng 精 and are the first to form in the body. They function as deep reservoirs from which the twelve main meridians can be replenished, and into which the latter can drain their excesses.

All extraordinary vessels are in some way connected to the kidney meridian system. This is considered to be the primary root of the physical body. It is akin to a tree with the kidneys as the roots, and with the extraordinary vessels forming the trunk. The twelve ordinary meridians correspondingly form the limbs and branches of the tree. We will extend this tree analogy when we move on to the subject of 站桩 Zhan Zhuang – Standing Like a Tree in a later chapter and which we liken to the sap. Ultimately, therefore, the functions of the vessels are used for the storage, circulation and distribution of energy.

The eight extraordinary vessels are 奇經八脈 qí jīng bā mài

1. 任脈 Rèn mài – Conception Vessel
2. 督脈 Dū mài – Governing Vessel
3. 冲脈 Chōng mài – Penetrating Vessel
4. 帶脈 Dài mài – Girdle Vessel
5. 阴维脈 Yīn wéi mài – Yin linking vessel
6. 阳维脈 Yáng wéi mài – Yang linking vessel
7. 阴蹻脈 Yīn qiāo mài – Yin Heel Vessel
8. 阳蹻脈 Yáng qiāo mài – Yang Heel Vessel

For 太极拳 tàijíquán, 氣功 qìgōng and Inner Alchemy practices, the four most important of the 奇精

八脉, qì Jing Ba mài are the the 任脉 Rèn mài – Conception Vessel, the 督脉 Dū mài – Governing Vessel, the 冲脉 Chōng mài – Penetrating Vessel and the 帶脉 Dài mài – Girdle Vessel. Traditional Chinese medicine concentrates on the the twelve primary channels and two of the eight vessels – the Governing,督脉 Dū mài and the Conception, 任脉 Rèn mài – Ren vessels.

The 奇經八脉 qí jīng bā mài Eight Extraordinary Vessels are divided into two components, each of four vessels: a primary set and a secondary.

The four primary ones are:

- 督脉 Dū mài – Governing Vessel – the sea of yang
- 任脉 Rèn mài – Conception Vessel – sea of yin
- 帶脉 Dài mài – Girdle Vessel/Belt Channel – holding the energy of the body
- 冲脉 Chōng mài – Penetrating Vessel/Thrusting Channel – the primordial channel of twelve primary meridians

The four secondary ones that traverse the arms and the legs are:

The four transporters:

- 阳蹻脉 Yáng qiāo mài – Yang Heel Vessel/Yang Bridge – The most yang
- 阴蹻脉 Yīn qiāo mài – Yin Heel Vessel/Yin Bridge – The second most yin
- 阳维脉 Yáng wéi mài – Yang linking Vessel – The second most yang
- 阴维脉 Yīn wéi mài – Yin linking vessel – The most yin

All of the Eight Vessels can be utilised and the energy circulated through 气功 qìgong and 太极拳 tàijíquán practice. The Eight Extraordinary Vessels work in conjunction with the three 丹田 dān tián, or 3 Dantians.

The functions of the Eight Extraordinary Vessels or Meridians are as follows:

Serving as 气 qì Reservoirs

Using water as a metaphor, Biǎn Què 扁鹊,[51] who was a famous physician and who lived around 500 BCE, stated that the twelve organ-related 气 qì channels or meridians can be described as rivers, and that those of the eight extraordinary vessels can be considered to be reservoirs. This is especially so as regards the 督脉 Dū mài and 任脉 Rèn mài, both of which perform the function of redistributiing 气 qì, and absorbing any excesses and thereupon distributing where the body is deficient.

All eight of the vessels contribute to this activity of rebalancing the 气 qì. This store of 气 qì can be accessed through either the practice of 气功 qì gōng or 太极拳 tàijíquán working with the Eight Extraordinary Vessels or by using acupuncture or 推拿 tuī ná (a form of massage) to directly connect the eight vessels to the twelve channels.

Regulating the Changes of the Life Cycles

Chapter 1 of The 黄帝内经 Huáng dì nèi jīng or the Yellow Emperor's Classic of Medicine states that the Thrusting Conception vessels regulate the changes of the life cycles of men and women, which are seven and eight years long respectively. This is through the circulation in the 任脉 rèn mài 督脉 dū mài described below.

In his discourse with Qíbó 岐伯,[52] the latter stated

In general, the reproductive physiology of woman is such that at seven years of age, her kidney energy becomes full, her permanent teeth come in, and her hair grows long. At fourteen years, the

---

[51] According to legend, he was the earliest known Chinese physician. He died around 310 BCE
[52] A mythological doctor, employed as a minister by Huáng dì

tian kui, or fertility essence matures, the ren/conception and chong/vital channels responsible for conception open, menstruation begins, and conception is possible. At twenty-one years, the kidney energy is strong and healthy, the wisdom teeth appear, and the body is vital and flourishing. At twenty eight years, the bones and tendons are well developed and the hair and secondary sex characteristics are complete. This is the height of female development. At thirty five years, the yangming/stomach and large intestine channels that govern the facial muscles begin to deplete, the muscles begin to atrophy, facial wrinkles appear, and the hair begins to thin. At forty-two, all three yang channels Taiyang, shaoyang and Yangming are exhausted the entire face is wrinkled, and the hair begins to turn grey. At forty-nine years, the ren and chong channels are completely empty, and the tien kui has dried up. Hence, the flow of the menses ceases and the woman is no longer able to conceive.

In the male, at eight years of age, the kidney energy becomes full, the permanent teeth appear, and the hair becomes long. At sixteen years of age, the kidney energy is ample, the tien ku is mature, and the jing is ripe, so procreation is possible. At twenty-four years, the kidney qi is abundant, the bones and tendons grow strong, and the wisdom teeth come in. At the thirty-second year, the body is at the peak of strength, and the functions of the male are at their height. By forty, the kidney qi begins to wane, teeth become loose, and the hair starts to fall. At forty-eight the yang energy of the head begins to deplete, the face becomes sallow, the hair grays, and the teeth deteriorate. By fifty-six years, the liver energy weakens, causing the tendons to stiffen. At sixty-four, the tian kui dries up and the jing is drained, resulting in kidney exhaustion, fatigue and weakness. When the energy of the organs is full, the excess energy stored in the kidney is excreted for the purpose of conception. But now, the organs have aged and their energies have become depleted, the bones and tendons have become frail and stoff, and movements are hampered. The kidney reservoir becomes empty, marking the end of the power of conception.[53]

---

[53] The Yellow Emperor's classic of medicine. Trans. Maoshing Ni

Circulating 气 qì to the Entire Body, particularly the Five Ancestral Organs

One of the most important functions of the eight vessels is to deliver 气 qì to the entire body, including the skin and hair. Hence, in traditional Chinese medicine, the lustre of the hair and the sheen of the skin are considered to be a sign of the health of the kidneys.

气 qì is also transported to what is known as the five ancestral organs. These are the brain and the spinal cord, the liver and gall bladder, the bone marrow, the uterus and the blood system.

Guarding Specific Areas against Evil 气 qì

卫 气 wèi qì or Guardian 气 qì protects the body from outside pathogens and diseases. The Thrusting Governing Conception vessels all have major roles in guarding the abdomen, the thorax and the back from disease.

河 图 Hétú known as the River Map and Luò shū 洛书 diagrams

In association with Lǎozǐ's verse 42 from the 道德经 Dào dé jīng mentioned earlier is the Hétú 河图, known as the River Map, and 洛书 Luò shū diagrams. These diagrams are representations of energy combinations, based on yin and yang, and underpinned by numerical calculations. Hence, all methods are ways to describe the manifestation of energy, whether by prose or by diagrams or by numbers or in combinations thereof. The 河图 Hétú and 洛书 Luò Shù are prominent with regard to the 奇經八脈 qí jīng bā mài, Eight Extraordinary Channels and are believed to have gained prominence from the Yi Shu Yin Pictures by Liu Mu during the Song dynasty.

To remind ourselves, Verse 42 states the following:

The Dao produces the One,

The One gives birth to the Two,

The Two generate Three,

The three gives birth to the Ten Thousand Things,

The Ten Thousand beings hold yin on their backs,

Embrace yang to their chest,

And amalgamate the two qi to achieve harmony.

Therefore, we can observe the numerical theme to expressions of energy and its relationship to 內丹, nèi dān, Internal Alchemy, which is based on the expansion of binary numbers. We shall examine this further.

The 河图 Hétú is arranged as:

Six black and one White on the Bottom,

Two Black and Seven White on the Top,

Three Black and Eight White on the Left,

Four Black and Nine White on the Right.

Fu Xi drew his Eight Trigrams based on this arrangement.

The 奇經八脈 qí jīng bā mài are arranged together as follows:

South – Fire – 2 Yang dots; 7 Yin – (7-2=5) –陽蹻脈 yáng qiāo mài and 督脈 dū mài

West – Metal – 4 Yang dots; 9 Yin – (9-4=5) –陽維脈 yáng wéi mài and 帶脈 dài mài

North – Water – 1 Yang dot; 6 Yin – (6-1=5) –衝脈 chōng mài and 陰維脈 yīn wéi mài

East – Wood – 3 Yang dots; 8 Yin – (8-3=5) –陽蹻脈 yáng qiāo mài and 任脈 rèn mài

Centre – Earth – Primordial – 5 dots (5)

All combinations of the numbers relate to 5.

The numbers and relationship to the Five Phases are illustrated below:

Table 2-5: Five Phases and the Vessels

Numbers	Direction	Element	Primary Vessel	Secondary Vessel
1 and 6	North	Water	冲脉 chōng mài	阴维脉 yīn wéi mài
2 and 7	South	Fire	督脉 dū mài	阴跷脉 yīn qiāo mài
3 and 8	East	Wood	任脉 rèn mài	阳跷脉 yáng qiāo mài
4 and 9	West	Metal	带脉 dài mài	阳维脉 yáng wéi mài
5 and 10	Centre	Earth		

Figure 2-18: Luò Shù and Eight Extraordinary Meridians

阳维脉 yáng wéi mài South East 4	带脉 dài mài South 9	阳跷脉 yáng qiāo mài South West 2
阴跷脉 yīn qiāo mài East 3	无极 wú jí Centre 5	督脉 dū mài West 7
任脉 rèn mài North East 8	阴维脉 yīn wéi mài North 1	冲脉 chōng mài North West 6

The difference between each number is always 5, which corresponds with the 5 in the centre.

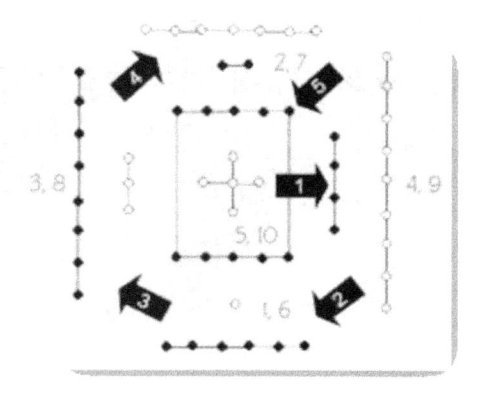

Figure 2-19: Flow of the numbers combinations

As we move in a clockwise direction starting from the centre, which is Earth, we can observe the flow of the reproductive Elements Phase. That is:

Earth [5,10] produces Metal [4,9]

Metal produces Water [1,6]

Water nourishes Wood [3,8]

Wood nourishes Fire [2,7]

Fire produces Earth.

This continues in an endless virtuous cycle.

For each and every combination, there is one yang number that is odd. This is represented by white dots. One yin, as an even number, is shown as black dots. For every opposite yang number, there is a corresponding yin number and, of course, the opposite is true. This diagram depicts the mutual interdependency of yin and yang and their effect on the other in an endless self-generating and sustaining cycle.

On the 河图 Hétú diagram, the odd numbers are used to denote yang or male and the even numbers that of yin or female. Hence, the numbers are arranged oppositely to produce a synthesis from the central number 5. This number 5 denotes the human being at the centre – between Heaven and Earth.

The Luò shū 洛书 River Map, also known as the Magic Square, was created by the ruler Yu the Great and is believed to be the oldest of its type in existence. Studying the diagram, you will notice that all of the three numbers of any line add up to 15, and hence this is the source of the term 'Magic Square'. The Magic Square in ancient times was used to produce calculations of daily and seasonal cycles.

Both diagrams are mentioned in the 易经 Yì jīng commentaries and the Magic Square is referred to by Zhuāng zi with regard to its application to 內丹, nèi dān, Internal Alchemy. These cosmic number diagrams were later converted into the 易经 Yì jìng Early Heaven 先天 Xiān tián and Later Heaven 后天 Hòu tián trigrams. This provides our link to the 易经 Yì jìng.

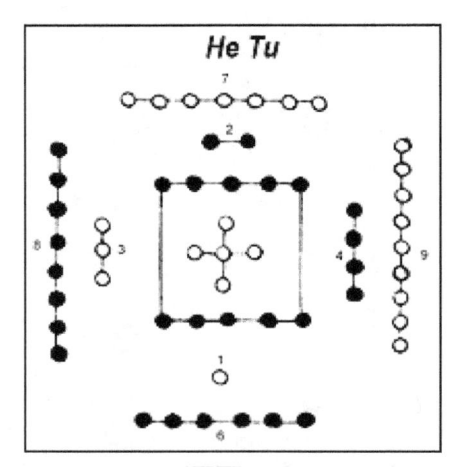

Figure 2-20: The 河图 Hétú Arrangement

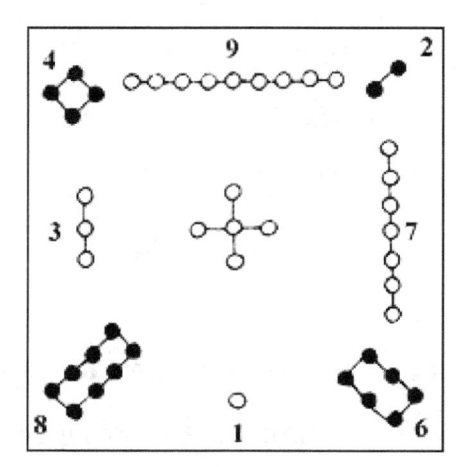

Figure 2-21: The 洛书 Luò Shù Arrangement

The pairings based on the diagram above show the classic pairings of the 奇經八脈 qí jīng bā mài Eight Extraordinary Channels depicting the energy network within the microcosm (the human body) as aligned with the macrocosm (the Universe), moving from 无极 wú jí to yin and yang to the Four Forces and the Eight Trigrams: 2 moving to 2^2 moving to 2 to 4 to 8.

1. The 陰維脈 yīn wéi mài is placed in the North
2. The 陽蹻脈 yáng qiāo mài is placed in the South West
3. The 陰蹻脈 yīn qiāo mài is placed in the East

4. The 陽維脈 yáng wéi mài is placed in the South East
5. The centre is the base for all of the vessels and is empty
6. The 衝脈 chōng mài is placed in the North West
7. The 督脈 dū mài is placed in the West
8. The 任脈 rèn mài is placed in the North East
9. The 帶脈 dài mài is placed in the South

Once the 奇經八脈 qí jīng bā mài Eight Extraordinary Vessels are superimposed upon the Luò Shù based on the sequence listed, they are paired according to the 河图 Hétú combinations. Note the 河图 Hétú combinations form the classic coupled pairings of the 奇經八脈; qí jīng bā mài Eight Extraordinary Vessels.

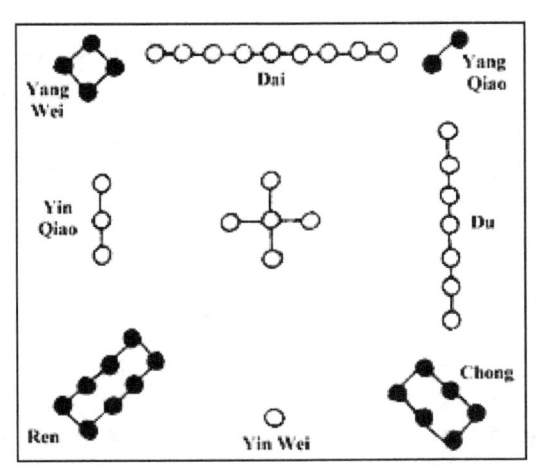

Figure 2-23 Luò Shù and 8 Extraordinary Vessels

There are specific points – referred to by their acupuncture names – that can be used to stimulate the energies of the Eight Extraordinary Vessels. These are listed below:

- Lu7 opens up the 任脈 rèn mài
- Kd6 opens up the 陰蹻脈 yīn qiāo mài
- SI3 opens up the 督脈 dū mài
- Sp4 opens up the 衝脈 chōng mài
- Ub62 opens up the 陰蹻脈 yīn qiāo mài

- Pc6 opens up the 陰維脈 yīn wéi mài

- Sj5 opens up the 陽維脈 yáng wéi mài

- Gb41 opens up the 帶脈 dài mài

The following acupuncture points are connected as pairs:

- SI3 and UB62
- LU7 and KD6
- PC6 and SP4
- SJ5 and GB41

Now let us explore in more detail each of the Extraordinary Vessels.

1. 督脈 dū mài – The Governing Vessel -兌 duì ☱

The Trigram relating to this vessel is the symbol of a lake or a marsh and represents the emotion of Joy. The 督脈 dū mài is also known as the Sea of all Yang Meridians. It is also sometimes referred to as the confluence of all the yang channels, of which it is the governor. Hence, it is most usually translated into English as the Governing Vessel. This vessel is one of the two meridians, and is paired with the 任脈 rèn mài to form the Small Heavenly Circle or Microcosmic Orbit. This is a movement of energy in a circle at the back and front of the upper torso.

The function is to maintain the fire inside the body and it controls the loss of body heat. It is inextricably linked with the fluids that flow outside of the channels, in the skin and flesh. The 督脈 dū mài has 27 points. As the governor of all of the yang channels, it is able to increase the yang energy of the whole body.

The 督脈 dū mài commences within the lower abdomen or 下 丹田 xià dān tián and moves downwards through to the 会阴 huì yīn point or perineum before ascending inside the spinal column on to the nape of the neck. This is the acupuncture point called 风府 – fēng fǔ, wind. It then enters the brain and moves on to the vertex across the forehead and then descends to the nose bridge on to the lips and enters the labial frenulum inside of the upper lip. The 会阴 huì yīn is a major energy point for our practices. It relates to the perineum between the anus and the genitals and corresponds to the root chakra in Indian yoga esoteric systems. A slight contraction in the 会阴 huì yīn prevents energy leakage and acts as a pump to circulate the cycle through the Microcosmic Orbit, also known as the Small Heavenly Cycle.

It represents the fire of the body as it is yang 气 qì. It governs the back, which is the area most abundant in 卫气 wèi qì. It circulates the 卫气 wèi qì in order to guard against external pathogens or diseases. With regard to the circulation of 卫气 wèi qì, the process commences at the acupuncture point known as Fēng fǔ 风府 (Gv-IG), and descends to the Governing Vessel to the 会阴 huì yīn (LI-I), the perineum. It takes a monthly cycle of twenty one days from the flow from Fēng fǔ 风府 to the 会阴 huì yīn, and nine days from the 会阴 huì yīn to the throat.

The 气 qì flows through a pathway up the midline of the back and through all of the yang meridians. This excludes the stomach meridian, which flows along the front,

There are four major points along the vessel. These are:

The 尾闾 wěi lǘ that is located on the tail bone. It is the beginning of the 督脉 dū mài channel, and is known as the 'Rising Yang Point'. Once it is opened, the yang 气 qì will rise and nourish and heat up all of the body.

The Jia ji is located in the middle point of the twenty-four joints of the spine. It connects the upper and lower spine. Once this point in energetically opened, a person can enjoy longevity. Hence, it is also known as 神道 shén dào – the spiritual path.

The **Yu zhen** are two points that connect the spinal cord with the brain. Once these have been opened, the body and the Shen will be unified.

The **Ni wan** is also known as the called Tian men or the Gate of Heaven, and once opened communication outside of the body is possible. It enables the Shen to travel outside of the body.

The master point of the 督脉 dū mài is the small intestine point 3. It is coupled with the 陰蹺脈 yīn qiāo mài, whose master point is the bladder point 62.

The Governing Vessel is also responsible for nourishing the five ancestral organs, as described previously. These include the brain and spinal cord. Hence, it is stated in traditional medicine that the kidneys control the brain.

As the 督脉 dū mài Du meridian runs through the back of the body, the five Zang organs, the Six Fu-organs and the twelve ordinary meridians are all dependent on it. This means that it needs to be operating in a relaxed state for it to be functioning optimally. The 内经 Nèi jīng, the Internal Classic, states that: 'To treat one hundred kinds of illnesses, start with treating the back'.

The 督脉 dū mài, together with the 任脉 rèn mài, is considered to be the one of two most important 气 qì channels to be trained in the commencement of 內丹, nèi dān, Internal Alchemy cultivation. The objective in 太极拳 tàijíquán is to fill these two vessels with 气 qì so that circulation through both vessels can be achieved. This is known as the Small Heavenly Circulation. This is how energy circulated when we were in our mother's womb. This process is achieved by converting the Jing stored in the kidneys into 气 qì, and then circulating the 气 qì in the 督脉 dū mài and 任脉 Rèn mài channels. By leading this 气 qì to the head the Shen (spirit) will be nourished.

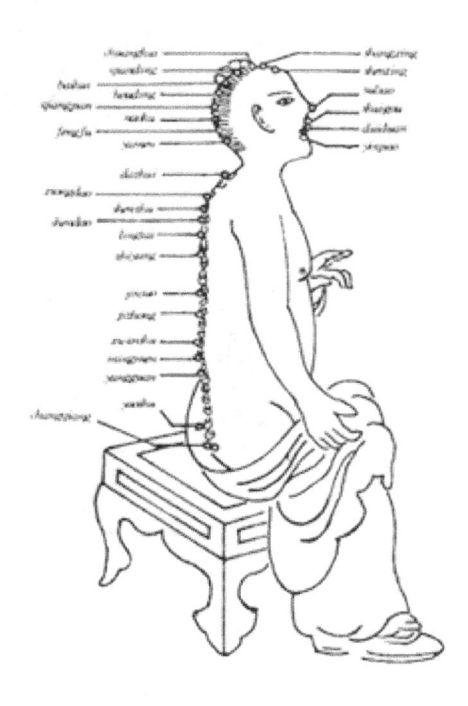

Figure 2-24: The 督脈 dū mài

2. The Conception Vessel – Rén 脈 mài 离, lí ☲

The corresponding trigram is 離 Lí, known as the Clinging. The 任脈 rèn mài is the Sea of Yin. It monitors and directs all of the Yin Channels, plus the stomach channel. The 任脈 rèn mài flows from the 会阴 huì yīn – perineum up along the front mid-line of the body and ends in the lower mouth. Together with the 督脈 dū mài vessel, it forms the Small Heavenly Cycle, described above.

The 任脈 rèn mài at the 会阴 huì yīn perineum before ascending to the public region. It then moves along the internal abdomen and ascends through the 關元 guān yuán to the throat and there ascends further to circle around the lips, passing via the cheeks and emerging into the infraorbital region.

As previously mentioned, the 督脈 dū mài and 任脈 rèn mài control the life cycles of men and women of seven and eight years respectively. The changes in these two vessels over time cause the ageing that occurs in our lives.

The master point for the 任脈 rèn mài is the acupuncture point Lung 7 and, therefore, is associated with respiration. This is coupled with acupuncture point Kidney 6, on the 陰蹻脈 yīn qiāo mài vessel. It receives and transports the 气 qì of all the yin meridians. It regulates the female functions, such as menstruation, menopause and pregnancy.

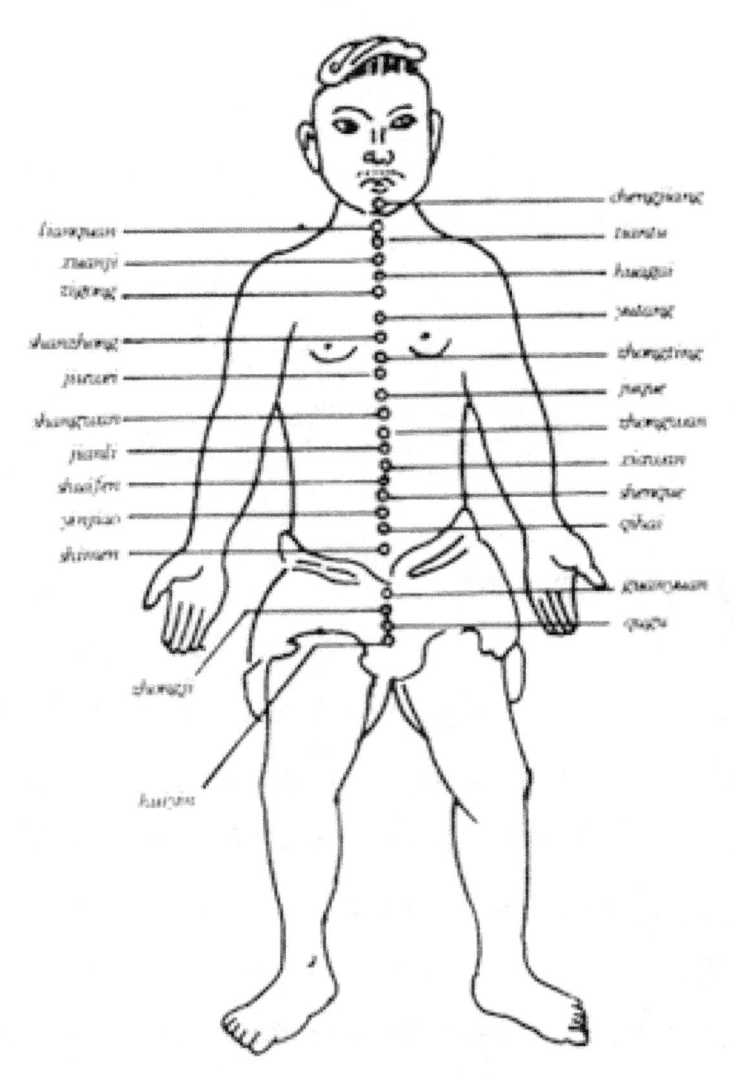

Figure 2-25: The 任脈 rèn mài

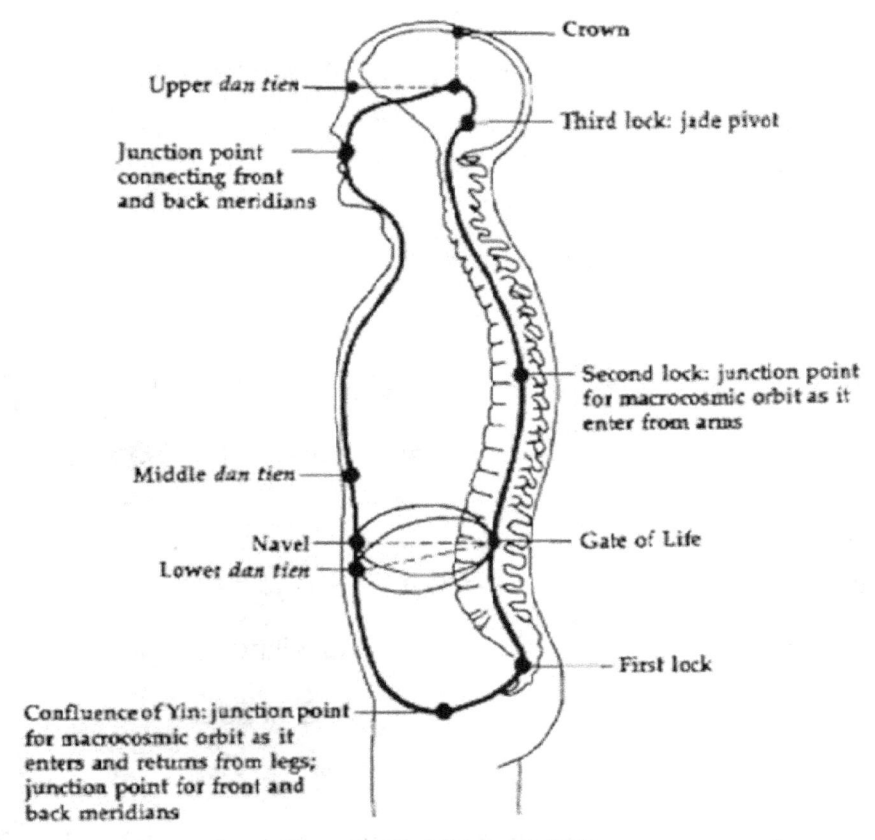

Figure 11.1 The Microcosmic Orbit

Figure 2-26: The Microcosmic Orbit or the Small Heavenly Cycle

The Small Heavenly Circle or Microcosmic Orbit comprises eleven points and courses up the back 督脈 dū mài vessel and down the front, down the 任脈 rèn mài Vessel.

The 任脈 rèn mài points are as follows:

1. **下 丹田 Xià dān tián – the lower 丹田 dān tián**: It is located about three finger-widths down from your navel and resides within the inside of the body. It is at the level of your centre of gravity, which is between the fourth and fifth lumbar vertebrae. It is also known as the hara in Japanese and is a well-known energy centre within other traditions.

2. **会阴 Huì yīn – perineum**: The point is situated on the pelvic floor between the anus and the genitals. It is the point at which the 任脈 rèn mài, 督脈 dū mài and 衝脈 chōng mài vessels cross. It is a point that can be activated by gentle pressure led by the mind.

3. 尾闾 Wéi Lǔ: This is situated at the DU Meridian Point 2. The point is located within the space between the sacrum and the coccyx.

4. 命门 Mìng mén – Gate of Life, situated at the DU Meridian Point 4: This is located at the midline, between L2 and L3, at the inflection point of your lower back. This is a key point to open during the 太极拳 tàijíquán practise. It is crucial for the development of 坎 Kǎn and 離 Lí that we covered earlier as exchange of water and fire to produce steam that will activate the 气 qì in your body.

5. 至阳 Zhì yáng situated at the Du Meridian Point 9: This is located on the spine, at the level of the inferior angle of the scapula when you are standing or siting, below the spinous process of the seventh thoracic vertebra, level to the inferior angle of the scapula

6. 大椎 Dà zhuī, the big bone of the lower neck: This is situated at the Du Meridian at point 14. This is located at a point that is just interior and below the spinous process of C7 at the base of the neck, approximately at the level of the shoulders in the depression below the spinous process of the seventh cervical vertebrae.

7. 风府 Fēng fǔ – the Jade Pillow, situated at DU Meridian Point 16: This is located on the midline, in the depression just below the occiput at the base of the skull. This is a point that is difficult to open and can impede the flow of qi up the Du channel.

8. 百会 Bǎi huì known as the One Hundred Meetings, situated at Du Meridian Point 20: The hundred denotes many rather than being a literal number. It is the highest point at the top of your head. It also corresponds to the crown chakra in Indian yoga.

9. 印堂 Yin Tang known as The Hall of the Seal: This is located at the midpoint between the medial ends of the eyebrows.

Bridging point:

10.   **Que qìao – known as the Magpie Bridge, transition point**: This is located on the hard palate, where the tip of the tongue touches the roof of your mouth with least effort. It is an important point as it joins the 督脈 dū mài and 任脈 Rèn mài and in 站桩 zhàn zhuāng will cause the saliva to develop. It contains important enzymes that should be swallowed into the abdomen.

The last 任脈 Rèn mài Point

11.   膻中 Shān Zhōng – known as the Heart Centre, on the 任脈 Rèn mài meridian point is located on the sternum midline, in the depression level with the fourth rib at the level of the nipple.

Figure 2-27: The main points on the Microscopic Orbit

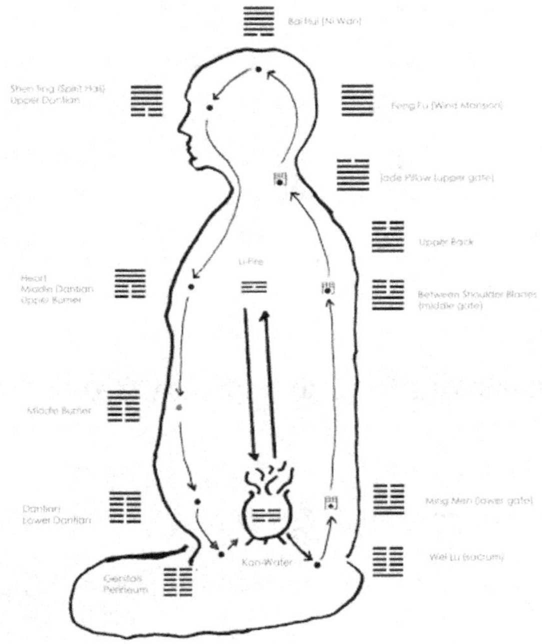

The Waxing and Waning of Qian and Kun (Yin and Yang)
Qian - The Opening of the Gates - Inhalation
Kun - The Closing of the Gates - Exhalation

Key points:

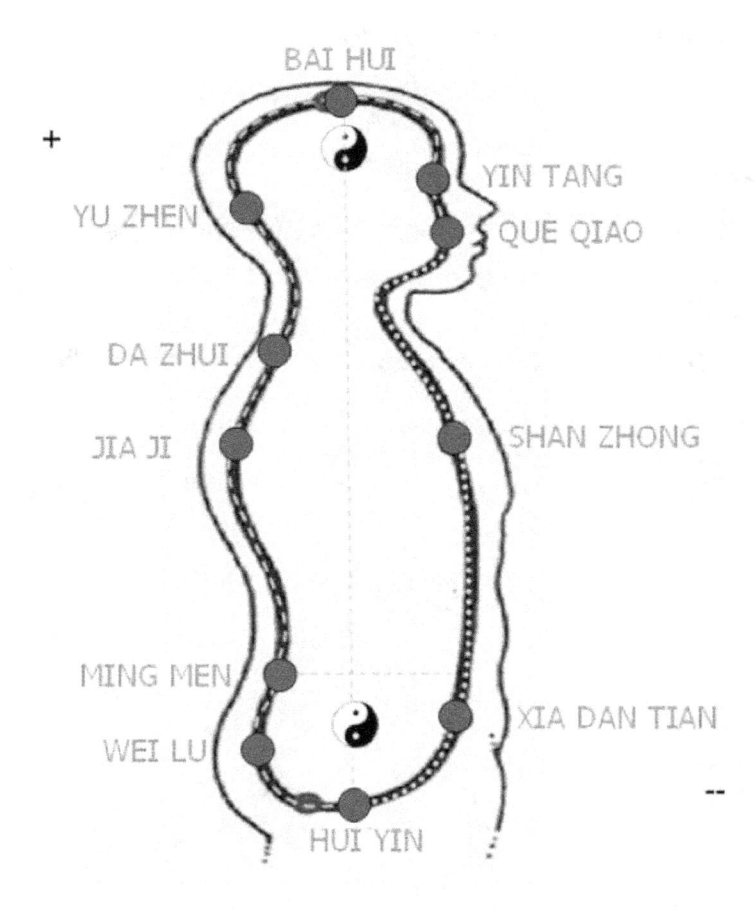

Figure 2-28: Circulation of energy through the 督脈 dū mài and Rén vessels

3. The Thrusting Vessel –衝脈 [chōng mài] - ☰, 乾 qián

This is known as the sea of blood, arteries and the twelve primary channels, yuán 气 qì, and is the blueprint of life.

The Trigram is Heaven. This is the most Yang of the Trigrams. The 衝脈 chōng mài connects to and communicates with the 任脈 Rèn mài, and both mutually regulate the 气 qì in the kidney channel. The kidneys, one of the most vital Yin organs, store the 元气 yuán qì - the Original 气 qì. The 衝脈 chōng mài flows vertically deep within the body, along the front of the spine. The 衝脈 chōng mài has a close resonance with Sushumna Nadi, described in Hindu yogic traditions. It is our energetic core.

The 衝脈 chōng mài commences in the uterus, branching along three channels: the first goes along the posterior wall of the abdominal cavity, ascending within the spinal column; the second ascends to the umbilicus along the anterior wall of the abdominal cavity and circulates in the chest, upwards to the throat and around the lips; the third branch descends to the perineum along the medial aspect of the thigh, where it stops at the big toe.

The master point of the 衝脈 chōng mài is Spleen 4, which is combined with Pericardium 6 on the *yin wéi 脈 mài* channel. This pair affects the heart, chest and stomach. This 'vital passage' regulates the flow of 气 qì and blood in the twelve regular meridians, and is significant in gynaecological disorders, digestive issues, prolapses and problems with the heart. Energetically, it relates to intergenerational patterns, issues arising from abuse and cellular memory. Psycho-spiritually, an imbalance in the 衝脈 chōng mài negatively impacts our self-acceptance and self-love.

The Thrusting vessel is considered one of the most important and decisive vessels in successfu 气功 qìgong training, especially in Marrow Washing. There are many reasons for this. The first reason is that this vessel intersects two cavities on the Conception vessel: Huiyin (LI-I) and Yinjiao (LI-7). Huiyin means 'meeting with Yin' and is the cavity where the Yang and Yin 气 qì are transferred. Yinjiao means 'Yin Junction' and is the cavity where the Original 气 qì (Water 气 qì, or Yin 气 qì) interfaces with the Fire 气 qì created from food and air. The Thrusting Vessel also connects with eleven cavities on the kidney channel. The kidney is considered the residence of Original Essence (Yuan Jing), which is converted into Original yuán qì 元氣.

The second reason for the importance of the Thrusting Vessel in 气功 training is that this vessel is connected directly to the spinal cord and reaches to the brain. The major goal of Marrow Washing i 气功 is to lead the 气 qì into the marrow and then further on to the head, nourishing the brain and spirit (Shen).

And finally, the third reason is found in actual 气功 practice. There are three common training paths: Fire, Wind and Water. In the Fire path 气功 qìgong, the emphasis is on the Fire or Yang 气 qì circulating in the governing vessel and therefore strengthening the muscles and organs. The Fire path is the 气 qì training in muscle/tendon changing (Yi Jin Jing). 气功 qigong. However, the Fire path can also cause the body to become too Yang, and therefore speed up the process of degeneration. In order to adjust the Fire to a proper level, Marrow Washing 气功 qigong is also trained. This uses the Water path, in which 气 qì separates from the route of the Fire path at the Huiyin cavity (LI-I), enters the spinal cord, and finally reaches up to the head. The Water path teaches how to use Original 气 qì to cool down the body, and then to use this 气 qì to nourish the brain and train the spirit. Learning to adjust the Fire and Water 气 qì circulation in the body is called Kan-Li, which means Water-Fire. You can see from this that the Thrusting vessel plays an important role in 气 qìgong training.

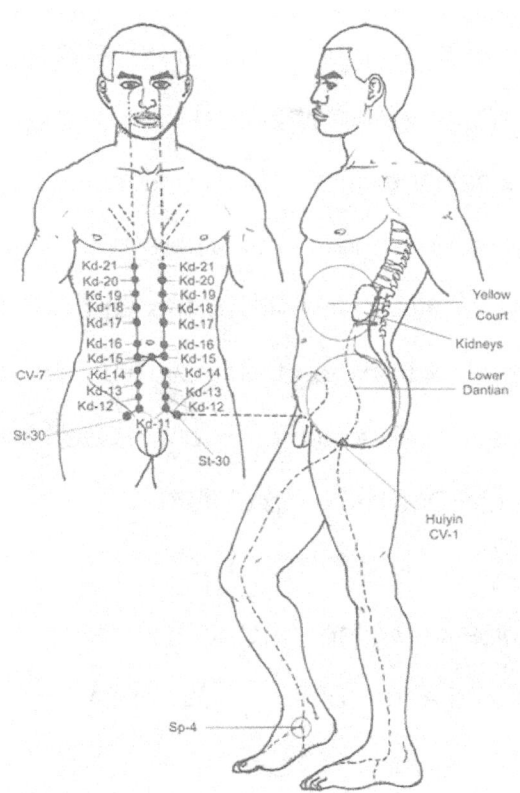

Figure 2-28:

4. The Girdle Vessel – Dai 脈 mài - 巽 xùn ☴

The Trigram symbolises the Wind or the Gentle. The purpose of the 帶脈 dài mài is to harmonise and control all of the channels, and it influences the 督脈 dū mài and 任脈 Rèn mài. The 帶脈 dài mài circles the waist and flows horizontally, and is unique amongst the eight 脈 mài in so doing.

The 下丹田 xià dān tián is located within the region of the 督脈 dū mài. It is responsible for the strength of the waist and for maintaining the kidneys' health. If the 气 qì is flowing fully smoothly within the 督脈 dū mài, then problems with the back will not occur. In order to lead **yuán qì** 元气 from the kidneys to the 下丹田 xià dān tián dān tián, must be healthy and relaxed. This means that the 气 qì flow in the waist area must be smooth. The master point of the 帶脈 dài mài is Gall Bladder 41, paired with Triple Heater 5 on the 陰維脈 yīn wéi mài.

The 帶脈 dài mài creates the 气 qì's horizontal balance and hence centres you physically and mentally. It also regulates the 气 qì of the gall bladder.

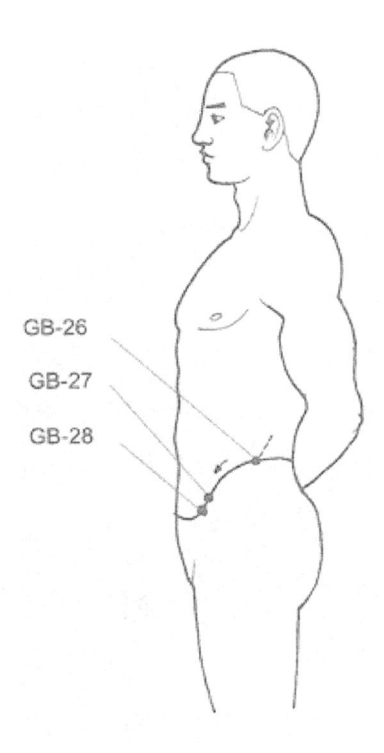

Figure 2-30: The Dai Mai

5. The Yang Heel Vessel 陽蹻脈 yáng qiāo mài ☵, 坎 kǎn

The Trigram is ☵, 坎 kǎn, Water or The Abysmal. The 陽蹻脈 yáng qiāo mài is located in the trunk and legs. It is connected to the 督脈 dū mài. It controls the Yang of the left and right-hand sides of the body and regulates the urinary bladder, the gall bladder, the small intestine and the large intestine.

The Trigram is ☵, 坎 kǎn, Water or The Abysmal. The Yang 气 qìao 脈 mài is located in the trunk and legs. It is connected to the 督脈 dū mài. It controls the Yang of the left and right-hand sides of the body and regulates the urinary bladder, the gall bladder, the small intestine and the large intestine.

Practising 太极拳 tàijíquán, which requires strong legs, converts the fat stored in them, and this is led upwards to nourish the Yang channels.

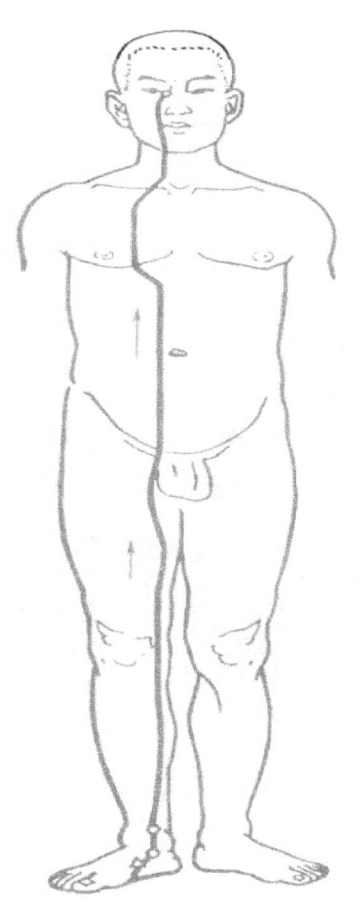

Figure 2-31: The Yang Heel Vessel

6. The Yin Heel Vessel –陰蹻脈 yīn qiāo mài, 坤 kūn

The trigram is ☷, 坤 kūn. It is the most Yin of the Trigrams and represents Earth, the Receptive. The Yin Heel Vessel –陰蹻脈 yīn qiāo mài is paired with the 任脈 rèn mài]. The Yin Heel Vessel –陰蹻脈 yīn qiāo mài is connected with the two cavities of the kidney channel. It is used in the 內功 nèi gōng practice of Bone Marrow Washing. As the kidneys are classified as inner and outer, the testicles, representing the latter in men, are stimulated so as to increase hormone production and increase the conversion of 精 jīng into 气 qì. This can then be led to the head to nourish the brain and Shen in order to reach the Dao or enlightenment.

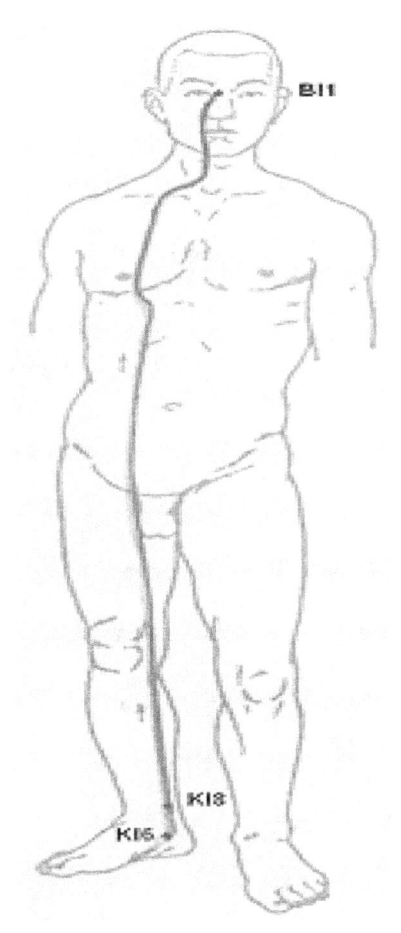

Figure: 2-32: The Yin Heel Vessel

7. The Yang Linking Vessel - 陽維脈 yáng wéi mài – 震 zhèn

The Trigram is ☳, 震 zhèn, the Arousing. The 陽維脈 yáng wéi mài regulates the 气 qì mainly in the Yang channels: the urinary bladder, gall bladder, Triple Burner, small intestine and stomach channels. It is connected with the governing vessel at Yamen (Gv-l5) and 风府 Fēng fǔ (Gv-l6). The opening point of the 陽維脈 yáng wéi mài is Triple Heater 5, which is coupled with Gall Bladder 41 on the 帶脈 dài mài. This combination dominates the exterior of the body.

8. The Yin Linking Vessel 陽維脈 yáng wéi mài - 艮 gèn

The Trigram is ☶, 艮 gèn or the Mountain. The Yin Linking vessel has connections with the kidney, spleen and liver Yin channels. The Yin Linking vessel also communicates with the 任脈 Rèn mài at two cavities. The 陰維脈 yīn wéi mài is paired with the 衝脈 chōng mài as previously stated. This combination permits energetic access to the Neiguan, the inner gate to the self. Psycho-spiritually, this vessel is concerned with the meaning that we derive from life, and helps us to respond to life with clarity and compassion.

The Yinwei channel binds together the six yin channels as it joins with the 任脈 Rèn mài channel. It starts from the medial aspect of the lower leg and runs along the medial aspect of the thigh where it ascends to the abdomen and meets with the Foot-Taiyin channel from where it moves through the chest and joins with the 任脈 Rèn mài channel at the neck.

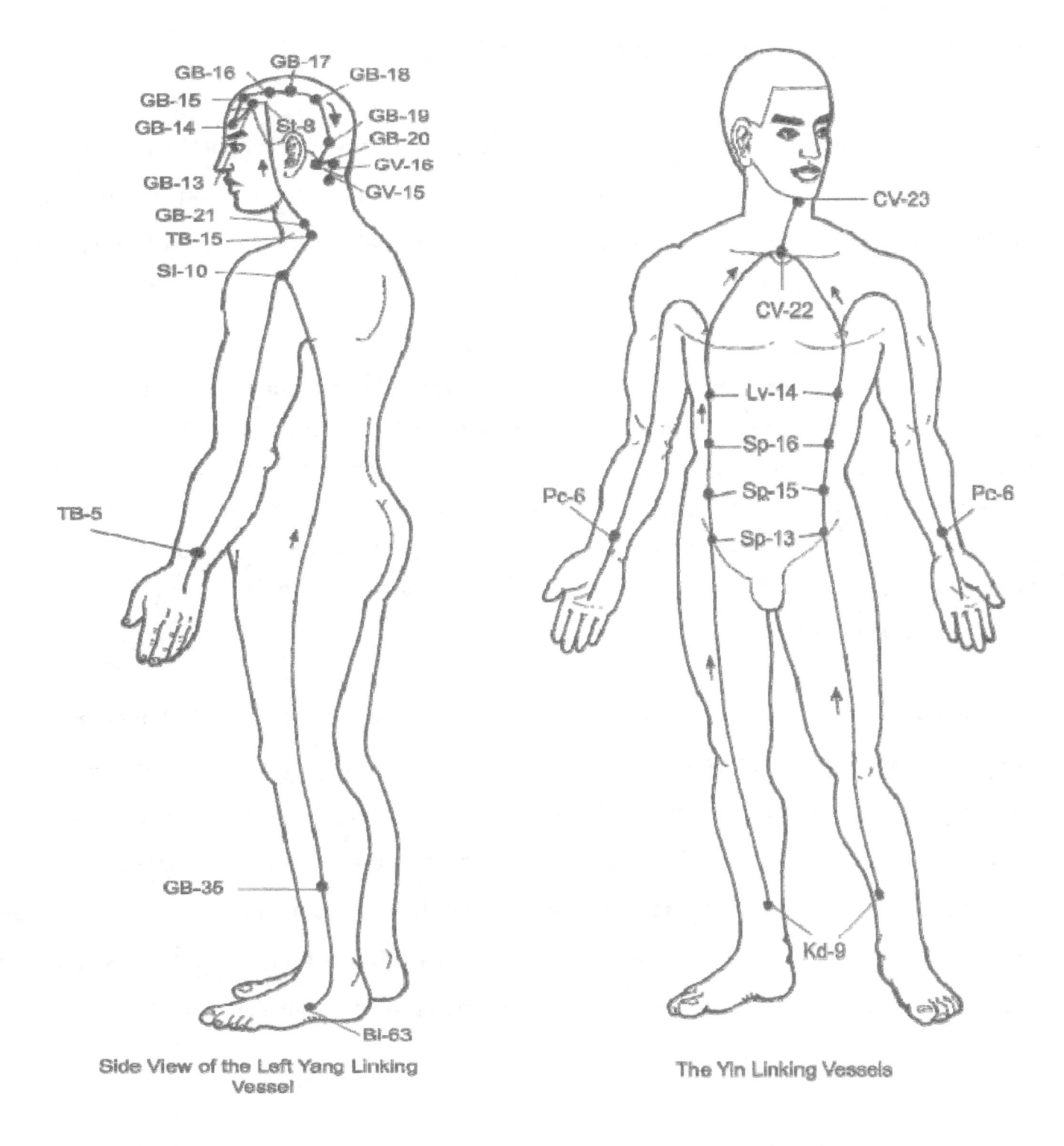

Figure 2-33: Yin Linking Vessels

Figure 2-34: The Yang Linking Vessel

The Kinship of the Three, According to the Book of Changes

魏伯阳 Wei Boyang's seminal work, the 周易参同契 *Zhōu yì Cāntóng qì*, is translated as *The Kinship of the Three, According to the Book of Changes*. It is a fundamental source for our studies into 內丹, nèi dān, Internal Alchemy. Wei Boyang wrote this around 142 AD. The oldest edition in existence is from the Ming dynasty in the sixteenth century.

Wei Boyang uses the terminology of the 易经 Yì Jīng in terms of the trigrams to explain the workings of the process of cultivation and that of the refinement of the three treasures of 精 Jīng, 气 Qi, and 神 Shén. It needs to be noted, however, that the use of the symbols is only relevant within the context of the 內丹, nèi dān, Internal Alchemy process he is describing. In other words, the trigrams are used differently to those within the work of the 易经 Yì Jīng. The terminology also makes use of the metaphor of External Alchemy and the substances and processes relevant therein, for example furnace, vessel and chemicals.

The main terms used from External Alchemy are 炉顶 lú dǐng, which literally translates as the furnace top, and it is symbolised as the reaction vessel to describe the microcosm in the human body and the 药物 yào wù representing the medicine. In the metaphor, this represents the 气 qì of the yin and yang of the original 易经 the Yì Jīng hexagrams that we have encountered previously within our discourse. That is, those of ☵, 坎 kǎn and 离 Lí ☲, the water and fire, retiring and advancing. As this process cannot be seen by the eye or taken out and evaluated like a tangible entity, it is only the adept, the 真人 zhèn rén who is able to intuit this through his higher consciousness.

This is a profound and difficult book to comprehend and is worth studying in its own right. Some of the book will be used to further emphasise the theory and how it relates to practical use.

Chapter One sets the scene:[54]

---

[54] *The Kinship of the Three, According to the Book of Changes*. Trans. Zhou Shiyi

乾 qiàn ☰ and 坤 Kūn mean the two gates for yi or the parents of the family of trigrams, or the walls of 坎 Kǎn ☵ and 离 Lí ☲.

The working of 坎 Kǎn ☵ and 离 Lí ☲坎 is like that of a hub which spins the wheel and holds the axel in place.

The microcosm of man's body might be likened to the Chinese double-acting piston bellows.

The other four trigrams, male (☳ ☶) and female (☴ ☱), symbolise the cylinder of the bellows with 乾 qiàn ☰ and 坤 Kūn ☷ as its two valves.

This describes the movement of yin going and yang coming in through 坎 Kǎn ☵ water and 离 Lí ☲ fire – the yào wú or chemicals circulating through the body.

In Chapter 27, we see the concept of 无为 wú wéi and the eschewing of some of the more contrived Daoist practices. It is a reminder that the earliest Daoist teachings were aimed at achieving a state of simplicity – letting go, of reducing, rather than adding to. The highest levels of 內丹, nèi dān, Internal Alchemy are at the Pre-Heaven Xian Tian level, where cultivation takes place through inner observation, stillness and awareness watching awareness.

It should also be noted that 精 Jing, 气 Qì and 神 Shén exist at both pre-heaven and post- heaven levels. At Pre-Heaven levels there is in fact no separation – even as perceived. Most of 內丹, nèi dān, Internal Alchemy is at the post-heaven level and uses the physical body. Yet students need to start from the point at which they are at, and so ascending to the higher levels is like climbing a mountain. Few are able to ascend in one go. This alluding to cultivation at the level of pre-heaven is stated here:

I am not in favour of the following methods for longevity when;

- *Attention is turned to the organs within man's body, and thoughts are made to throng upon the mind.*

Paces are made in accordance with the movement of the Great Dipper during certain periods such as 六甲 liù jiǎ

- 六甲 liù jiǎ *(six periods of time).*

- *Sexual techniques such as 'Methods of Ninety One' (the so called Left Hand Path of Daoism) are used and the mystic embryo is thrown into disorder.*

- *Air is swallowed only causing noises inside the intestines and stomach, but the vital* 气 *qì will be exhaled and the evil inhaled. Sleepless day and night, the body will be tired out, giving an appearance of idiocy. All kinds of pulses will stir and boil so as to drive away peace of mind and body.*

- *A temple is erected, and pious offerings are morning and night at the altar. Then ghostly things appear and the operator will be inspired and led to great joy, thinking that he is sure to attain longevity. But all of a sudden he is seized by an untimely death and his corporeal remains are left to decompose.*
  The above-mentioned operations go against the order of nature and will give rise to the loss of the vital axis. Numberless methods like these are all against the doctrines laid by Huanglao (The Yellow Emperor and Lǎozǐ) and will violate the process of juidu.

We have covered some of the ground of 內丹, nèi dān, Internal Alchemy, which is a subject that could cover several books to do it full justice, but it is really to be practised rather than theorised. We will return to this topic later in the book when we explain the practical dimensions of learning the 太

极拳 tàijíquán form. This will enable you to understand not just the terminology used by your teachers but also how the positions held relate to the map of energy flow within the body.

Figure 2-35 Found in Mǎwángduī manuscript depicting exercises performed in ancient times.

Chapter 3 – The Origins of Chen Style 太极拳 tàijíquán

太極

坎 kǎn

天下莫柔弱於水.

而攻堅強者, 莫之能勝.

其無以易之.

弱之勝強, 柔之勝剛, 天下莫不知, 莫能行.

是以聖人云受國之垢, 是謂社稷主.

受國不祥, 是謂天下王.

正言若反

Nothing in the world is more flexible and yielding than water.

Yet when it attacks the firm and the strong, none can withstand it, because they have no way to change it.

So the flexible overcomes the adamant, the yielding overcomes the forceful.

Everyone knows this, but no one can do it.

This is why the sages say those who can take on the disgrace of nations are leaders of lands;

and those who can take on the misfortune of nations are rulers of the world.

True sayings seem paradoxical.

Chapter 78 道德经 Dào dé jīng – translated by Thomas Cleary, 1991[55]

We have by now most probably overtaxed your brains and depleted your Shén in so doing by taking you on the journey, albeit fleetingly, through the landscape of Daoist principles and associated esoteric language. It is appropriate at this point to move on to our next topic. That is the subject of the origins of Chen style 太极拳 tàijíquán. Unfortunately, as we stated earlier, much of China's heritage and historical documentation was destroyed by the Communist government in the first 30 years or so of its rule. Hence, the historical narrative is somewhat sketchier than it could have been otherwise, which is a great pity. Nevertheless, we shall endeavour to provide you with the key facts available.

First of all, let us get an idea of the physical geography that we are dealing with. Chen style 太极拳 tàijíquán originated in the village of 陈家沟 Chénjiāgōu, which is literally translated as Chen Family Village. The name of the village is a composite of three words: 陈 Chen is the family surname,[56] 家 jiā means family and 沟 gōu is a gully or ravine. The name is apt as the village lies in a gully not far from the Yellow River. 陈家沟 Chénjiāgōu is situated in the county of 溫县 Wēn Xiàn in 河南 Hénán Province, eastern central China.

---

[55] *Further Teachings of Lao-Tzu: Understanding the Mysteries*. Trans. Thomas Cleary.
[56] In Chinese, the surname is placed first in the opposite way to English.

河南 Hénán province is located slightly to the east of the centre of China. The name 河南 Hénán is comprised of two words, 河 or hé, which means river, and 南 or nán, which means south. Hence, 河南 Hénán means south of the river. The river which Hénán lies to the south of is in fact the famous 黄河 Huáng Hé or Yellow River that runs for over 700 kilometres through 河南 Hénán. Most of the province lies to the south of the Yellow River.

河南 Hénán province is considered by many to be the birthplace of Chinese civilisation, and it is where China's first dynasty, the 商朝 Shāng cháo (shortened to the Shang), came into existence after the clan had conquered the local tribes in around 1700 BC. Modern 河南 Hénán Province is densely populated, with around a hundred million people living there. It has mainly a rural character and is the third most populated province in China. It is the location of the famous 少林 Shàolín[57] temple, the birthplace of the famous martial art named and the spiritual home of the Chan Buddhist monks.

Our history of Chen style 太极拳 tàijíquán starts in the fourteenth century. In 1367, an army largely composed of rebel farmers from 安徽省 Ānhuī Shěng, Anhui Province crossed the Yellow River, overthrew the 大元 Dà Yuán, Yuan Dynasty[58] and founded the 大明帝国 Míng Cháo Ming Dynasty. During that period, there was a famous battle in Huai qìng City, near the site of the present day 陈家沟 Chénjiāgōu. The recently enthroned Emperor Ming believed that the local population would rebel in support of his predecessor and so engineered three massacres that led to the decimation of the region's population. This resulted in the land being left to lie fallow.

Just five years later, in 1372, at the beginning of the Ming dynasty epoch,[59] the Emperor published an order to the people living in Hong Dong town in 山西省 Shānxī province, which borders 河南 Hénán immediately to the north, to move and resettle in Hai qìng City. It appears to have been more of a request than a compulsory re-location.

---

[57] Now world famous for its Kung Fu.
[58] The Yuan dynasty lasted from 1271 to 1368.
[59] The Ming dynasty lasted from 1368 to 1644

Amongst those who moved to 河南 Hénán was a youth whose name was 陈仆 Chen Bu.

陈仆 Chen Bu moved with his family to a small village about 20 miles from Wen Town, which was part of Huan qìng City. In those days, the village was known as 常阳村 Cháng yáng cūn or Changyang village, a hybrid name that reflected the preponderance of the two families of Chang and Yang. Sometime later, the name 陈家沟 Chénjiāgōu was given to the village. 陈仆 Chen Bu was a skilled and widely recognised martial artist at that time and later became known as the first generation of Chen family martial artists.

Martial arts continued to be practised by the family until the time of 陈王庭 Chen Wang Ting, who was both an outstanding scholar and a martial artist. It was he who created the Chen family system of 太极拳 tàijíquán, as it is known today. He is known as the ninth generation of the Chen family martial artist and the first generation of 太极拳 tàijíquán. Hence, this indicates that there was a martial arts lineage pre-dating the 太极拳 tàijíquán form. He based his new system on the family's own martial arts and incorporated into it the Daoist metaphysics and energy principles that we covered in the previous chapter.

Figure 3-1 Map of China

Figure 3-2: Statue of 陈王庭 Chen Wang Ting in modern-day Chénjiāgōu village.

Figure 3-3: Statues of 陈王庭 Chen Wang Ting, Chen RuXin and Chen Suo Le.

Figure 3-4: 陈长兴 Chen Changxing and Chen Qing Ping 陈清萍.

Due to his martial arts prowess, 陈王庭 Chen Wang Ting was recruited by the government of the 清朝 Qīng Cháo, the Qìng Dynasty, to help solve the persistent problem that the country had with marauding bandits.

As a result, he became very famous. Following this service, he became disenchanted with his treatment by the government (it seems like things don't change much!) as he believed that his talents were not being fully utilised. Hence, he resigned his position and concentrated on further developing his martial arts skills. He invented a kind of martial art that had special features, incorporating combinations of the soft with the hard and fast and slow movements. This is now known as 太极拳 tàijíquán. Note that this creation of 太极拳 tàijíquán was not something totally new but was an evolution of traditional Chinese martial arts infused with Daoist principles.

The Daoist metaphysical principles that were incorporated in the art include Yin and Yang theory, the universal principles of the 易经 Yì Jīng or the Book of Changes, the system of internal energy channels and meridians of the 经络 jīng luò theory, and the arts of dào yīn 导引[60] combined with the breathing of 吐纳 tǔ nà.[61]

To describe it in this way, however, makes it appear as though entirely separate systems were amalgamated into some form of scientific formula. This is not the case. The principles used to perfect the martial form as we have seen are complementary and interrelated. Only during a process of conceptual analysis do these strands appear to be separate or standalone. Movement, breathing and meditation are applied consistently.

As we know, the symbol of the 太极拳 tàijíquán, with its famous depiction of the black and white yin-yang emblem, is quite common to most Westerners, even to those who have had little exposure to Daoist metaphysics. Hence, emblems and symbols used to advertise 太极拳 tàijíquán classes or flyers, posters or websites usually display the yin-yang symbol to represent the derived source of the arts taught.

Also incorporated into 陈王庭 Chen Wang Ting's art were elements from 戚继光 General Qì Jiguang,[62] Jixiaoxinshu 继效新书, a book of techniques derived from different schools, and that of the 黃帝內经 Huang Di Nèi Jing, the Yellow Emperor's Canon of Chinese Medicine, which we have discussed previously. According to research conducted by Gu Liu Xin[63] on historical records during the middle of the sixteenth century, Qi Jiguang developed a martial arts set known as the thirty-two forms of boxing.

Each of these contributions to the development of 太极拳 tàijíquán will be explored further in the following chapters. The thirty-two forms in General Qi Jiguang's set had some similar names but different meanings to the ones that we use today. Often these older sets had one name for an entire

---

[60] A set of stretching and breathing exercises similar to Hatha Yoga.
[61] Known today as 气功 qì gōng.
[62] Qì Jiguang (1528–1588) was a military general and national hero during the Ming Dynasty.
[63] Gu Liu Xin (1908–1991) was a martial artist and prominent historian, who undertook research into 太极拳 tàijíquán and the martial arts.

section or set. Like the Chen forms, Qi Jiguang's form included Lazy About Tying Clothes[64] and Single Whip.[65] Qi Jiguang supposedly took the forms and knowledge of sixteen different martial arts schools to develop his form.

陈王庭 Chen Wang Ting further developed a structure for the 太极拳 tàijíquán practices and, in doing so, produced seven training forms. These consisted of five sets of 太极拳 tàijíquán movements, one set of 炮捶一路 pào chuí yī lù or canon fist, which is the most martial of the forms, and one set of 一百零八势長 cháng quán or Long Fist, which consisted of 108 movements. In addition to this compendium of routines were the Thirteen Postures, Shi San Shi, in addition to the sets of weapon forms and the two-person combat form called 推手 tuī shǒu, Push Hands.

陈王庭 Chen Wang Ting used the experience that he had gained from his martial arts together with his battle experience. He then incorporated what he had learned from his Daoist studies to combine the best aspects of other martial arts traditions to blend with the system developed by 陈仆 Chen Bu. He succeeded in creating the unique system of martial arts now known as Chen style 太极拳 tàijíquán.

陈王庭 Chen Wang Ting perfected the art and in doing so created five sets of 太极拳 tàijíquán and five sets of fist boxing, Wu Tao Chui, as well as Canon Fist, 炮捶一路 pào chuí yī lù. Also included were weapons sets, including spear and broadsword. From these origins came other styles of 太极拳 tàijíquán, such as Yang, Sun, Wu (Jianquan), Wu (Yuxiang) and Zhaobao. Each of these latter styles further developed their own changes.

After the downfall of the 明朝 Ming Cháo dynasty, which had lasted from 1368 to 1644, the political scene was volatile and not for the first or last time the whole of Chinese society was reduced to a state of turmoil. This was the last 汉人 Han ruling dynasty of China.[66] It had gained power following

---

[64] Each posture in 太极拳 tàijíquán has a name, some of which sound strange when translated literally into English.
[65] This is a repeat posture in the form and is common to all 太极拳 tàijíquán styles.
[66] The Han Chinese are an ethnic group native to China. Other ethnic groups have ruled China, such as the Mongols and the Manchurians.

the collapse of the Mongol-led Yuan Dynasty. (See Table 1 in the Introduction for a list of Chinese dynasties.) The Ming dynasty had been one of the greatest eras of orderly government and social stability in China and a time of flourishing spiritual and cultural endeavours.

Following his successful martial arts and scholarly career, 陈王庭 Chen Wang Ting decided to withdraw from public life and retired to live in Chénjiāgōu village. Just prior to his death, he wrote:

Recalling past years, how bravely I fought to wipe out enemy troops and what risk I went through. All the favours bestowed on me are now in vain! Now old and feeble, I am accompanied only by the book of Huang Ting Jing,[67] 黄庭内景经 a classic on 气功 qigong. Life consists in creating actions of boxing when feeling depressed, doing field work when the season comes and spending leisure time teaching disciples and children so they can be worthy members of society.

Development and recent history

The 14th generation Chen martial artist, Chen Changxing 陈长兴 (1771–1853) further synthesised 陈王庭 Chen Wang Ting's style into two routines that came to be known as 老架, lǎo jià, Old Frame. His family relative and fellow martial artist, Chen Youben 陈有本, also of the 14th Chen generation, is credited with starting a mainstream Chen training tradition that differed from the one created by 陈长兴 Chen Changxing. It was originally known as New Form 新架 xin jia as opposed to 陈长兴 Chen Changxing's 老架, lǎo jià, old frame or old form. It gradually became to be known as xiao jia 小架 or small form. He simplified the First and Second Routines of Chen Changxing. His style was called Xiao Frame, Small Frame, Xin Jia. From then on, people called the form developed by Chen Changxing Old Frame or 老架 lǎo jià

.

---

[67] Huangting Jing (*Scripture of the Yellow Court*) is part of the Daoist Canon that includes Huangting Neijing Jing and the Huangting Waijingjing. It is said that Huangting Jing was written by Wei Huacun, a famous female Daoist in the Western Jin Dynasty (265–317).

Figure 3-5: Chen Changxing

陈长兴 Chén Chángxīng (1771–1853) was the fourteenth generation descendant and sixth generation Master of the Chen family. He wrote some important books such as *Ten Important Items of 太极拳 tàijíquán*, *Useful words of 太极拳 tàijíquán* and *Chapter of 太极拳 tàijíquán – Application*. As there were so many repetitions of the routines, he adapted the five routines into one. His form was called Old Frame or Big Frame. His students, such as his family relative 陈耕耘 Chen Gengyun, became very famous too. His other prominent student was Yang Luchan, the founder of Yang style 太极拳 tàijíquán.

He famously stated what is a key principle of 太极拳 tàijíquán movement, that is, that the three parts of the body, the top, middle and the bottom, should be harmonised with the breathing in co-ordination as one breath. He also stated that the body, hands and legs should move as if bound by a rope.

In the late 1920s, 陈发科, Chén Fākē (1887–1957) and his nephew broke with Chen family tradition and became pioneers by moving to Beijing and openly teaching Chen style 太极拳 tàijíquán to 'outsiders', defined as those who were from outside the Chen family village. He remained in Beijing and taught public classes for several years. He famously was quoted as saying: 'There are three steps to learn 太极拳 tàijíquán: first to learn the correct moves, then to practice often, and finally to understand the details.'

Around that same time, a 16th generation family member Chen Xin 陈鑫 (1849–1929) wrote the first published book on the style, 太极拳图说. *Tàijíquán illustrated 太极拳 tàijíquán Chen Shi Tu Shuo.* This was not published until after his death three years later, in 1932, as *Illustrated Explanation to Chen Family 太极拳 tàijíquán*. This book describes movements that are practised within the Small Frame system. In the book, he describes his motivation for writing it. Primarily, he was a Daoist practitioner rather than a martial artist.

Chen Xin did not practise martial arts himself and was asked how he could write a book on martial arts if he was not a practitioner. He stated that 太极拳 tàijíquán is based on the Yijing, the *Book of Changes*, and that as he was himself was a scholar in that field, it was no problem for him to do so, as indeed proved to be the case.

His motivation was clear. He was worried that the theory would be lost to posterity if he did not put it in writing. The book covers only 64 postures rather than the 74 postures one would expect. This is believed to be due his wish to concentrate on what he believed to be the main postures. And, of course, this is the magic number of the total number of hexagrams contained within the 易经 Yì Jīng. Excluded are the opening and closing forms and some transitional movements.

To the reader, it may seem odd to exclude these and thus, seemingly, not achieve completeness. The answer may be obvious to you by now, as correspondence with the magic number of the sixty four hexagrams is the obvious reason.

Whilst Chen Xin's book is a veritable gold mine of useful information for the Chen practitioner, it is also very difficult to understand for the vast majority of students. This is due to it being aimed at fairly advanced practitioners: it is written in Traditional Chinese and the grammar and punctuation are not standard.

There is also information missing in the sense that there is an incompleteness to it. This is considered to be due to the tradition that prevailed in China of not fully providing all of the answers but to leave gaps to be filled by direct transmission by a teacher.

There are explicit references made to the philosophy of the 易经 Yì Jīng and Internal Alchemy. It is stated clearly that the practice of internal alchemy is the main driver of the movements of the 太极拳 tàijíquán forms.

Between 1930 and 1932, Tang Hao, a well-known and respected martial arts historian, was commissioned by the government of the time to establish what exactly the origins of 太极拳 tàijíquán were. He visited 陈家沟 Chénjiāgōu three times to undertake research into the Chen family records. He concluded that 太极拳 tàijíquán originated from 陈王庭 Chen Wang Ting of 陈家沟 Chénjiāgōu in the middle of the 17th century.

陈发科 Chen Fa Ke

The story of 陈发科 Chen Fa Ke is a very illuminating and encouraging one for 太极拳 tàijíquán students. His great grandfather was Chen Chang Xing (1771–1853) who was the teacher of Yang Lu-chan (1799–1872), the founder of Yang style 太极拳 tàijíquán.

Until the age of about fourteen, he was weak and sickly. Until that time, he had never really practised his family's 太极拳 tàijíquán. Upon realising that the family's tradition was always passed on through the generations and that he was palpably much weaker than the other boys of his generation, he decided to embark upon a regime of diligent practice.

After 陈发科 Chen Fa Ke arrived in 北京 Beijing, 陈照丕 Chen Zhao Pi (1883–1972), who was a student of his father, Chen Deng-Ke, together with Chen Yan-Xi and Chen Pin-San, travelled to Nanjing to teach at the behest of the mayor of 南京市Nanjing. From 1928 to 1958, 陈发科 Chen Fa Ke also taught in 北京 Běijīng, 西安 Xī'ān, 兰州 Lánzhōu, 洛阳 Luòyáng, 开封 Kāifēng and 郑州 Zhèngzhōu. In 1958, he returned to 陈家沟 Chénjiāgōu village amidst the revolutionary fervour of the new Communist government. Not very many people practised 太极拳 tàijíquán after that period, due to the years of civil war between the Nationalists and the Communists and the repudiation and oppression by the latter of all ancient and traditional Chinese arts.

Chen Fa Ke started to teach again in his home village in the nineteen fifties, concentrating on the younger generation; especially children.

This was a difficult time for everyone in China, due to the 文化大革命 Wén huà Dà gé mìng Cultural Revolution.[68] 太极拳 tàijíquán was considered to be part of the old traditions that the new government repudiated and wished to eliminate. This antagonism (a euphemism for the outright cruelty of the Communist government) also included attacks on religion, not just Daoism, but Buddhism and Confucianism[69] and anything that did not conform to Marxist ideology.

---

[68] The Chinese Cultural Revolution, which took place between 1966 and 1976, was a period of widespread social and political upheaval during which anyone suspected of not being fully loyal to the Communist dogma was persecuted. Many innocent people were murderd and anything that was tainted with being deemed of the old order, that is, pre-Communist brought down the full force of the government. It was launched by Mao Zedong, the chairman of the Communist Party of China.
[69] Confucius: *Kǒng zǐ* 孔夫子; Kǒng Fūzǐ; literally 'Master Kong', 551 BCE–479 BCE) was one of the foremost philosophers and thinkers and has schools in existence to this day that follow his teachings.

Figure 3-6: 陈发科 Chen Fa Ke

To either teach or practise 太极拳 tàijíquán at this time was a very dangerous undertaking. Reflecting on that era, Chen Ke-Sen, 陈照丕 Chen Zhao Pi's son, wrote:

During the Cultural Revolution, my father was persecuted and subjected to public 'struggle sessions', but during the still of night, Chen Zheng-Lei and several others of his prized disciples secretly went to study under him. My father, demonstrating that he was not afraid of persecution, bravely carried on with his teaching of 太极拳 tàijíquán.

陈照丕 Chen Zhao Pi died in 1972, aged seventy-nine, but was still active right up to his death. After 陈照丕 Chen Zhao Pi moved to Nanjing, he did not see his uncle 陈发科 Chen Fa Ke for thirty years. Afterwards, when he moved back to 陈家沟 Chénjiāgōu, he taught the next generation of grand masters: Chen Xiao Wang, Chen Zheng Lei, Wang Xian and Zhu Tiancai. These four are known as the Four Buddha Warriors. So, the forms taught by 陈照丕 Chen Zhao Pi and 陈发科 Chen Fa Ke are different. Nowadays, the forms practised in 陈家沟 Chénjiāgōu are derived mostly from 陈照丕 Chen Zhao Pi.

Figure 3-7: Chen Fa Ke

Since the mid 1990s, Chen style 太极拳 tàijíquán has spread throughout the world. Previously, only the other styles had branched out throughout China and abroad. This expansion of Chen style 太极拳 tàijíquán happened for a variety of reasons.

Firstly, there was the initial stage of expansion when the teaching moved out of 陈家沟 Chénjiāgōu village and spread throughout China during the 1920s, as explained above. Hitherto, the family art was regarded as a closely guarded secret. Following the oppression of the Cultural Revolution in the 1960s and 1970s, things opened up considerably. This was at a time of increasing interest in the West for authentic Daoist arts and teachings. This thirst for knowledge enabled top masters, among whom are Grandmaster Wang Hai Jun, and his master, Grandmaster Chen Zheng Lei, to visit the US, the UK and other Western countries. The dissemination of information via the internet has also enabled Chen style 太极拳 tàijíquán to develop further to the extent that its presence is becoming firmly established around the world.

Figure 3-8: Young people practising in Chénjiāgōu village.

Figure 3-9: Grandmaster Wang Hai Jun in Diagonal posture in 陈家沟 Chénjiāgōu village.

Figure 3-10: Grandmaster Wang Hai Jun in Lazy About Tying Coat posture.

Grandmaster Chen Zheng Lei

Grandmaster Chen Zheng Lei was born in 陈家沟 Chénjiāgōu Village in May 1949. He is the son of Chen Zhao Hai, who died just a year after he was born. He has been practising 太极拳 tàijíquán for more than fifty years. His teachers include his famous uncles, 陈照丕 Chen Zhao Pi and Chen Zhao Kui, masters of the 18th generation in Chénjiāgōu.

In Grandmaster Chen Zheng Lei's own words:

My family was involved in 太极拳 tàijíquán for generations, and my whole family practises 太极拳 tàijíquán. Since I was a kid, I was influenced by 太极拳 tàijíquán and my family style only – I had no contact with other martial arts. Since I was eight I studied 太极拳 tàijíquán with my uncle, Chen Zhao Pei. He started teaching in Beijing in 1928, and taught for the Nationalist government during the 1930s. Later, in 1938, the Japanese invasion occurred, and he moved back to 河南 Hénán province. He was part of the Flood Control Committee, and worked to control the Yellow River and taught Chen style. In 1958, he retired from the Flood Control Committee and moved back to Chen village.

Figure 3-11: Grandmaster Chen Zheng Lei

His early childhood was very difficult. His father died young having been cruelly persecuted by the government. The family was placed in the category of what was contemporaneously and somewhat euphemistically termed 'struggled against', a form of cruel physical and psychological persecution meted out by the government. This meant that it was very difficult for the family to survive and food was never sufficient to abate hunger. He started work at the age of fourteen, doing agricultural work, and his education was curtailed.

Practising tàijíquán offered him an escape from the terrors of the Cultural Revolution. Together with his cousin, Grandmaster Chen Xiao Wang, he was banned from competing in martial arts competitions. The situation changed for him for two reasons. One was the death of Mao Ze Dong and the accompanying reduction in the persecution of that political era, which was gradually replaced by a more open policy and an easing of the suppression of traditional Chinese culture in general. The other factor came down to the 'Tiger Dance'. The Chen family has a long tradition of performing a dance known as the Tiger Dance. This is similar to the Lion Dance, known as 老虎 lǎo hǔ in the south of China, which is popular at Chinese New Year festivals and is performed by ethnic Chinese in the West. The Tiger Dance is performed as the story of the tiger and the tiger fighter. It has series of six sets. The tiger fighter demonstrates his martial arts skills to fight and defeat the tiger. There are lots of theatrical and acrobatic movements and rings of knives and fire to pass through.

The Lion Dance is a tradition that had been passed down from each generation but had been neglected due to the suppression by the government of anything that was regarded as being of the traditional ways. Chen Zheng Lei learned it from the older generation, which, due to their age and associated physical conditions, had to explain, rather than demonstrate, the very vigorous movements of the form. In 1972, he started to perform the Lion Dance in Chénjiāgōu village. He was the only one skilled enough to go through the two sets of fire and knife rings. His skill in performing this dance and the Tiger Dance provided him with a pivotal moment in his life.

In 1974, the party secretary and officials of the National Competition came to visit Chénjiāgōu. Chen Zheng Lei performed the Lao Hu, the Tiger Dance, and they were so impressed that they asked him to perform the dance as a representative of 河南 Hénán Province in the National Competition. Hence, knowing the Tiger Dance enabled him to compete at twenty-four years of age. Today, Chen Zheng Lei is the only surviving person who knows the entire Tiger Dance.

Chen Zheng Lei is known as 太极拳 tàijíquán Jingang, that is one of the top four practitioners of his generation in China. He received his degree in physical education from 河南 Hénán University in 1985. He has twice won the 太极拳 tàijíquán Grand Champion at the National 太极拳 tàijíquán Competitions, and has won more than ten gold medals at the 河南 Hénán Martial Arts Competition.

In 1987, he was selected as a National Martial Arts First Class Judge, and his team won three team first places from 1989 to 1991 at the National and 河南 Hénán Martial Arts competitions. In 1995, he was awarded the title of Da Shi or Grandmaster.

Grandmaster Chen is in great demand both in China and abroad as a teacher. He has trained over 15 of the coaches for the Beijing and Wuhan Physical Education Institutes, and is an advisor to 太极拳 tàijíquán organizations in Japan, France, the US and Italy. Master Chen currently teaches in 郑州 Zhèngzhōu City in 河南 Hénán Province.

He was just eight years old when he started to study with his uncle, who also taught the remaining three Buddha Warriors or Three Tigers: Chen Xiao Wang, Zhu Tiancai and Wang Xian. He studied with his uncle for fifteen years until the latter's death in 1972. He learned all the family's routines: 太极拳 tàijíquán, 太极拳 tàijíquán sword, double sword, double broadsword, 13 staff, spear and kuan dao, and other weapons.

His training also included learning the philosophical aspects of the Dao. Following his uncle's death, he continued his studies with another uncle, Chen Zhao Kui, who was himself the grandson of Chen Fa Ke. He further developed his 太极拳 tàijíquán skills, comprising both the theoretical and the practical aspects. He has commented that he would sweat so much during practice that his socks would fill up with water. Chen Zheng Lei remembers that after his uncle died, it was the spirit of Chen Zhao Pi that helped him carry on, even though his family suffered very much.

From 1974 to 1988 he competed in, and won, national championships many times, being awarded titles in staff, spear, and sword. In 1976, the Cultural Revolution was practically over and he went to work in a factory. The owner was very interested in 太极拳 tàijíquán and this persuaded him to recruit Chen Zheng Lei. This was not easy even then, and he was labelled as a 'Rightist' and the owner was equally criticised for employing him.

The reforms of the 1980s, when 太极拳 tàijíquán was allowed to be opened up to both China and the rest of the world, resulted in big changes for Chénjiāgōu as it became a location for foreign 太极拳 tàijíquán enthusiasts to visit. One of the first major groups of people that came were the Japanese.

太极拳 tàijíquán was very popular in Japan in the 1970s, more so even than in China where it had been suppressed for decades by the government. In 1981, a 太极拳 tàijíquán association in Japan visited 陈家沟 Chénjiāgōu and this grabbed the attention of the media as foreign visits were not that common, especially for the arts and culture. This raised the profile of both the village and the martial art of 太极拳 tàijíquán.

The government was then developing tourism facilities and thus encouraged visits from foreign 太极拳 tàijíquán enthusiasts. At the same time, in 1983, martial artists from Chen village began to tour China and the outside word, teaching 太极拳 tàijíquán.

Chen Zheng Lei and his cousin, Chen Xiao Wang, were recognised by the government due to this international interest. He finally left his job in the factory and went to the 武术 Wǔshù Institute to teach 太极拳 tàijíquán and also wrote some books. In 1994, he was appointed Vice Director of the 武术 Wǔshù Institute, to manage and develop not just 太极拳 tàijíquán, but other styles as well. He is the leader of 400 schools in 河南 Hénán province. He has written several 太极拳 tàijíquán books, produced DVDs and has travelled to many countries in Europe, Asia, the Americas and Australasia. He has also produced several books, including *Ten Phases of 太极拳 tàijíquán Practice*, *A Collection of Chen's Taichi Boxing and Weapon Routines*, *Chen's Taichi for Health and Wellness* and *The Wonders of Taichi Kungfu*.

Chen Zheng Lei is a happily married man with three children: two daughters and one son. All are involved in 太极拳 tàijíquán. He still travels extensively throughout the world to teach.

Figure 3-12: Children assembled in Chénjiāgōu.

Figure 3-13: 陈家沟 Chénjiāgōu village

Chen Family 太极拳 tàijíquán Genealogy

1st Generation: 陈仆 Chen Bu Founding Ancestor 1368–

2nd to 9th Generation: 陈王庭 Chen Wang Ting (1580–1660)

10th Generation: 陈汝信 Chen Ruxin

11th Generation: 陈大鹏 Chen Dakun

12th Generation: 陈善通 Chen Shantong

13th Generation: 陈秉旺 Chen Bingwang

14th Generation: 陈长兴 Chen Changxing (1771–1853)

15th Generation: 陈耕耘 Chen Gengyun

16th Generation: Chen Yan Xi

17th Generation: Chen Fa Ke

18th Generation: Chen Zhao Pei

19th Generation: 陈正雷 Chen Zheng Lei (1949–present)

Chapter 4 – 站桩 zhàn zhuāng

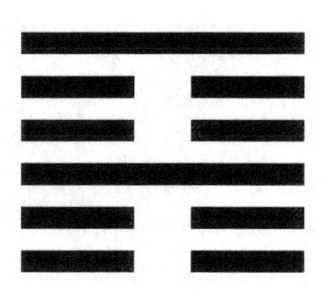

艮 gèn

Standing Meditation is the single most important and widely practised form of qigong integrating all elements of posture, relaxation, and breathing previously described. It improves alignment and balance, stronger legs and waist, deeper respiration, accurate body awareness, and a tranquil mind.

Kenneth S. Cohen, *The Way of 气功 qìgōng*, p. 133

有物混成.

先天地生.

寂兮.

為兮獨立不改.

周行而不殆.

可以為天下母.

吾不知其名.

字之曰道.

強為之名曰大.

大曰逝.

逝曰遠.

遠曰反.

故道大, 天大, 地大, 王亦大.

域中有四大, 而王居其一焉.

人法地.

地法天天法道.

道法自然.

There was something undefined and complete, coming into existence before Heaven and Earth.

How still it was and formless, standing alone, and undergoing no change, reaching everywhere and in no danger of being exhausted!

It may be regarded as the Mother of all things.

I do not know its name, and I give it the designation of the Tao.

Making an effort to give it a name, I call it The Great.

Great, it passes on in constant flow.

Passing on, it becomes remote.

Having become remote, it returns.

Therefore the Tao is great, Heaven is great, Earth is great, and the sage king is also great.

In the universe there are four that are great, and the sage king is one of them.

Man takes his law from the Earth.

Earth takes its law from Heaven.

Heaven takes its law from the Tao.

The law of the Tao is its being what it is.

Translated by James Legge (1891), Chapter 25

We have already described the Daoist principles upon which the practice of 太极拳 tàijíquán is based. Now let us analyse these principles in more detail by examining how they manifest themselves in terms of the practical application of 太极拳 tàijíquán. These applications must first be learned and then correctly put into practice to achieve the maximum benefits from the practice of 太极拳 tàijíquán.

In order to properly practise 太极拳 tàijíquán and to be able to embrace the fundamental Daoist principles, it is vital to understand the basic premise, that is, that we know how to move with mind and body in full harmony whilst at the same time maintaining a correct physical alignment. The practice and physical posture of this can be described somewhat simply as Standing Still. This is a very practical method of achieving this objective.

This is in one sense an easy and yet in another a quite difficult skill to achieve. Why is this so? Well it is easy because from a physical perspective, we are stationary and thus do not have to engage the mind and body by absolving any notion of the dynamics of movement. And yet in another sense, it is more difficult inasmuch as we have to engage our minds much more as it is not focussed on the external stimulus of movement. In the balance of things, it is nevertheless a sound principle to use the practice of standing. This is so as to be able to fully teach our mind and body to be in a state so that later it can effortlessly move in the harmony of mind and body when required.

Although learning how to stand and practising thereafter need not precede learning the movements, it is much easier to first learn the principles that must be incorporated into our movement practice by adopting a static pose. This process provides a solid foundation and allows for the embedding of the key principles that can then be incorporated into the movements. It is of course useful to practise both together and incorporate them into a daily regimen.

In recent times, 站桩 zhàn zhuāng has become both a physical exercise and a meditation in its own right and is taught to students who may never wish to practise 太极拳 tàijíquán or any other of the martial arts. It may not appear to you at first instance as anything physical – what is difficult about standing still? Once having tried it, you will soon realise that it is not so physically easy, notwithstanding how much you will notice your wandering mind as you do so.

In the Chen style 太极拳 of the tàijíquán tradition of teaching, movements from the form were adopted as static postures in the middle of practising the forms. This was undertakn to develop the strength and concentration required, as well as opening the energy channels of the body, and it allowed the teacher to adjust the posture. Typically, two of the most popular postures were 懒扎衣 Lǎn zhā yī, Lazy about Tying Coat and 单鞭 Dān biān, Single Whip with 70% of the weight on one leg, making the postures very powerful if held for a significant time. Whilst learning the form, these postures would be held, and the master would align the body posture and allow the student to feel the differences that small adjustments would bring in terms of feeling the 气 qì and feeling more stable and grounded in the postures. In more recent times, the Wuji posture has been adopted as a foundation for 站桩 zhàn zhuāng as a separate exercise.

The practice of standing is known as Zhàn Zhuāng 站桩 or, in English, *Standing like a Tree*, or, as it is sometimes referred to, *Stance Training*. It is also known by the terms Standing 气功 qìgōng and Standing Meditation. The Chinese term 站桩 zhàn zhuāng is used in both martial arts and in 气功 qìgōng schools. The word *Zhuāng* has a few meanings, and although one of them is 'stakes', the more appropriate term within the context of the exercise is 'stances'. 站桩 Zhàn Zhuāng is most aptly translated therefore as 'standing in stances', but Western students mostly use the Chinese pinyin term without fretting over the English translation.

In essence, 站桩 zhàn zhuāng is a generic term used to describe a 气功 qìgōng exercise where the practitioner remains stationary in one chosen posture for a length of time, which may be anything from a few minutes to a few hours. In the rest of the chapter, we will use the Chinese term 站桩 zhàn zhuāng rather than one of the other terms. There are very many types of 站桩 zhàn zhuāng with different feet and hand positions, weight ratios applied and other additions and variations, just as there are many types of movement: walking, jogging, sprinting and variations of standing. It may surprise you to learn that different stances and hand positions have very different and nuanced effects on the practitioner.

The aim of the exercise therefore is to remain stationary in posture. This is termed as a static position, especially as viewed from the outside. From the inside, it is not so much so. Internally, it allows the energies to move, circulate and build up. It also aligns the body in a stance that is beneficial for health and provides the correct alignments for the movements and postures that are required for 太极拳 tàijíquán. It is the perfect practice to enable the 小周天 xiǎo zhōu tiān and 大周天 dà zhōu tián, Small and Big Heavenly Cycles of energy that we covered in Chapter Two, to circulate through the 奇经八脉; qí jīng bā mài, Eight Extraordinary Vessels.

The 站桩 zhàn zhuāng form of 气功 qìgōng was a secret practice in times past and would not be taught outside of the immediate circle of students in martial arts schools. 王薌斋 Wang Xiang Zhai, a famous martial arts master, made this technique the basis of a new internal martial art that he called 意拳 Yiquan, or Mind Boxing. He used to say: 'The immobility is the mother of any movement or technique'.[70] It is also practised in other internal martial arts styles, such as Bàguà Zhǎng 八卦掌 and 形意拳 Xíng yì quán, as well as some external ones, such as 少林寺 Shàolín Kung Fu.

站桩功 zhàn zhuāng gong, is also a description of practising in order to develop and harmonise the 三宝 sān bǎo, or Three Treasures that we covered in the previous chapter. As well as being an important component in the training for 太极拳 tàijíquán, 站桩 zhàn zhuāng is also an excellent exercise in its own right. As a standalone exercise, it may be classified as a form of 內功 nèi gōng. There are some variations in the positions for the stances, with different hand and feet positions within the various martial arts and 气功 qìgōng schools, but the principles remain the same. All students learning internal martial arts should practise 站桩 zhàn zhuāng in order to make any substantial progress.

---

[70] Victoria Windholtz, *Standing Like a Tree*

We will now explore the principles of the practice and explain why it is such a good 內功 nèi gōng exercise in terms of the accrued benefits for health and the strengthening of the body, and also explain to you how it is a fundamental supplement to 太极拳 tàijíquán practice.

At the physical level, 站桩 zhàn zhuāng integrates all of the essential elements that we need, that of correct posture, moving towards a relaxed state and incorporating deep, slow and whole body breathing. It will develop for you a better structural alignment and balance. It will make your legs and waist grow stronger and will allow your breathing to settle into a much deeper and natural rhythm until eventually the body breathes by itself without any forced effort. It is a form of internal aerobics and develops 气 qì in the body. It is also a way of progressing slowly without straining yourself and allows the progress that is made to be sustainable.

站桩 Zhàn Zhuāng is also known as Standing Meditation because one adopts a meditative state whilst standing. Meditating in a standing posture has some very clear benefits. It helps the mind to concentrate, as it needs to remain clear and alert in order to remain in the balanced posture. You are constantly aware of the body and the mind. Both the body and your mental awareness are maximised when meditating in a standing posture and consequently your energy and blood circulation are much improved. Indeed, the paradox is that the more relaxed you are the easier it is to stand for longer periods. Only when you are not doing anything does something happen!

As an advanced form of Daoist meditation, the exercise develops stillness which firstly allows you to reduce the external, and then, latterly, the internal, distractions. This allows us to listen to our body and gain knowledge and an increasing awareness of its state. Externally, there is no obvious movement, contrasting with the internal state where there is energy circulation. The internal movement is the circulation of the 气 qì and the optimum working of the body's energies. Indeed, you will quite often hear your digestion working quite loudly and, if you are unlucky enough, also sense it through the olfactory system.

The effectiveness of 站桩 zhàn zhuāng as a 气功 qìgōng or nèi gōng practise is widely accepted:

If I had to choose one qigong technique to practice, it would undoubtedly be this one. Many Chinese call standing meditation 'the million dollar secret of qigong'. Whether you are practising qigong for self-healing, for building healing ch'i, for massage or healing work on others, standing is an essential practice. Acupuncturists feel that by practicing standing meditation they can connect with the ch'i of the universe, and be able to send it through their bodies when they hold the acupuncture needle ... Standing is probably the single most important qìgōng exercise. One of the reasons that standing is such a powerful way to gather and accumulate fresh ch'i in the body is that during the practice of standing the body is in the optimal posture for ch'i gathering and flow.

Kenneth S. Cohen

The distinction to be made between that of movement and stillness is, however, not one of an absolute dichotomy. More accurate is to say that it is the distinction made from different reference points. This means that nothing is either entirely still or entirely moving in extremis.

In our minds, we distinguish as opposites between movement that is perceived as explicitly active (yang), and standing, which is perceived as wholly being static (yin). As we have shown from observation of the 太极 tàijí symbol, yin and yang are relatives, with an element of each contained within the other. Hence, the stillness of the body as it remains fixed, allows for internal movement of the energies within the meridians of the body. And when you have practised for a reasonable length of time, you will find that your body can move without any appreciable volition, especially the arms and hands. Your legs at first of course – but that is probably shaking through fatigue.

From a relative perspective, during the 太极拳 tàijíquán movements, the body moves externally and yet internally remains still as the mind settles and breathing settles to a rhythm that is long and slow. The practice of 站桩 zhàn zhuāng may be described as akin to the action of the peeling of an onion in the sense that there are layers and layers to it. This means that as you become more relaxed, yet more layers of tension will be discovered below and yet even further ones discovered beneath that.

There is always more relaxation so it is easy to assume that you have cracked it and have peaked at the absolute relaxed state. This is not the case, however, and further states of relaxation and energy circulation are always achievable, even after many years of practising.

The inherent danger for students, therefore, is that many stop practising when they are able to stand for twenty minutes or more, thinking that they have mastered the art and it has become too easy. This is a profound mistake. You can only realise how tense you were in the past, not how tense you may still be. If you stop practising you are preventing yourself from achieving further levels of relaxation and awareness. Hence, there is a constant and ongoing relaxation and stillness to be achieved, although nothing is actually strived for. Also, the practice is not just one that exists at a physical level. Although the body is able to remain still, this does not mean that further progress cannot be made – in other words, the 气 qì continues to accumulate and the mind may go further to reach an even calmer state.

The way that 站桩 zhàn zhuāng adheres to the principles of 太极拳 tàijíquán can be explained quite simply. In order to move, it is necessary to do so as one unified structure. To do this, we need to stand and to get the body trained in posture and balance. This provides not just an optimum way to move but it aids in maintaining balance. Also, if the legs are not strong enough, the body starts to bend to meet the challenge of the movement and the sinking, so at the physical level, getting stronger legs is essential for the practice of 太极拳 tàijíquán.

Like much in life, the simplest things are often the best and often, equally, the hardest to do or to attain. 站桩 zhàn zhuāng is in principle the simplest thing of all to do. Just stand. But the mere act of just standing and preventing the body from tensing and not allowing the mind to wander is much more difficult in practice. Most people are unable to remain motionlessly relaxed. They tense their muscles and this creates stress and leads to a non-relaxed state.

In contrast to 站桩 zhàn zhuāng, Western style callisthenics exercise systems are based upon vigorous movement that are used to achieve the benefits of physical conditioning. These exercises are fine. There is no criticism of these exercises at all. They are not full mind, body and spirit exercises, however, and they also are more difficult to continue with as one ages. 站桩 zhàn zhuāng is a composite system that harmonises all levels of the person and allows you to grow steadily and to feel more energised at the conclusion than at the outset.

Even physically fit people who are beginners to 站桩 zhàn zhuāng may initially experience severe muscle fatigue and subsequent body trembling until sufficient stamina and strength have been developed. The difference is in the emphasis on the use of tendons rather than on the muscles. Many exercise systems pay attention to the muscles and neglect the tendons, ligaments and the bones, and this can cause injury due to the body being put in positions where it is out of balance.

站桩 zhàn zhuāng will enable the martial arts practitioner to develop a state known as 中定 zhōng dìng or central equilibrium, as well as sensitivity to specific areas of tension in the body. It is not only the physical centre, it is a state and a feeling of centeredness. It needs to be present at all times so that you are always rooted and stable. This state is essential for effective 太极拳 tàijíquán.

Standing in a perfectly aligned posture is a powerful 气功 qìgōng exercise because the body is in the optimal posture for gathering 气 qì and letting it flow through the body. It also removes any energetic blockages.

If you can stand in perfect alignment and are relaxed, your centre of gravity will naturally sink to the 下丹田 xià dān tián. For most people in their daily lives, it is very difficult to stand in perfect alignment. This means that there is tension in the body due to the inability to stand upright and an associated absence of complete relaxation. This is due to bad posture, which is caused by tension and modern lifestyles and the adoption of bad habits.

Once the correct posture is achieved and maintained, you then need to be emotionally and mentally relaxed. When you have acquired the ability to maintain the physical position and you do not have to worry about it anymore, you can focus on your mind. This may be either on one point within the body, such as the lower 下 丹田 xià dān tián, or on nothing – that is no one thing except your own awareness.

wish to attain the boundless state known as 无极 wú jí. Emotions are a major factor for influencing the mind and the body. Negative emotions affect the internal organs and cause disharmony within the mind. There are sounds and colours and meditative exercises that can be performed to alleviate these negativities, and these will be covered in greater detail in a subsequent book.

Western medicine and physiology seek to analyse the body by dividing it into component parts or functions – the blood circulation, and the muscular, skeletal, lymphatic and nervous systems. Of equal importance for our purposes in viewing the body from a holistic perspective is to understand the fascia system. This is the composition of the connective tissues that connect the bones, ligaments, muscles and tendons. The pathways of the fascia map to the meridians and the patterns of movements 缠丝劲 chán sī jìn that we will cover later in Chapter 7. This concept of fascia has recently been discovered in the West; within Rolfing[71] and other body health systems, it has been discovered that everything is interconnected through the fascia that links to every muscle in the body. As everything is interconnected, the weakest point adversely affects the whole like a link in a chain.

站桩 zhàn zhuāng fixes the body in a perfect alignment to gravity so it avoids having to exert energy to counteract it. Hence, energy is not wasted when you remain upright and the muscles are maintained in a balanced state. To attain this state, it is essential that the body be relaxed, as this will allow the internal organs to settle. Only then can the blood flow freely, which will allow the muscles, tendons and ligaments to strengthen. When the body is as fully relaxed as possible, the 神 Shén is working at optimal levels. This allows you to breathe in a fully optimised respiratory manner as the abdomen and diaphragm are relaxed.

---

[71] Rolfing is an alternative treatment developed by Ida Pauline Rolf, who was born in New York in 1896.

Not only does 站桩 zhàn zhuāng develop a solid stance, but consistent training also delivers internal power. The system has been developed by masters through the centuries and in terms of martial arts effectiveness, it is based on the principle of 无为; wú wéi. As we know from this principle of effortless activeness, reducing the movement to the minimum will be accompanied by a concentration of the energy and the mind.

Studies have shown that the average person has more than sixty thousand thoughts every day – a meagre total in comparison to the number generated in writing this book! The vast majority of these thoughts in fact are almost all fruitless. The thoughts are actually thinking the person until a person becomes a self-hypnotised robot. The mind flows like a river of thoughts, which increase in number and intensity by our paying attention to them. In actual fact, these thoughts do not belong to what you perceive as 'you' and are not 'you'. Hence, by observing this rootless master that should instead be an obedient slave, you will thereby gain great insight into your day-to-day actions and what drives them.

When emotion is added to the mix, further energy is provided to our thoughts, which creates a vicious circle of anxiety. This type of incessant anxious thinking drains you of your energy. In order to be able to practice 太极拳 tàijíquán properly, we need to have a relaxed, concentrated and calm mind. Practising 站桩 zhàn zhuāng helps us achieve this.

Do not make the mistake, however, of trying to drive the thoughts away as this will provide further attention and increase both the number and intensity of them. The more relaxed you are, the more you will notice your mind's chattering. This is very difficult for students and beginners to do. But don't worry, this is the state of all people. The thoughts are not the real 'you', and they will slow down and eventually stop if you persist. Adopt the state of 自然 zìrán, of naturalness as it is.

The still mind of the sage is the mirror of heaven and earth, the glass of all things. Vacancy, stillness, placidity, tastelessness, quietude, silences, and non-action – this is the Level of heaven and earth, and the perfection of the Tao and its characteristics. 莊子 Zhuāngzǐ – Legge version

In terms of the 气 qì, you can let this flow naturally and accumulate. A chattering mind will dissipate the spirit. Once you have achieved a harmony of 精 jīng 气 qì and 神 Shén, this total integration of the whole person can be applied, not just to your 太极拳 tàijíquán movements, but also to any other forms of movement and stillness.

Standing and movement make use of different muscle groups or systems, the latter making use of the central nervous system and fibres. In addition, even when the same muscles are used, they will be performing differently for standing and movement activities. Those muscles used in the standing posture do so in a non-volitional way. That is, we need to use our mind and intend to control the muscles. This is the term 意 Yi that we encountered earlier, which means Mind intent. Ultimate movement is through intent and free flowing 气 qì moving the muscles and tendons of the body in a completely uniform exercise.

Do not under any circumstances feel that you should put your body under pressure to adopt a lower stance or to stand for longer than you can through gritted teeth. This is an exercise in 气 qì development and mental and body relaxation.

Imagination is also very important in your practice. It is stronger than willpower. You may combine the uplifting feeling of the relaxed state to do some visualisation using the imaginative power of your relaxed mind. In reality, the body is never completely still as the heart is always working, as are the lungs and the rest of the body. The position of stillness is, therefore, more allegorical than literal.

There are many forms of the 站桩 zhàn zhuāng stance. Different schools have adopted variations of feet and hand positions. All retain the same principles, however.

Postural and Phasic Muscles

The terms *postural* and *phasic* muscles were first introduced by Vladimir Janda,[72] a doctor who treated patients who suffered from muscular imbalances. He categorised muscles into two types, postural and phasic. These were differentiated on the basis of their functions and properties. Examples of postural muscles are the calves, hip flexors and spinal muscles. These are mostly involved in maintaining our upright posture. Phasic muscles include the abdominals, deltoids, glutes and triceps.

Postural muscles are used to maintain an upright posture and for countering gravity. When these muscles are put under any stress, they shorten and become more hypertonic. They respond well to relaxation and lengthening techniques, such as Hatha Yoga, 导引 dǎo yǐn and stretching. Phasic muscles are used for movement and respond to strengthening but are generally prone to weakening under stress.

The key difference between the two muscle types is that the postural muscles are designed so as to become hyperactive, and tight. This can lead to some possible pain and discomfort, especially where there is over exertion. Phasic muscles in contrast are prone to becoming inhibited and weak. The postural muscles are, therefore, more prone to becoming tight or painful or both. In contrast, the muscles that are difficult to strengthen are the abdominals, the gluteus maximus and the lower trapezius, all of which are all of the phasic type. If the body is not properly balanced or aligned optimally and it is not exercised in the right way, muscle imbalances leading to injuries will occur.

Before you make any movement in 太极拳 tàijíquán, there is a short pause whilst the body prepares. What is happening is that the lumbar muscles of the lower spine are being activated. This occurs through the use of the deep abdominal postural muscles. For beginners, there will be little if any

---

[72] Vladimir Janda 1928-2002. At the age of 15, he contracted polio and was paralysed and unable to walk for two years. Later, he recovered some walking function, but developed post-polio syndrome and needed the use of a walker until the end of his life.

awareness of this happening. This lack of awareness will disappear once your 站桩 zhàn zhuāng practice develops. The postural muscles movement will harmonise with the 气 qì activation in your 下丹田 xià dān tián. Further development will result in the activation of the power of 發勁 fā jìn from the release of energy when the body is held so that your postural muscles become engaged.

The postural muscles co-ordinate across the whole body and so ensure that the body moves as one unified unit. Training of the postural muscles and their associated strengthening enables the 太极拳 tàijíquán practitioner to rely ever less on the phasic muscles. This will also increase the speed at which the body can move when practising. This proves essential as a reactive movement when self-defence is required. Many of the problems that occur with our muscular-skeletal system are due to imbalances between these two main muscle types of postural and phasic.

The states of being in both static posture and movement may be regarded as something quite separate if viewed in an abstract way. That is how we usually analyse and describe things conceptually. And yet this need not be the case. As elucidated in the 道德经 Dào dé jīng, these are not really opposite states but are apparent differences only when viewed from the relative perspectives, rather than from the whole. From a 太极拳 tàijíquán aspect, static posture embodies internal 气 qì movement and movement embraces stillness.

Postural muscles react against the force of gravity and are controlled indirectly through intent and during the transition between two postures.

In a situation where you do not utilise the principles of 太极拳 tàijíquán when you face a situation where someone pushes you and you automatically resist in a tense way, the phasic muscles will be used in reaction, using strength and tensed muscles. A seasoned 太极拳 tàijíquán practitioner will, however, use the postural muscles to absorb the push by adopting a natural relaxed alignment. Hence, we are getting to the nub of the appliance of Daoist principles and their application for practical martial purposes.

In Daoist philosophy, you will see the statement, 'Where the Mind directs, the 气 qì will follow. The mind Intent 意 Yi will guide the postural muscles without feeling like there is any effort.

To again quote Kenneth S. Cohen:

I call the ancient, natural way of standing 'the Paleolithic Posture'. In the Paleolithic Posture, the knees are slightly bent, the spine is straight, and long, the breath is deep and quiet, and the eyes are open and alert. The body feels like a tree with deep roots for balance and tall branches for grace. Although we usually think of a 'posture' as a static pose, it includes our carriage in movement as well. Since a straight and tall stance confers the greatest balance, sensitivity, awareness, and alertness, we see it in a scout standing still on a mountain lookout or walking through camp to a council meeting.[73]

Stances

Double weighted

This is a conventional stance that has the weight evenly distributed between the two feet in a 50%-50% balance. An example is the Wuji stance, where the hands are at the side of the body.

Unevenly weighted

This is where more weight is placed on one leg than the other. Examples are the 三体势 Sān Tǐ Shì stances used in 形意拳 Xing yì quán and the 单鞭 dān biān Single Whip and 懒扎衣 lǎn zhā yī Lazy About Tying Coat stances used in Chen style 太极拳 tàijíquán, where the weight can be distributed in 70%-30% or 60%-40% ratios.

Single leg weighted

This is where all of the weight is placed one leg and the other is off the ground. These stances obviously are more difficult to hold than the double or single leg weighted

---
[73] Ken Cohen, *Honoring the Medicine*, p. 240

Now let us examine the precise details of how to stand properly. We will take the double weighted stance as an example.

How to stand

Step out

We step out in order to move the legs into the correct position. To do this we can opt to follow the first movement of Chen style 太极拳 tàijíquán. (Each of the Chen routines adopts this same initial movement.) Start by raising the left foot and move it sideways so that the distance between the feet is equal to that of the width of the shoulders. This is a shoulder width measured as the distance between the inside, not the outsode of the feet which would be too narrow a stance. Please note, however, that we are all different sizes and therefore the most important factor is that you be both centred and aligned. Within the parameters of these guiding principles, simply adjust your posture to feel centred and grounded.

See Figures 4-1 to 4-6 below

Your weight should be evenly distributed over your feet, with not too much on your toes, and equally not too much on your heels. You may gently rock on your heels until you feel that both of your feet are balancing the weight and gripping the ground. Your feet may be aligned facing straight ahead or you may feel more comfortable with them pointed out slightly. To emphasise the point again, the important thing to consider is that you are aligned and feel so. It is an easy trap to fall into that, either at the one extreme, you are spending so much time thinking about the precise bodily adjustments that you become tense and, at the other, you become floppy and badly aligned. Use your sensitivity rather than a ruler or spirit measure!

Sink as though you are about to sit down on a chair that has been placed behind you. Do not bend the knees as an isolated movement, but just release the hip and waist area: that will allow the knees to move forward slightly. One way to get the right position is to tense the knees and then relax them. Your knees will then rest in the correct position due to the sinking of the hips rather than by a deliberate bending of the knees operating in isolation.

Sink the body from the bottom. This will open the area known as the inguinal, known in Chinese as the 胯 kuà. This will naturally flatten out the arch in your lower back and the coccyx will hang down naturally. Check that your body weight is still distributed evenly. You should feel balanced over your feet. Relax your arms and shoulders. Relax the 下丹田, Xia Dān tián. Do not tuck in the tailbone too far in an exaggerated way as you would, for example, when expanding and releasing the energy in a movement, but release the tension in the lower back by ensuring the correct alignment of the top of the body.

Figure 4-1: Your weight should be distributed evenly over both of your feet

Stand straight with your feet together and prepare by relaxing and clearing your mind. Look for any tension within. Relax and breathe soft and naturally.

Figure 4-2: Raise the left knee while balancing on your right leg.

Maintain the weight on the right leg and slowly raise the left. Make sure that you remain balanced and avoid your body rising. Perform this movement slowly and maintain your balance.

Figure 4-3: Raise your knee until it is almost at a 90-degree angle.

Raise the knee to an almost 90-degree angle from the ground. Keep your weight on the right leg and maintain the vertical height of the body. Do not raise the body or over compensate by leaning over to the right.

Figure 4-4: Lower the left leg, placing the toes on the floor.

Slowly lower the left leg and place the toes on to the floor. This movement is performed slowly, with the weight maintained on the right leg

Figure 4-5: Place all of the left foot on the ground and balance your weight evenly between the feet.

Place the whole of the left foot on to the floor by lowering the heel. Balance your weight evenly.

Figure 4-6: This is the correct posture to begin the zhàn zhuāng exercise.

This is the posture for you to commence the 站桩 zhàn zhuāng exercise. You should feel balanced and relaxed. The knees are slightly bent as the 胯 kuà is sunk. The spine is stretched upwards and the weight is spread evenly over the feet. Relax and smile down internally to relax the internal organs. Breathe slowly and take your time in feeling adjusted.

Aligning the body

Aligning the body properly is of crucial importance. If this is correctly performed, then a vertical line through the vertebrae is maintained and the least possible stress is placed on the muscles, tendons and ligaments as you are aligned perfectly and without tension. This is the optimum state for the fascia and allows the 气 qì to flow most easily. This will also allow the line of gravity to pass through the centre of the vertebrae that should align as though one vertebra at a time has been stacked on top of the others and balanced.

It is important to surrender the body to the pull of gravity as though you do not need to exert any effort to stay perfectly upright. Normally, muscular tension is required to remain upright and it feels challenging and unnatural to relax and just trust to gravity.

The head should be held is as if it was suspended from above by a cord attached to the crown. This will allow the 百會 bǎi huì point to open and allow the yang energies of heaven to enter the body. The neck will then be relaxed and free of tension, and it will feel as though it is lengthening. The lower jaw should be relaxed to prevent the neck muscles from tightening up. The jaw should move slightly backwards and downwards to allow the neck to move slightly backwards. Do not strain to raise the head as though it is a competition or act like the time when you were a child and stretched against the wall to see how much taller you had become since the last measurement. Remain relaxed. Relax, relax and relax again!

In *Illustrated Chen's 太极拳 tàijíquán*,[74] Chen Xin is quoted as follows:

The head is the governor of the six positive (yang) meridians and it controls postures and movements of the whole body. The whole body moves according to the position of the head.

---
[74] As quoted in *Taichi Old Frame One & Two* by Chen Zheng Lei, page 41

Chen Xin also alludes to the importance of the 百会 bǎi huì acupuncture point.

In the Alexander technique,[75] emphasis is similarly placed on the importance of the head. Relax the small of the back and the buttocks, and let the pelvis settle into a comfortable position, so that it feels connected to both the knees and the ankles. The eyes may be either open or closed. The mouth is almost open. Remember to avoid tightening up the body as the mind tries to remember all of the individual body positions. It is a bit like learning to drive a car where you initially feel overwhelmed by all of the individual instructions. To start off, you are trying to remember everything all at once and find it difficult to grasp, until eventually everything fits into place and your body relaxes into a natural state and perfect alignment follows.

The tongue should rest very lightly on the upper palate. To find the correct place, put the tip of your tongue just behind the upper teeth and curl it towards the back of the mouth. At a certain point, it will rise up a steep incline. Usually, this is somewhere around the point where the incline starts. It may feel like it is more sensitive or it may tickle more than in other places. This is the place to rest the tongue. There are different points at the top of the mouth that are more yang or yin. Use your intuition and experiment and you will reach the right position.

This is known as the Magpie Bridge and connects the 督脉 dū mài and 任脉 rèn mài – the yang and yin meridians that we covered earlier. There are, in fact, five different positions that the tongue may be placed in to stimulate different types of energy, based on the energy cycle phases that we discussed earlier.

The shoulders should relax downwards, the opposite state to that of being hunched up. Widen the shoulder blades and let them relax both backwards and downwards. The shoulders are places where much tension is stored and tightness occurs. This is often caused by postures adopted when

---

[75] The Alexander technique was created by Frederick Matthias Alexander and is based on teaching the body to refrain from using excess amounts of muscular and mental tension.

using computers, sitting at a desk too much, driving for long periods of time or slaving over the laptop for the writing of this book!

The chest should be relaxed and kept slightly concave, not sticking out like the exaggerated military stance so beloved by sergeant majors. The stomach should be full and relaxed, not pulled in. Again we can quote Chen Xin from the *Illustrated Chen's 太极拳 tàijíquán,* as quoted in Chen's *Taiji Old Frame One & Two.*[76]

The chest needs to be tucked and hollow. Relax the chest. Once the chest is relaxed the whole body is relaxed.

Depending on how far the arms are raised, the movement should always be performed very slowly to avoid lifting the shoulders.

As the arms are raised, check for tension, especially in the shoulders and the neck. Avoid using the shoulders to raise the arms. In high arm positions, the wrists are never lower than the elbows.

Lift the arms with the wrists rather than the elbows. As your practice develops, your arms will rise with little effort as the body becomes more relaxed. The arms should be held as though someone is placing a fist between your armpit and the upper arm. Keep checking to make sure that your posture and weight do not shift. Observing the body is a key component of the practice. Once you reach a particular level, the arms will raise seemingly on their own slowly through the 气 qì movement. Strange, but it is true. Then you know that the 气 qì is full and moving properly.

The arms should be held so that you look like you are embracing a big balloon in front of your chest. Keep the fingers open and maintain a small open distance between them. The eyes can be half closed, keeping a soft focus, or closed. When everything is aligned correctly, the body's energy system is open and 气 qì can flow freely.

---

[76] Chen Zheng Lei, *Chen's Taichi Old Frame One & Two,* page 43

The posture is most suitable for those without any particular illness to strengthen the constitution, prevent illness and promote health into old age.[77]

The abdomen should be relaxed. Tuck in the chest and keep it closed in towards the abdomen.[78]

Sensations

The first sensations you will experience will vary in intensity and variety. Much depends on the current state of your body and mind. Those feelings that manifest themselves are of physical and mental discomfort. This discomfort is natural and must be differentiated from pain. Physically, you will feel where the tension lies in your muscles, joints and tendons. You will mistakenly feel as though it is the exercise itself, rather than the current tense state of your body that is the source of your discomfort. Just as when you first start to meditate and you notice how loud and chattering your mind is, rather than meditation being the actual cause, so it is that you will notice blockages of energy in your body.

As the energy circulates, the blockages in the energetic body become increasingly noticeable and the negative sensations will gradually ease. This process of peeling away the layer of tension whilst feeling unpleasant is nothing to worry about.

Initially, and even for some time thereafter, when the body has become relaxed, the mind will seem like it is much noisier and ceaseless in its repetitive chattering. This is because your attention is focussed inwards and has no external object upon which to attach.

Your mind will chatter away like the monkey it so resembles. Thoughts will seem to appear more often and you will feel like something is trying to annoy you when you want to remain undisturbed –

---

[77] J. P. C. Moffett, Wang Xuanjie, *Traditional Chinese Therapeutic Exercises: Standing Pole.* Foreign Languages Press, May 1994. ISBN: 7119006967, pages 49-52
[78] Chen Zheng Lei, *Chen's Taichi Old Frame One & Two,* page 46

just like those buzzing mosquitos in 陈家沟 Chénjiāgōu village that treat themselves to your body as if it were five-star cuisine. This feeling is due to your awareness of the mind and the fact that it is 'not doing anything'. It has always been thus. It is just that now you are becoming more aware.

Eventually this situation will improve and, with regard to your body, you will become more aware of what is going on all the time and be able to distinguish between the cause and the effects. The mind acts like a sports commentator, constantly running through external and internal events, wanting to trap you into engaging and, hence, increasing the mental noise. One way to overcome this is to imagine that you are perched on a bridge over the top of a motorway and you are watching the cars flying past below in both directions. The cars are your thoughts, but you can watch them with insouciance knowing that you cannot be knocked over by them.

As the body becomes naturally aligned and the mind more relaxed, a whole range of sensations will manifest themselves. These include a feeling of warmth in the hands and feet, the body feeling light and the muscles feeling empty and floating. There will be a state of euphoria, centeredness and balance and a feeling that the energy is holding your body in place, rather than your muscles. Your body may sway but if your practice is mature, then it is the energy moving to align your body, distinguishable from the early stages when your muscles are shaking because they are not used to the exercise.

The Chinese word for relaxation is 松 sōng. It means to completely relax both mentally and physically and thus includes the mind as well as the body. 松 sōng is a fundamental requirement for 太极拳 tàijíquán practice. Every 太极拳 tàijíquán master will keep repeating this. Without 松 sōng, 太极拳 tàijíquán is just a physical movement with no substance. Absence of 松 sōng will interfere with the smooth flow of 气 qì and blood, preventing smooth movements. 松 sōng creates more 气 qì and allows the body to be supple.

The transition through various levels of the exercise from one of pure physical endurance to a state of meditation will occur naturally if you persevere. You will be surprised at how quickly your body will become still where once you thought that this was impossible. Your negative emotions will rise but later subside. The real challenge will be, as previously mentioned, the chatter in your mind. Remember it is like an internal commentary. It is not the real 'you' but an incessant loop that has always been there. You are just becoming more aware of it as your body relaxes and your mind has no outward phenomena to latch on to. You will reach a stage where it is the chatter that stops your practice rather than the physical demands on your body.

Please persevere as the rewards are so great, and it is a gift to anyone who sticks at it. And what is more engaging than standing around doing nothing? Remember that you are using your own body to cultivate yourself and who knows more about their own body than the occupant? Also, do not forget to train on a regular daily basis, so that you can grow the way a tree grows, slowly but inexorably, and with incremental strength and progress. Also, focus on relaxing physically, emotionally and mentally, and do not think that you have reached the end. Eventually you will look forward to it and feel as though you are being recharged like a battery. You will notice when you walk how much easier it is to take strides than before.

It is a state obtained through a process, not an event. Please persevere, persevere and persevere. The rewards are worth it.

Push Hands in relation to standing

The practice of Push Hands, which we will cover in another volume of this series, is predicated upon the adoption of an aligned posture, achieved through the practice of 站桩 zhàn zhuāng and the correct use of the postural muscles. This occurs, for example, when your Push Hands partner pushes you and you are able to stay rooted and balanced. If the temptation to use raw muscular power is overcome, the Yi can lead the 气 qì in harmony with the balance and rooting of the postural muscles to respond.

Rooting

Rooting is achieved through the practice of 站桩 zhàn zhuāng and is incorporated into your practice. Without rooting, 太极拳 tàijíquán fails as a martial art. Without the body being rooted and balanced and relaxed, and the 气 qì centred in the lower 丹田 dān tián, there is no power. With rooting, the upper body is empty and the lower body is full.

Without rooting, movement has no balance and power is lost moving from one position to another. You feel balanced even when the weight is transferred from one leg to the other.

Postures

There are a number of postures that can be practised for your 太极拳 tàijíquán training. The first is the Wuji Zhuāng. This is where the arms are held by the side of the body. It is the stillness position before any motion begins. It is a state of emptiness, the primordial condition. It precedes the emergence of Yin and Yang. It is Wu Ji giving birth to Tai Ji. The emptiness transforms into the cyclic dualities.

- Ping Bu Cheng Bao Zhuāng – embracing the tree
- Single Whip
- Lazy about Tying Coat.

Meditation

The 站桩 zhàn zhuāng postures have many health benefits and are a very sublime form of meditation in their own right. The benefits of meditation are now well known from an esoteric perspective, as well having been proved by scientific studies. Aside from feeling more relaxed, having your stress levels reduced and avoiding the ailments that arise as a result, meditation also produces an increased level of awareness and a higher state of consciousness that feed the spirit.

Perseverance

The secret to the practice of 站桩 zhàn zhuāng, aside from the simple rules that need to be adhered to, is a simple one. It is that of perseverance. Whilst it must be acknowledged that in earlier times, there were fewer distractions and it was easier to live in a more natural environment, nevertheless life was much harder. Thus, it is fair to say that there are always advantages and disadvantages to living in any particular era. What the earlier generations respected was the allocation of time and effort to learning a skill. This is an attitude we must also adopt.

Modern studies show that it is how much you practise that counts, rather than your initial level of skill. Perseverance leads to repetition, which causes the myelin insulation to increase in your nervous system. Myelin is a dielectric or electrically insulating material that forms a layer that is known as the myelin sheath, usually around only the axon of a neuron. It is essential for the proper functioning of the nervous system. Myelin is also the primary substance affect in Multiple Sclerosis sufferers.

This will create circuits that allow you to speedily activate what you have learned and practised. It is akin to programming a computer. Hence, regular practice will enable you to achieve all of the advantages that this practise can bestow upon you.

Figure 4-7: The Wu Ji posture – Embracing the Tree

Figure 2-8: Side View of the Wu Ji posture – Embracing the Tree

Figure 4- Side view of Wu Ji – Embracing The Tree

Figure 4-8: Embracing the Tree, showing the alignment

Figure 4-9: Single Whip – Facing Posture

Figure 4-10: Single Whip rear view

Chapter 5 – Principles of 太极拳 tàijíquán

坤 kūn

天地不仁,以萬物為芻狗.

聖人不仁,以百姓為芻狗.

天地之間,其猶橐籥乎.

虛而不屈.

動而愈出.

多言數窮.

不如守中.

Heaven and earth do not act from any wish to be benevolent;

They deal with all things as the dogs of grass are dealt with.

The sages do not act from any wish to be benevolent;

They deal with the people as the dogs of grass are dealt with.

May not the space between heaven and earth be compared to a bellows?

'Tis emptied, yet it loses not its power;

'Tis moved again, and sends forth air the more.

Much speech to swift exhaustion lead we see;

Your inner being guard, and keep it free.

道德经 Dào dé jīng – Translated by James Legge, 1891, Chapter 5

In the last chapter, we covered the principles relating to stillness – that of the practice of standing. These principles that you have just learned and which relate to your posture, relaxation and balance equally apply to those of the 太极拳 tàijíquán movements. Learning how to stand symmetrically, balanced, relaxed and being aware of our bodies and energies – will provide the training that enables these to be maintained whilst we are practising the movements. Stillness and movement are yin and yang, the opposite yet complementary forces that exist in our Manifest World and within the microcosm that is our bodies.

There are also further aspects, however, that we must take into consideration, such as appreciating the importance of not succumbing to hubris, that is, not falling into the trap of underestimating the need to first fully master the basics. Although many people spend years studying 太极拳 tàijíquán, unfortunately their progress is slow. And their aim of acquiring the skills they so earnestly strive for is inhibited, tempering their zeal and making their development seem all too slow and frustrating. Part of the reason for this may be due to insufficient or inappropriate practice. But a significant element is down to a basic misunderstanding. That is, there is a core set of skills that has first to be mastered in order to achieve real and sustainable progress.

For some students watching a master practising the forms, there is also a natural tendency to wish to immediately imitate. It is not possible to commence 太极拳 tàijíquán training, or, in fact, much else, by starting at the highest, rather than at the most basic level. A master practising the movements will make it look easy in the eyes of the observer. The paradox is that the easier it looks, the higher is the accomplishment of the practitioner. As you start off practising yourself, you will feel clumsy and yet gradually you will refine your energy until you arrive at a point where what you are doing really is meditation in motion.

Taking education as an analogy, it would be like trying to start one's studies at a PhD. level. The natural route, of course, would be to first undertake primary education, then secondary, and to continue through to undergraduate and postgraduate studies. This is equally true for 太极拳 tàijíquán. Without a good mental and physical understanding of the basic skills that make up the foundation level of 太极拳 tàijíquán, higher levels will not be reached. It is not magic, but merely the result of consistent training in a correct and progressive manner.

Learning 太极拳 tàijíquán consists of applying a set of core principles that appertain to each and every movement. This holds true regardless of each individual movement appearing totally different when you observe it through the eyes of an inexperienced learner. Just as when the Unmanifest expresses itself in its myriad forms through the Manifest, the 太极拳 tàijíquán movements have the same characteristics of self-unfolding. The movements as expressions of energy are inter-related, inter-connected and flowing and are derived from the same source. For example, the 缠丝劲 chán sī jìn silk reeling exercises that we cover as a separate topic in Chapter 7 are all embedded within the movements, as are the principles of the 站桩 zhàn zhuāng, as we have just illustrated.

You have probably have heard of the maxim, 'Be still as a mountain, yet move like water'. It is important to remember this. Being still as a mountain was described during the 站桩 zhàn zhuāng chapter, and moving like water is what you will learn through the 缠丝劲 chán sī jìn and 18 movement sets, covered in the ensuing ones. As your practice develops, this statement will resonate and become a living part of your being.

Movement starts from a solid base in the rooting. We have covered the importance of rooting in the previous chapter, and it is difficult to over-emphasise the need for this. Movement starts with the feet and is controlled by the waist. This leads to an energy release from the legs and an ensuing movement through the back and on through to the arms and the hands, where it is finally expressed by and through the fingers. This is the skill of moving effortlessly, and the energy movement is like that of sap moving through the roots of a tree and ascending upwards through to the branches and leaves. This is achieved when 松 sōng exists so that the body is relaxed and not stiff.

When you move, whether it be to the left or to the right, or if it is forwards or backwards, you do so by pressing downwards then lifting up the foot. This action is akin to that of compressing a spring. The force of 劲 jìn is derived from the feet and thereby directs the waist. When you pivot on the heel, the movement must be opposite to the direction that the heel is pivoting in so that the hips follow the movement. This will lead the body in its movement.

The body moves as one in a co-ordinated and aligned way. The upper and lower parts of the body are co-ordinated in what is known as 上下相随 shàng xià xiāng suí – above, below – mutually dependant. Breathing takes place through the lower abdomen and harmonises with the movement, inhaling and exhaling in harmony with the movement. This will all occur naturally as the body moves in the correct way.

The lower 下丹田 xià dān tián moves a fraction of a second before the upper body, although both parts move as a unified whole. This is known as 相连不断 xiāng lián bù duàn. This maximises the power and prevents the breakdown of energy flow. There is also a continuous movement. Although each move has its own name and description, there is a seamless inter-connected flow between each. Hence, the movements are performed smoothly and evenly with continuity. This means that each of the individual movements do not stop and start. And there is no discernible gap between them. This moving with ease and poise is known as 动中求静 dòng zhōng qiú jīng – seek stillness within movement.

So, although we have chosen to break down the movements into 18 individual parts, there is continuity and perpetual movement throughout – without pause. In other words, there are no breaks when one part of the form starts and another one finishes. The principles of 缠丝劲 chán sī jìn that are embedded within the movements should resemble that of reeling silk from a cocoon, neither too fast nor too slow. This ensures that there is no risk of either breaking or slackening the thread. Eventually your form will move from being external to internal so that you will use your mind rather than physical force to effect the movements. This is known as 用意不用力 yòng yì bù yòng lì – use intention, not strength.

When you first practise and for some time after, you feel the same way as you do when you first learned to drive a car. Can you remember how complicated it seemed to learn the different hand and foot movements and to be aware of the traffic at the same time? Eventually through the passage of time and lots of practice, the mechanical aspects of driving the car became ingrained, and then you were able to become more aware of the externals – the behaviour of the other drivers and the surrounding environment. Judging by the way some people drive, you may think that this is an inappropriate analogy!

The hips should avoid swaying from side to side and the centre of gravity needs to be kept low, with your body at a constant height so that it does not bob up and down. This requirement will be easy for you to fall foul of. The natural tendency for most beginner students is to make this mistake of rising and falling. Your weight should not be balanced equally over both legs in a 50-50 ratio. This should be avoided. Rather the weight should be held over the legs in a 70-30 ratio. This helps with your balance and movement from one position to the other. When you are kicking, there will be

100% weight on one leg (not the one that kicks – despite what you see in the Hong Kong films!). As you move, there should not be any loss of balance as you migrate from one position to another.

All of the movements are spiralling and are performed in the form of a circular movement, with the arm following the body. The body should keep relaxed at all times and turn at 90 degrees but whilst at the same time ensuring that it does not extend too far. Feel as though you have something in reserve. This is the principle of always having the yin within the yang and the yang within the yin, as depicted in the 太极拳 tàijíquán symbol. Remember that you are not performing callisthenics or performing a gym workout and that you must not use force.

The 气 qì will flow through the body from the 下丹田 xià dān tián to the hands from the energy stored there through the meridians. When you turn the body to one side, the other takes the weight and the hands follow naturally. We must now provide a more precise explanation for the use of the lower 下丹田 xià dān tián in 太极拳 tàijíquán. We have previously covered in some detail the mechanics of the body in terms of rooting and alignment. This is akin to building the best chassis for a car, and can be recognised as akin to your physical body. Now we need to concentrate on the engine and see how it can make optimum use of the fuel. For this, we need to examine the role of the lower 下丹田 xià dān tián in some more detail.

There are two main principles to discuss with regard to use of the 下丹田 xià dān tián. The first is with regard to breathing and the second, to rotation. Both are difficult to learn, and at this stage you should just be aware of these rather than try to learn them from this book. Full instruction from a qualified teacher is required. This is because of the dangers of engaging in incorrect or forced breathing and the negative effects it could have on the body.

The practice of co-ordinating the lower 下丹田 xià dān tián with the rest of the body produces both 缠丝劲 chán sī jìn and, hence, the power of 太极拳 tàijíquán. The spiral force of 缠丝劲 chán sī jìn emanates from the rotation of the 下丹田 xià dān tián like an electric motor. The 下丹田 xià dān tián is like that of the bellows mentioned in the 道德经 Dào dé jīng. The hands express the energy. The 站桩 Zhàn Zhuāng Standing Postures will strengthen your lower 下丹田 xià dān tián considerably. Hence, 站桩 Zhàn Zhuāng will, in addition to strengthening the lower 下丹田 xià dān tián, open the meridians that will enable the mechanics of the body, the tendons, ligaments, and muscle, to co-ordinate with the 气 qì to provide an integrated internal and external wrecking ball to one's opponent.

The lower 下丹田 xià dān tián is also similar to the operation of a fuel storage depot. The fuel needs to be pumped into the car for propulsion and for the ensuing speed to occur. The mind is like the driver and the 气 qì moves in a yang fashion from the 下丹田 xià dān tián to the hands and back to yin in the reverse situation. Once your practice is advanced, you can thereafter concentrate fully on the opponent. This is known as achieving the harmony of the internal and external, or 内外相合 nèi wài xiāng hé. The easiest way to think about this is to consider the situation where you have become an experienced car driver and you are then able to concentrate on the other traffic and conditions on the road. This is an advance on the situation when you focused primarily on the mechanics of driving the car. Breathing must be natural and not forced; otherwise, it will inhibit the flow of 气 qì through the 奇经八脉 qí jīng bā mài – Eight Extraordinary Vessels.

Breathing

There are many types of breathing exercises. These are used by Daoists, Buddhists, Yogis and many other cultivators. Each technique has a different purpose. Forcing yourself to breathe is not only contradictory but may be dangerous if taken to its extremes. You can spend a lifetime learning all of the different techniques. It is essential as always that you are relaxed when breathing and not forceful because, as stated, inappropriate breathing techniques may cause you harm. The highest level of breathing is actually natural, slow and embryonic. 莊子 Zhuāngzǐ described the ancients breathing through the heels. This is the ultimate natural breathing, through the pores of the skin and key energy points of the body. It also absorbs energy into the bones and marrow.

Daoists consider the lower 下丹田 xià dān tián to be a very important part of the body and a source of great physical energy. Other cultures too consider it important. To the Japanese, it is known as the hara. In yoga systems from India, the lower abdomen is the crucial centre in the body for full and proper breathing. Not only does breathing through the lower 下丹田 xià dān tián provide for good health and contribute to a more youthful countenance, but for martial artists, it provides explosive power. Unfortunately, most people inhale in a shallow way using only the upper lungs, where only the chest expands.

There are two main types of 下丹田 xià dān tián or abdominal breathing. In the first method, the abdomen expands upon inhalation and contracts upon exhalation. In abdominal breathing, when we inhale, our stomach, as well as our chest, expands. This type of breathing expands the abdomen on all sides – front and back and sideways, rather than restricting movement to just the front, as is often supposed. When we exhale, the carbon dioxide is expelled and our chest and abdomen contract. This is known as normal abdominal breathing.

As we progress with our 太极拳 tàijíquán training. we use reverse abdominal breathing, together with a rotation of the 下丹田 xià dān tián. In reverse abdominal breathing, the lower abdomen contracts when we inhale and expands when we exhale. Hence. it is the reverse of normal abdominal breathing – as is implied by its name. Both methods are very useful for health and the strengthening of the body and to enable you to gather energy. The use of reverse abdominal breathing will enable you to generate and distribute the 气 qì and be able to use 发劲 fā jìn or release of energy.

Reverse abdominal breathing builds up a good store of 气 qì and packs it into the bones. During inhalation, the diaphragm will move downwards as the belly contracts inwards. This creates a pressure in the abdomen, and the breath is packed with the resultant 气 qì and then stored in the tissues and organs. During exhalation, the diaphragm relaxes and moves upwards, and so the abdomen also relaxes. The internal organs are also massaged during this type of breathing and the mind becomes centred within the body, rather than remaining stuck in the head.

Reverse abdominal breathing is more difficult to learn and then master than normal abdominal breathing. This is simply due to it appearing to be against the natural way – which is that of expanding the breath upon inhalation. To practise reverse abdominal breathing, adopt either a sitting or 站桩 zhàn zhuāng stance. Inhale slowly through the nose, evenly and smoothly. Slowly contract your abdomen in and up. Inhaling through the nose is important because nostril hair works to filter the pollution in the air and also warms the air as it enters the body.

You should slowly contract your perineum or at the 会阴 huì yīn point as you inhale. This should be done gently rather than using too much muscular power. As you practise more and become adept, you can use your mind to invoke the contraction. The perineum is the point between the anus and the lower edge of the pubis at the front of the pelvis. It is known as 会阴 huì yīn, meaning union of the yin. It is akin to being the floor of the ocean, the very lowest point of yin. It is the point at which the 任脉 rèn mài and 督脉 dū mài converge with the 冲脉 chōng mài.

It is also an important point in yogic traditions and corresponds to the root chakra. The upper chest will expand naturally as oxygen fills up your lungs. Next, slowly exhale through the nose, at the same time relaxing the 会阴 huì yīn down and gently pushing the abdomen out and down. Much practice will create a rhythm and feeling of unified and relaxed breathing.

During exhalation, it is important to ensure that the 下丹田 xià dān tián expands on all sides: to the front, by ensuring that the lower abdomen protrudes, to the back, by expansion through the míng mén, and sideways, to the left and to the right. The 命门 mìng mén is a point located between the kidneys, at the level of the second lumbar vertebrae. It has a Fire and Water relationship with the kidneys and 元气 yuán qì. It can be translated as the Door of Life. It is a very important point for 太极拳 tàijíquán practice.

Many times you will not find reference to the expansion of the left and right sides mentioned, but the belly should expand on all sides. This is also a good exercise to massage the kidneys. It is worth mentioning that in here the rotation of the 下丹田 xià dān tián should be accomplished in a very relaxed manner and not with force so that no harm can occur. It will be easier if you start with the reverse abdominal breathing first and move into the rotation of the 下丹田 xià dān tián later on under the guidance of your teacher.

In Chen style 太极拳 tàijíquán, the 下丹田 xià dān tián is likened to a ball, as viewed from both physical and energy perspectives. The primary rule of movement is that the 下丹田 xià dān tián must move first before the body does. The 下丹田 xià dān tián also moves in a spiral and circular way, the same as the body.

If you throw a punch and use just the arm to do so, the strength is limited. If you use the whole body, generating power from the breath and by the turning of the 下丹田 xià dān tián, the power is multiplied significantly. This power is known as 發勁 fā jìn, or releasing energy. This is the main focus of all of the training and is what makes it such an effective martial art.

The 胯 kuà

The 胯 kuà is of crucial importance to body mechanics, relaxation and smooth circulation of 气 qì. It is located in the inguinal groove, from the ball joint at the top of the thigh bone. There is no direct translation of the term 胯 kuà in English. It is as much a function as a location, more verb than noun. It can be described otherwise when there is a more expansive explanation of the movement within the form. Without proper relaxation and movement of the 胯 kuà then, the lower and upper body cannot act in co-ordination or move in an optimal state.

So, in order to understand the subject properly, we must describe its relation to the rest of the body and how it operates within the entire body. The 胯 kuà controls the movement of the waist and is like a control motor that unites the waist, the legs and the upper body right through to the fingertips. To do so, the 胯 kuà must be properly aligned. It must be both strong and flexible, and attaining the optimum state takes practice and sensitivity. To use yet another metaphor, it moves like a raft in the water, adjusting to the waves and yet in control, ensuring that you do not fall overboard. Hence, the waist must be moved just right; not too much and not too little, not too stiff and not too loose. So for beginners, after much practice, the knee movements become much less as small movements of the 胯 kuà can be used to create larger ones in the whole body. The lower 下丹田 xià dān tián works in tandem with the 胯 kuà, which rests above it. Practice of both the 站桩 zhàn zhuāng and 缠丝劲 chán sī jìn exercises will enable you to appreciate the use of the 胯 kuà in a practical sense. Without proper use of the 胯 kuà, real progress will be inhibited.

The psoas muscle

This is a crucial muscle to relax and keep flexible. It connects the hips, the spine and the legs and is crucial to the smooth movement of 太极拳 tàijíquán. Sitting around all day (a slight exaggeration we know) on chairs shortens the psoas muscle and, accordingly, the body becomes tighter. Many Western students are unable to maintain a squatting position or even get there in the first place. This is due to a tight psoas muscle and corresponding restricted flexibility in the hips.

As we have hitherto expounded, the 松 sōng, which roughly translates as relaxed but not flaccid, is the state to which you aspire (but do not aspire too hard!). It is a relaxed state but within a structure. It is more of a verb than a noun, inasmuch as it is dynamic and needs adjusting when circumstances change. In other words, it is a state of being.

There is a term used to describe the release of 发劲 fā jìn. This process is to maintain the energy as soft until it becomes as hard as iron at the last moment. The body should be like a bow stretched and ready to release an arrow using the legs, waist, shoulders, elbows and the wrists and hands.

Earlier on, we covered the eight trigrams of the 易经 Yì Jìng and the 五星 wǔ xīng five phases of energy change. Here, we now map the basic techniques of 太极拳 tàijíquán, as set out below.

13 Basic Techniques

There are 13 basic techniques. This number is an aggregate of the eight energies, depicted by the trigrams of the 易经 Yì jìng and the five phases.

The Eight Energies

掤 pěng – warding off

捋 lǚ – rolling back

挤 jǐ – pushing

按 àn – pressing

采 cǎi – taking

挒 liè – spreading

肘 zhǒu – elbowing

靠 kào – leaning

The Five Directions or the Five Steps 五步 wǔ bù

进步 - jìn bù – forward step

退步 - tùi bù – backward step

左顾 - zǔo gù – left step

右盼 - yòu pàn – right step

中定 - zhōng dìng – central position, balance, equilibrium

The relationship of the Eight Movements and the Yì jìng:

The Eight Movements relating to the Yì jìng				
Movement	Trigram	Name	Translation	Image in Nature
1. 掤 - pěng – warding off	☰	乾 qián	Creative, Force	Heaven, Sky 天
2. 捋 - lǚ – rolling back	☱	兌 duì	Joyous, Open	Swamp, Marsh 泽
3. 挤 - jǐ – pushing	☲	离 lí	Clinging, Radiance	Fire 火
4. 按 - àn – pressing	☳	震 zhèn	Arousing, Shake	Thunder 雷
5. 采 - cǎi – grabbing	☴	巽 xùn	Gentle, Ground	Wind 风
6. 挒 - liè – spreading	☵	坎 kǎn	Abysmal, Gorge	Water 水
7. 肘 - zhǒu – elbowing	☶	艮 gèn	Keeping Still, Bound	Mountain 山
8. 靠 - kào – leaning	☷	坤 kūn	Receptive, Field	Earth 地

Relationship of the Five Steps and the Five Phases

The Five Steps		
Step	Direction	Phase
进步 - jìn bù – forward step	West	Metal
退步 - tùi bù – backward step	East	Wood
左顾 - zuǒ gù – left step	North	Water
右盼 - yòu pàn – right step	South	Fire
中定 - zhōng dìng – the central position, balance, equilibrium	Centre	Earth

The Eight Energies

The eight trigrams that we encountered when we covered the subject of the 易经 Yì jìng are used as an expression of the energies used in the 太极拳 tàijíquán movements. These are known as the Jìn 劲 that we mentioned earlier. Of the eight, four are considered direct forces, and four indirect. The direct forces are: 掤– pěng relating to 乾, qián; 捋– lǚ relating to 兑 duì, 挤– jǐ relating to 离 lí, and 按 àn relating to 震 zhèn. The indirect forces are -采– cǎi relating to 巽 xùn, 列 – liè relating to 坎 kǎn, 肘 – zhǒu relating to 艮 gèn and 靠 – kào relating to 坤 kūn. These eight energies apply equally to all of the forms, weapons, push hands and 缠丝劲 chán sī jìn.

In the book *The complete verse of 太极拳 tàijíquán*, Grandmaster Chen Wang Ting said:

Practice Peng, Lu, Ji, and An seriously. When fluent it is very difficult to challenge. Even if the opponent charges with tremendous power, I can apply four teals of force to redirect and neutralize a thousand pounds.

It is advisable for beginners to learn and practise the four direct forces first before moving on to the four indirect ones. Practice enables you to balance the posture and remain stable whilst directing the energy. These are developed through the 缠丝劲 chán sī jìn exercises that we cover later in the book. The principles that we learned in 站桩 zhàn zhuāng are maintained through the movements. The movements consists of firm balance and rooting and circular arcs, co-ordinating the upper and lower limbs.

The four indirect forces deliver power and consist of three elements. The first is neutralisation, the second is generation of energy and the third is unleashing the power. This requires co-ordination, and fluid movement.

For the eight forces or movements, attention needs to be paid to the hands, and in particular, the five fingers, which should be held close to each other and relaxed. Then then you will have a feeling of the 气 qì reaching there. The hands lead the movement, working through the elbow, with the shoulder following, rotating with the waist; the navel and the nose are aligned. The eyes will follow the hands as they move.

The Four Direct Energies

掤 Péng – ward off

掤 劲 péng jìn is the primary and direct force of 太极拳 tàijíquán. It is one of the four direct energies. Quite simply, without it, there is no 太极拳 tàijíquán. 掤 劲 péng jìn exists within all of the remaining seven 劲 jìn. 劲 jìn is the expression of your 气 qì or energy. It is an intuitive release of energy that is appropriate for the situation and movement.

掤 劲 péng jìn can be likened to a tyre filled with air or also that of a raft floating on water. It is both flexible yet resilient and is controlled from the centre. 掤 péng is used to absorb an opponent's energy and dissipate it so that your rooting is not disrupted. It requires a combination of the correct rooting, balance and a relaxed awareness. It can react to attacks on your unified energetic body from any direction. It is akin to the energy released from a coil of a spring, where the potential stored is released when compressed – quickly and powerfully. It has a twisting action that can be used effectively in defensive situations to block an opponent's attack and throw them off balance.

掤 Péng movement is expressed as that of an upward direction. Releasing 掤 péng requires the actions of stretching and extending. This requires the body to be cultivated so that there is sufficient flexibility in the tendons and the entire body is fully co-ordinated with the breath and that internal development so that the 奇经八脉; qí jīng bā mài, the Eight Extraordinary Vessels, are open. Thus, an unimpeded flow of 气 qì through the meridians is created. Peng should always be present throughout the body and be expressed throughout the movements of the entire form. The reasons that the body positions are adopted during the routines are to ensure that 掤 péng exists and that the energy is not broken. It is essential that the principles outlined in the chapter on 站桩 zhàn zhuāng be applied. The use of 中定 zhōng dìng by lowering the 胯 kuà and sustaining the posture is essential to attaining 掤 劲 péng jìn. Without this, the body will not move fluidly and there will be no proper rooting.

掤 Péng energy is not something that can be developed quickly. It is accumulated and can be expressed only after much practice and the appropriate amount of relaxation and internal energy awareness is attained. As always, this requires that there is co-ordinated breathing and concentration on the 下丹田 xià dān tián. Any of the 太极拳 tàijíquán movements that are forward or upward in direction are in a state of generating 掤 péng energy. This may be observed in the first movement, where the arms are raised.

捋 - Lǚ – Rollback

As with the other 劲 jìn that has 掤 péng as its basis, 捋劲 lǚ jìn is one of the four direct energies. It is an active force, yet a yielding one in the event of an attack by an opponent. It is an evasive energy that avoids the attack. It is most effective when the mind is in a state of relaxed awareness, a challenge when actually being attacked by an opponent. The body twists and turns and the opponent is rendered unbalanced. 捋 Lǚ energy may be expressed with either one or both hands. The force of 捋 Lǚ cannot be undertaken in isolation from, but needs to work with, that of 掤 péng. If this is not so, there will be no power.

挤 - Jǐ – Push

This is an attacking force, as the word 'push' suggests, and is used to throw an opponent off balance. You must remain rooted and use the waist for this type of energy delivery. As well as the arms and forearms, the shoulder, chest, elbow, hip or thigh may also be used to push your opponent. This depends on the precise circumstances of the attack, of course. This is a difficult release for energy and requires a sublime level of skill to deploy effectively. As with all of these forces, it is underpinned by and is, in fact, an extension of 掤 péng.

按- àn – Press

All of the usual attributes that constitute basic 太极拳 tàijíquán principles need to be applied here: rooting, use of the waist and legs, a relaxed awareness through a calm mind and use of the 下丹田 xià dān tián. This combination will neutralise an opponent and close down their offensive options. It makes use of the hands to push and press and feels like a wave of energy that is released through the palms. The whole body must move in co-ordination, between the upper and lower parts levels. Again, it is important to sink the elbows and ensure that the shoulders do not hunch, otherwise the energy cannot be fully expressed.

Four Indirect Forces

采- Cǎi – Grab

Almost any part of the body can be used to express 采- Cǎi energy. As the name conveys, it is a grabbing action. It is part of the martial art's 擒拿 qìn na – which means grab and hold techniques. We will cover this technique in some detail in a forthcoming book. It uses the grabbing techniques of 擒拿 qìn na. This technique, which exists in many martial arts, is that of holding an opponent in order to neutralise and throw them off balance. It requires sensitivity, and almost any part of the body may be used, but it is usually the hands. Although it is a grabbing technique, the principle of following one's opponent and reacting when appropriate still prevails. The opponent's muscles, bones and tendons are targeted. Some basic knowledge of the human physiology and energy points are required for 采 cǎi to be most fully effective.

挒 - Liè – Spreading/Break

This uses a shaking technique to force an opponent off balance. It is akin to a spinning and revolving action, like the waves of the ocean, to completely disrupt and attack your opponent. This energy can be applied in any direction and is often effected by engaging an opponent's wrists and elbows.

肘 - Zhǒu – Elbow Strike

This is a very fast, explosive and debilitating force to unleash on an opponent. It is expressed through the use of the elbow or shoulder and hence can cause major harm if it hits soft tissue. It is especially effective in close range combat. For the full effectiveness of 肘 Zhǒu, much depends on how you use your footwork. This needs to be precise and well timed and in conjunction with the co-ordination of the elbow, the waist and the hip.

靠 - Kào – Leaning

The method utilises the body between the shoulder and the knee to attack and is a very aggressive and potent skill to deploy.

The Five Most Important 太极拳 tàijíquán Skills for Beginners

Traditionally, the Chen family have compiled the following as the Five Most Important 太极拳 tàijíquán Skills for Beginners:

They are as follows:
- 放 松 Fàng sōng – Loosen the body by relaxing the joints
- 掤 劲 péng jìn – an outward supportive strength, the basic skill of 太极拳 tàijíquán

- 顶劲 Ding Jin– upright and straight
- 沉 Chen - rooted
- 缠丝劲 chán sī jìn – Reeling Silk Skill

These five basic skills should be considered the composite elements that need to be mastered in order to progress in your 太极拳 tàijíquán training. Without these basic skills being learned and embedded in the body knowledge through perseverance and training, you will not be able to achieve the benefits that accrue from practising authentic 太极拳 tàijíquán.

These five beginners' skills complement each other and are acquired slowly with patience and the persistence of practice. As we have previously stated, understanding the concepts at the same time as training the body is the best approach to learning. Nevertheless, this is not something to be grasped immediately but rather is something which slowly unfolds. It is neither an intellectual pursuit that the mind can grasp like solving a mathematical equation, nor a purely physical training where the mind can be absent. Rather, it is an amalgamation of mind-body discipline, and, hence, is to be grown slowly and methodically through ongoing cultivation.

These five basic skills are not learned in a linear fashion. Although it is easier to explain and understand these five skills when they are examined individually, they are integrated, and degrees of each intertwine with the others. For example, you would not completely understand 放松 fàng sōng before starting to understand 掤劲 péng jìn. Rather, they are collective, with progress in each skill acting as an aid to progress in the others.

It may be debated whether it should be mandatory to start with 掤劲 péng jìn as it is the first skill and is fundamental to all else. On the other hand, without developing 放松 fàng sōng, 掤劲 péng jìn cannot be learned. Either way, notwithstanding the merits of either premise, we commence with the subject of 放松 fàng sōng as a logical first step.

The Five Chen Style Basic Principles of 太极拳 tàijíquán[79]

1. 放 松 Fàng sōng – Loosen the body

放 fàng means to release, to free oneself, to let go, to let out. 松 Sōng means to loosen, to relax. Hence, the emphasis is on letting go. It is not something to be forced. Ironically, when people tell you to relax, you assume that you have to do something. In fact, you are not doing anything other than attaining a relaxed state. It is a state not an action. You cannot force relaxation. It is also not intended to foster a limp, flaccid body posture like you would imagine if you were lying on the sofa eating crisps and watching your football team yet again put on a disappointing display.

放 松 Fàng sōng is sometimes abbreviated to 松 sōng. Now that we have an idea of what the term means, we must provide you with more detail on how it relates specifically to 太极拳 tàijíquán. First of all, the joints must relax, but as a consequence, other parts of the body must work hard, particularly the legs. Loosening the joints is perhaps a better translation. The result should not be a body like a cooked bowl of noodles: rather it should be like a solid piece of rubber, strong but not stiff. The term 放 fàng can have two meanings. The first is about something remaining under control, being connected to both the mind and the body (i.e. in this case not going limp.). The second is to put something down, away from you. The combination of these two meanings provides the precise meaning.

For most people who practise 太极拳 tàijíquán, 松 sōng appears to them early on in their lessons. Unfortunately, most adults, and even many children are much stiffer than they realise. We do not know where we are tight, nor the degree of stiffness we generally maintain in our joints. In 太极拳 tàijíquán, 松 sōng describes the requirement of loosening the joints, relaxing the habitual stiffness from them, getting used to holding them without stiffness, then moving them without stiffness. This

---

[79] This was taken from an interview given to Nick Gudge. This was published in issue 36 of *Tai Chi Chuan & Oriental Arts*. It has been edited slightly for consistency and to conform with the style used in the rest of the book. Nick is a teacher in his own right and a student of Grandmaster Wang Hai Jun.

includes primarily the shoulders and hips, elbows and knees, the spine, and, in particular, the waist, the ankles and the wrists.

When a joint is loosened, it is free to rotate or turn without hindrance or resistance. It is this ability that is required in 太极拳 tàijíquán. The 太极拳 tàijíquán classics talk of even the smallest pressure of a feather or a fly causing movement, like a finely balanced and oiled ball-bearing, where even the lightest touch causes it to rotate.

How do we know when a joint is stiff? Well, initially we do not know that it is stiff, but as a learning tool, it is probably more effective to say that as adults they always are. Usually this is to a much greater degree than we realise. Whilst this is a sobering thought, it is necessary to face up to it. A good teacher helps a student to see where their stiffness lies. The student needs to be shown repeatedly where a joint is stiff. This is because the student neither knows that the stiffness is there nor how to loosen it. Their habit is to move with this stiffness. With practice, the joints become looser, and deeper structural stiffness becomes apparent. As the shoulders loosen, the arms feel heavier. As the hips loosen, the legs work considerably harder. So, for the beginner, heaviness in the arms and the legs working very hard are good indicators that the skill of 放松 fàng sōng is being developed.

Stiffness is difficult to recognise, but the effects of stiffness are easier to see. As the joints stiffen, they rise up. As they are loosened, the body, particularly the hips and shoulders, sink down. For a beginner, it is easy to confuse bending the knees for relaxing the hips (松胯 sōng kuà), and lowering the arms for relaxing the shoulders. One of the many reasons why 太极拳 tàijíquán is called an oral art is that it requires a teacher who understands these principles to show the way. Most people need to be shown the way repeatedly before they understand it in their mind, and then be corrected repeatedly before they understand it in their body. Much practice through this process is required for it to make sense and take hold. Many people get the basic idea in their mind but do not practise enough to realise it in their body.

There is a method or order to ensure progress. The forms of 太极拳 tàijíquán are the framework on which the method is hung. Within the forms, each posture offers an opportunity to understand the various levels of loosening the body.

Around each joint is a structure of muscle. For all the joints that rotate, we can initially consider them as having top, bottom, front and back sections. Each part needs to be trained to loosen before the joint will open properly. As an example, let us look at 放松 fàng sōng in the hips. This process is called 松胯 sōng kuà – loosening the hips. In the hips, usually the stiffness in the top section is most prominent. Once this is loosened, the front becomes more prominent and can be paid attention to. After this, attention can be paid to the back of the hip and then the underneath part. The student needs to be shown where to relax many times until they catch the idea, then practise until the loosening takes place in the body without the need of attention. Once this skill is gained in one part, the mind can be used to address the next area of stiffness. Each student is a little different, but the process is the same.

As each part of a joint is loosened, other parts of the body assume the workload of holding the body. In the beginning, this is mostly felt in the legs. Loosening the hips a little brings a significant additional workload onto the thighs. Until the legs become used to doing this extra work, no more loosening of the hips can be learnt. Loosening the hips a little makes the legs work much harder. Practising with this extra work in the legs makes them stronger. When they have been strengthened in this way and are used to this extra work, more loosening can take place. There is a saying that to gain 太极拳 tàijíquán 功夫 gōng fu, go to bed with tired legs and wake up with tired legs. In other words, loosen the hips so the legs work so hard that even in the morning they are still tired.

Once the joints are loosened, they will be free to rotate properly and to transmit rotation to and from other parts of the body. This is a fundamental requirement of 太极拳 tàijíquán. Any impediment to the joints rotating freely will result in a diminishing of 太极拳 tàijíquán skill. The more joints that are unable to rotate freely, or the greater the resistance within each joint, the less 太极拳 tàijíquán skill will be apparent. The more joints that are able to rotate freely, or the lesser the resistance within each joint, the greater the 太极拳 tàijíquán skill will be apparent. This is why loosening the body is the first, most basic skill in 太极拳 tàijíquán.

It is worth repeating that these are difficult ideas to formulate in words. It is difficult to gain an understanding of them. It is difficult to grasp their skill. They must be shown. The student must be led to them. They are not skills that lend themselves to be grasped intuitively. This is why most people do not get a good basic grounding in them. Many teachers do not have these skills, so naturally it is not possible for their students to gain them. Even today, after more than 20 years of high level 太极拳 tàijíquán and practitioners being available as visiting teachers, the level of these skills is not as high as it might be given the dedication, perseverance and effort of many students.

2. 掤 劲 Péng jìn – an outward supportive strength: the basic skill of 太极拳 tàijíquán

掤 劲 péng jìn is the core skill of 太极拳 tàijíquán. All other 太极拳 tàijíquán skills are based around this skill. It comes from loosening the body by (放 松 fàng sōng and stretching. In essence, 'stretching but not straightening' the joints. 掤 péng is not a natural or instinctive skill. It comes from a long period of correct practice. Without a good understanding of 掤 péng and considerable training to transform this understanding into this skill in every part of the body, it will not arise. 掤 péng will not be gained by accident. It is systematically trained into the body over time.

Whilst 掤 péng should be considered the most important skill, it is dependent on loosening the body. It is an effective argument that 太极拳 tàijíquán is 掤 劲拳 péng jìn quán because without 掤 péng, there is no 太极拳 tàijíquán. It is 太极拳 tàijíquán's essential skill. 掤 péng is always used when moving, neutralising, striking, coiling, etc. Through 掤 péng all other 太极拳 tàijíquán skills are utilised.

The phrase 掤 劲 péng jìn has been the source of some confusion. The two characters, 掤 péng and 劲 jìn, have several meanings in Chinese and specific meanings within the context of 太极拳 tàijíquán. 劲 Jìn is itself not simple to translate into English. There is no one effective word that can be used. It is translated variously as skill, strength and energy, and the term incorporates all of these meanings. 掤 péng is even more difficult to translate. It has been frequently translated as 'ward off energy'. The phrase 'outward supportive strength' will be used as a translation.

陈发科 Chén Fākē taught that there are two types of 掤 劲 péng jìn. The first is the fundamental skill or strength of 太极拳 tàijíquán. The second is one of the eight commonly recognised 太极拳 tàijíquán jìns that we have just covered. The first type of 掤 péng is the core element that is the foundation of these eight commonly recognised skills. It is perhaps best considered in English as a separate term from the 掤 péng that is listed as one of these eight skills. All eight jìns have their basis in 掤 péng, which is the fundamental skill. From the outside, 掤 péng has different appearances so it is sometimes called the bā mén 八門 eight gates (forces), but the heart of all eight is always 掤 péng as the fundamental skill. It is this fundamental skill or strength that we refer to when discussing beginners' skills. In the past 10 years, there has been much talk about 掤 péng. A student cannot simply demonstrate and use 掤 péng just because they will it. It requires external posture training combined with internal 劲 jìn training to be able to correctly express it. If you do not have 掤 péng, you do not have 太极拳 tàijíquán's 劲 jìn, and it follows that you also will not have any of the remaining 劲 jìns.

The fundamental skill of 掤 péng is when the limbs and body stretch or extend whilst maintaining looseness or 放 松 fàng sōng. Without looseness of 放 松 fàng sōng, the body is stiff and 掤 péng is lost. If the body is too loose or limp then 掤 péng is also lost. Without stretching, the body is not properly connected and 掤 péng is lost. If the limbs and body are over extended, they become rigid and 掤 péng is lost. So it is fairly easy to see that a balance must be 脈 maintained in order to retain 掤 péng. If any part of the body does not have 掤 péng, it is an error and must be remedied appropriately. Many form corrections are about regaining 掤 péng to various parts of the body, most commonly the knees and elbows. Typically, 掤 péng is lost or lessened because the body has stiffened or not been loosened sufficiently, most commonly the hips and shoulders.

For those who do not comprehend 掤 péng, it is difficult to observe the presence within the forms. On the other hand, if you are familiar with then, its absence is palpably noticeable. 掤 劲 Péng jìn is not something that you decide to use or not, like a choice. Whilst it is easy not to have it, once it is understood, its quality can be improved. Like any form of understanding, for example, learning a new language, it is quite possible not to understand anything in the beginning. In learning, there are many degrees of improvement or quality that can be sought and reached. From this understanding, it is quite easy to see not only the importance of looseness or 放松 fàng sōng as an integral prerequisite for 掤 péng – this fundamental skill of 太极拳 tàijíquán – but also that improving the appropriate looseness of the body will improve the quality or degree of 掤 péng skill.

It is very difficult to convey the idea of 掤 劲 péng jìn without hands-on correction. The student needs to be led to it after an initial degree of looseness is attained. It is not a 'step one leading to step two' type linear process. Rather it is a process of immersion that leads to understanding. Hands on, frequent correction of the body is required. Understanding the concept generally will not necessarily translate to a complete understanding in the body. It is a process. Wandering off can and does happen frequently. As an old Chen village saying states, what is required is a good teacher, good understanding and good practice. Without all three, 功夫 gōng fu will not be attained. It requires a teacher who understands and can see where the priority of correction is required to enable a student to understand the idea of 掤 péng in their body and not to become distracted or confused.

From inside the body, when 掤 péng is present, any pressure is transferred to the ground (rooted.) The stretching process connects the body in such a way that this happens without additional effort. It could be called a flexible structure inside the body. Consequently, when 掤 péng is present, the body becomes a little like a solid rubber object. It is not rigid, but loose and flexible, where pressure to any part is easily transferred across its whole structure.

When touching someone else, 掤 péng can be described as an audible skill because not only does it allow the detection of fine motions of an opponent (as if through the sense of hearing), it also allows determination of their structural weaknesses. When touching a person with 掤 péng, it becomes possible to know the best direction to attack them as well as being able to comprehend what the other person is doing and even intending to do. Listening skill 听劲 tīng jìn occurs through 掤 劲 péng jìn.

In 太极拳 tàijíquán, the emphasis of 掤 péng is on the leading and neutralising of an incoming force. When 掤 劲 péng jìn is present, there is the potential for rotation. With loose joints, the body becomes mobile and by stretching, it becomes connected. So any pressure on the body causes rotation or motion. 掤 péng is at the heart of 缠丝劲 chán sī jìn energy.

It is also the skill which allows and supports attack. It allows for a rapid response for rapid attack and a slow response for slow offense. In Push Hands 推手 tuī shǒu practice, the student is said to have crossed the threshold only when they have learned the meaning and skill of 掤 劲 péng jìn. Beginners often take years to accomplish this. While practising, when any part of the body, not just the hands and arms, comes into contact with the other person, a 太极拳 tàijíquán practitioner should make use of this outward supportive or warding force.

So, using 掤 péng, a skilled practitioner can not only detect what an opponent is doing, they can neutralise it, detect the direction of vulnerability and attack through it. When this understanding is reached, it is easy to see why it is considered the core skill of 太极拳 tàijíquán.

Where the joints are not loose, 掤 péng is lost. Where the limbs are limp and not stretched, 掤 péng is lost. So beginning with loosening the body, then adding stretching without becoming rigid, the skill that is 掤 劲 péng jìn becomes manifest in the body. Initially at the start and end of each posture, then continuously in the process of motion.

3. 顶 Dǐng – Upright and Straight

The meaning and understanding of 顶 dǐng is not difficult to grasp, though the practice of it takes much more time. 顶 dǐng means upright or straight and dǐng 劲 jìn means upwards pressing skill, strength or power.

When beginning to learn 太极拳 tàijíquán, loosening the body includes loosening the spine. If the body is not held upright, there will be excess muscular activity leading to stiffness. Most people do not know what it is to stand up straight. They have a habit of locking their knees, causing a tilt in the pelvis, which in turn causes their body to lean backwards. This creates significant stiffness around the spine and across the lower and mid torso and hips.

When the body is upright in a state of 顶 dǐng, it becomes possible to loosen the spine, the waist and then the hips. If 顶 dǐng is not present, it is very likely none of these can be achieved. When the student understands and maintains 顶 dǐng and consistently stretches without stiffening to produce 掤 péng in their body and movements, the circulation of 气 qì will become evident to them. As with all things in 太极拳 tàijíquán, this is a process, with consistent and lengthy practice producing results.

More importantly, this upward stretching without stiffening has the effect of lifting excess stresses off the various parts of the spine and allowing them to move freely, similar to the way traction in hospital can free the back from inappropriate strains and pressures so it can move freely. One additional result is that the circulation to the head through the neck is improved. Consequently, the movement of 气 qì around the body becomes more noticeable.

There is a famous 太极拳 tàijíquán saying 虚领 xū lǐng 顶劲 dǐng jìn, which can literally be translated as 'empty leading upward energy'. Its most common English meaning is 'the crown / top of the head is pulled upward as if suspended by a string'. Chen Xin's explanation uses the image of a string pulling upward the 百会 bǎi huì acupuncture point (at the rear of the crown of the head.) When the head is as if suspended or raised upward, the resulting position of the head enables it to turn freely and aid the balance of the body.

So to summarise, 顶 dǐng can be considered the principle that dictates stretching the spine upward to understand and 脉 maintain balance, reduce stiffness and understand and increase both 掤 péng and 放松 fàng sōng . This basic skill is frequently first grasped in standing exercises like 站桩 zhàn zhuāng. The lack of motion allows the student to focus more easily on gaining the correct balance and looseness in the body.

4. 沉 Chén – Rooted

沉 chén has two meanings in 太极拳 tàijíquán. This is in addition to the family name! The first meaning relates to how the body must 'sink' to connect to the ground. The second meaning relates to how the 气 qì must be trained to always be sunk. These two meanings refer to two separate but closely related skills.

The skill of 'sinking' the body is dependent on the skill of 放松 fàng sōng. The joints must remain loose but still be coordinated together. The body is allowed to compress, either using the force of gravity or a force applied from another person. This compression must be directed by the body down the leg without causing stiffness.

Training the 气 qì to remain sunk is more difficult to describe. The reference to 气 qì is difficult for many people to understand, and a direction to do something with 气 qì brings even more difficulties. The 下丹田 xià dān tián must first move freely. The 气 qì moves naturally initially. Then, as the 气 qì increases, it must be kept sunk and not allowed to rise out of control, for example, by getting excited or emotional. It must be sunk down to flood the legs and reach the ground. These actions can be felt clearly and unambiguously in the body and legs when the body and legs have been trained sufficiently and properly.

The combination of these two skills together produces a skill that may appear unbelievable, where a significantly bigger and stronger person is unable to push over a smaller, weaker person; where someone on their back leg can simply push backwards someone opposing them on their front leg. However, it is a basic skill that can be understood and developed with the correct teaching and considerable practice.

Through the skill of 沉 chén, incoming forces are directed down the legs to the ground and conversely outgoing forces are generally pushed from the ground. It requires a mobility in the hips and waist that is difficult to describe and comes from long and hard training in the correct manner. A student must be led to it by a teacher who not only understands the skill of it but also the skill of how to teach it. For strength to be connected to the ground, it must first sink to the ground. In this respect, 沉 chén is closely related to 掤 péng. Without a well-developed 掤 劲 péng jìn, including a mobility of the 下丹田 xià dān tián, a well developed 沉 chén or root will not be possible.

The outgoing force which pushes from the ground is not something mystical but the result of careful training, a coordination of 掤 劲 péng jìn, a certain type of leg and body strength and control. The body acts like a highly specialised and controlled spring. When it is compressed, the pressure goes to the ground, and when it is released, it pushes from the ground. A good root is essential to neutralise and release strength effectively using the internal method of Chen-style 太极拳 tàijíquán.

Understanding and developing a root is initially developed in standing in 站桩 zhàn zhuāng practice as we have covered in the previous chapter. Through correct standing practice, 掤 劲 péng jìn and 顶 劲 dǐng jìn are developed, along with balance and an understanding of 气 qì. Training the body and 气 qì to sink and to remain sunk under pressure is a major focus of 太极拳 tàijíquán forms. In the beginning, moderated and slow movement allows the quickest route to understanding and increased skill. Closing the body during practice – the elbows to the knees and the shoulders to the hips – helps develop the understanding of 沉 chén. Sinking at the start and end of each movement is part of the process of developing 沉 chén.

Practising with longer and lower postures and relaxing the upper body and hips develops the leg strength that is fundamental to developing a strong root. More importantly, loosening the body 放松 fàng sōng, particularly the hips, so your 气 qì naturally sinks to the legs and feet, helps develop a root. The intensity of practice and the strength required increases significantly as the 气 qì sinks more into the legs and a person's root develops. After the skills of 放松 fàng sōng, 掤 péng and dǐng are understood in the body and the mind, and specific rooting exercises can be used to aid the development of 沉 chén.

When rooted under pressure, the feeling is that the joints redirect in a downward direction and the joint itself may move down slightly. It should not be mistaken for crouching down or simply bending the joints. To crouch down low usually provides improved mechanical leverage and requires greater leg strength. The mistake of lowering the body is frequently mistaken for sinking. A lower stance will strengthen the legs but not necessarily develop the root. Initially, the skill of 沉 chén is trained in a more upright position as it takes more skill to be lower and be rooted than to be more upright and rooted.

5. 缠丝劲 chán sī jìn Silk Reeling Skill Energy Skills

The type of motion required in 太极拳 tàijíquán is called silk reeling energy. Although it is listed as the fifth most important skill for beginners to pay close attention to, without 缠丝劲 chán sī jìn, there can be no 太极拳 tàijíquán. 缠 Chán describes how the body must move to move the 气 qì, to maintain 掤 劲 péng jìn and to co-ordinate the constant opening 开 kāi and closing 合 hé of the outside and inside of the body that 太极拳 tàijíquán is composed of.

Again, it is not easy to describe or to understand in its entirety because it needs to be understood more by the body than by the mind. For the body to understand, it needs to be able to approximate 缠丝劲 chán sī jìn motion repetitively until its entirety is grasped. This is done through the continuous and repetitive practice of, initially, 缠丝劲 chán sī jìn exercises and then, more importantly, forms of 套路 tào lù.

For the beginner, 缠丝劲 chán sī jìn can be considered the early training of 太极拳 tàijíquán 身法 shēn fǎ or body mechanics in movement. By following the relatively simple choreography in a progression from simple to more difficult – first with one hand and then with both hands, first stationary then with steps – the beginner will find how the body moves in circles and spirals. For the beginner. the internal movement is not important. Paying attention to winding in 顺缠 shùn chán and winding out 逆缠 nì chán front circle 正面 zhèng miàn and side circle 侧面 cè miàn, and normal direction 正 zhèng and reverse direction 反 fǎn is sufficient. Try to move smoothly and without stiffness. Gain the skill of 放松 fàng sōng by removing the blockages caused by stiffness in the joints. Aim to get all parts of the body to move in a circle and spiral.

This form of spiral movement not only appears on the surface of the skin, but also appears inside through the whole body. It causes every joint and limb to experience motion. Through repeated coiling and stretching in the training for a prolonged period of time, the body will naturally attain resilient and elastic strength, 掤劲 péng jìn that is loose and yet strong at the same time. 缠丝劲 chán sī jìn is the method that the body uses to move so as to retain 掤劲 péng jìn.

In the mid-1980s, teachers from Chen village, notably Grandmasters Chen Xiao Wang and Chen Zheng Lei, among others, created silk reeling exercises, 缠丝功 Chán sī gōng. These exercises were derived from important movements in the training forms to aid in the development of 缠丝劲 chán sī jìn. It became well established quickly as a means of teaching larger groups the basic grasp of movement in Chen style 太极拳 tàijíquán, and is particularly helpful for those without regular access to direction and correction from a good teacher and whose practice time is too short to allow progress in understanding to be made immediately through the traditional forms. Although these sets of exercises may look different from teacher to teacher, they all involve training with the same set of principles.

In summary, we have described the five most important skills at the foundation of 太极拳 tàijíquán for beginners. Our hope is that those who wish to understand and gain the skills of 太极拳 tàijíquán will be aided by these descriptions. In Chen style, we say there are three requirements needed to gain 功夫 gōng fu. These are: a good teacher, good understanding and good practice. History has shown that all everyone who achieved a high level of skill had all three. This book aims to help to provide an understanding of what 太极拳 tàijíquán is and what its skills are. 功夫 gōng fu may be translated as skill, but the idea of time spent is a more useful translation.

In Chen style 太极拳 tàijíquán the 一路 yī lù, first form is the foundation route taken by all those who have reached the third level. While the 老架二路 lǎo jià èr lù weapons forms and push hands form an integral part of the training, too often students reach for these before the basic skills, which they need to use, are established. Without developing the basic skills of 放松 fàng sōng, 掤劲 péng jìn, 顶劲 dǐng jin, 沉 chén and 缠丝劲 chán sī jìn, progress will be limited.

These ideas should not just be explained in a theoretical way, but arrived at after long practice with regular correction and derived from experience. These are not ideas just to be discussed but principles that grow out of the practice and repeated correction from the teacher. It is important that the student understands this and not neglect their practice. In the beginning, training 太极拳 tàijíquán is like paddling upstream: as soon as you stop paddling you will move backwards. So train steadily and without a break with a good teacher and progress will come to you.

Health Benefits

As we have described, Chen style 太极拳 tàijíquán is a composite system of various routines and training exercises aimed at perfecting the martial art. So the question arises: how does this relate to health benefits, other than as a way of exercise like any other? To provide the answer, let us examine this question in some detail. As we have ascertained, the unified whole of a person as a microcosm of the universe is not just a compilation of the individual parts. Providing artificial demarcations and differentiations of different elements such as muscles, ligaments, tendons and blood for the sake of analysis fly in the face of Daoist principles. Nevertheless, if we adopt a granular approach to conduct an analysis, we may ascertain the health benefits within the orthodox categories used.

Although the Chinese have recognised the health benefits of 太极拳 tàijíquán for many centuries, only in the last 25 years or so has the Western medical community conducted scientific research to measure these benefits. These are covered below.

The movements make use of every muscle and joint of the body, which are stretched and also relaxed. Also, the tendons, which are often underappreciated – and where more emphasis is usually placed on developing the muscles – are both strengthened and stretched so that the body becomes strong and supple. The body not only strengthens, it also becomes more relaxed and starts to work in tandem with the mind and the spirit. The stretching comes from within rather than from without, which differs from ordinary stretching exercises. Much research has been undertaken into the benefits of meditation in itself. Slow, meditative practice helps to calm the mind and focus the senses on the present. All in all, the unit that is a human being is re-educated so that he works as one.

Breathing slowly and deeply improves the circulation of the blood and, accordingly, the lymphatic system becomes more efficient. By purifying the blood in this way, the elimination of waste products is also enhanced. Blood flows to the limbs and circulation is improved. This is a benefit that meditation and yoga practices from other cultures have also independently adopted and practised. You will find a universal consensus across these traditions, although they each place a different emphasis on individual practices. The breathing patterns and the 缠丝劲 chán sī jìn massaging of the internal organs, such as the heart, kidneys, liver, lungs, spleen and stomach, and the large and small intestines, provides many health benefits. The effect of the massage of the liver is that it is stimulated, which improves its cleansing function.

Practising the forms will make you stronger and also more supple. The spine is elongated when the body is held in an upright fashion. This allows the vertebrae to align properly and the internal organs attached to the spine to rest in the most optimum position. This works as an antidote to modern living where driving cars, using computers and sitting at desks (writing books like this) for hours at a time will have negative effects on the spine. Studies have shown that the cortisol levels in the body are increased, which are known to reduce mood imbalances. Also, in order to perform the movements, a great deal of concentration and co-ordination are required. This helps to keep the mind alert and makes use of both left and right brain hemispheres. This is especially important for older people in maintaining their mental and physical health. Practising 太极拳 tàijíquán will strengthen the bones and the connective tissue. Again, scientific studies have shown this to be the case. Blood pressure will decrease as a by-product of this exercise. One of the most exciting findings for older adults is the improvement of balance and coordination provided by these exercises.

太极拳 tàijíquán is also an excellent cardiovascular exercise that does not expose the joints to the sorts of pressure that other high impact exercises may do. Although many forms of 太极拳 tàijíquán practised in the West and, indeed, China no longer incorporate the martial arts aspect and, hence, are known as 太极 tàijí – omitting the word 拳 quán (fist), Chen style 太极拳 tàijíquán has low stances and also some fast as well as slow forms that are very rigorous, in particular the 炮捶 pào chuí or Cannon Fist forms. The deep breathing makes use of optimum lung capacity and increases stamina.

In the lower abdomen, glands that produce hormones exist. The rotation of the 下丹田 xià dān tián can stimulate these reproductive glands and make them work to their full potential and enhance the sex life. In China, some doctors use this breathing method as treatment for impotence with great success.

The 站桩 zhàn zhuāng postures, whilst providing all of the health benefits listed above, are a sublime form of meditation in their own right. The benefits of meditation are now well known, from both an esoteric perspective, as well as having been proven by modern scientific studies. Aside from helping one to feel relaxed, reducing stress and avoiding the negative medical ailments that accrue thereof, meditation also produces an increasing level of awareness, attention and higher states of consciousness that feed the spirit. Whilst it is easy to be calm, relaxed and stress free during the practice, you need something to take with you into the manic, stressful and nerve and energy sapping world that you live in. Holding to the principles that you have learned for your day-to-day living will help you not just to survive but to fully embrace life. Recently in the UK, Mindfulness meditation was adopted by some corporations and even members of parliament.[80]

The Composition of 太极拳 tàijíquán

The full range of 太极拳 tàijíquán consists of several individual components. These can be grouped together as follows: the Forms, the Weapons, Push Hands, 气功 qì gōng and combat. Please note that these groupings are not mutually exclusive and that elements of each is embedded within the others. The intention is merely to provide a framework within which to categorise and describe these for you. These categories are described in more detail below.

The Forms

The forms are sequences of movements with a precise order, based on the principles described in the previous chapters of posture and movement. The forms are derived from the lineage of the Chen family and some are developed by particular individuals within that family. As explained

---

[80] Mindfulness meditation is based upon Buddhist awareness techniques that aim to maintain a presence in the immediate moment.

earlier, the starting point was 陈王庭 Chen Wang Ting's codification of the previously more loosely structured practices into a corpus of seven training forms.

By the 14th generation of the family, the forms started to differ into two slightly different routines. These are known today as big frame or large frame, also known as traditional and small frame. Yes it is slightly confusing. 陈王庭 Chen Wang Ting's composition of seven sets was embodied within both the big and small frame traditions, albeit in a slightly modified fashion. The Big Frame sets preceded those of the Small Frame which, confusingly enough, was originally also known as New Frame! Another variable in the equation is that some schools of thought within the 太极拳 tàijíquán fraternity argue the opposite. That is, that Small Frame preceded Big Frame. Apart from a purist's approach to historical enquiry and the love of intellectual pursuits, it need not delay us any further. It may have been easier to have called them Bill or Fred!

The words 'big' and 'small' should not lead the reader to assume that the former implies a set of large, external movements in comparison to the latter's closer, more compact inner movements. The Big and Small Frame traditions are very similar in terms of training methods and the inclusion of the 內功 nèi gōng principles that we have previously covered. It is only the outward appearance that, in the eyes of a less experienced observer, leads to the conclusion that there is a great deal of difference between them.

Big Frame is a composite of the methods taught by 陈照丕 Chen Zhaopei, which was described as 'Old frame' by 陈家沟 Chénjiāgōu village students in the 1970s. Due to the government suppression of 太极拳 tàijíquán and the inability to produce books or any other formal method of communication, 陈照丕 Chen Zhaopei's form was the only one that they were familiar with. This form thus only became known as 'old' when 陈照奎 Chen Zhaokui returned to the village to teach as there was a distinct possibility that the art would die out following the death of 陈照丕 Chen Zhaopei. Hence, the villagers of that generation were exposed to the methods of 陈发科 Chén Fākē through his son 陈照奎 Chen Zhaokui for the first time and ascribed the term 'new' to describe the form.

Anyway let's now move away respectfully from the wish to extract absolutes from relatives and explain about the forms themselves.

The forms are comprised of:

Big Frame or Traditional:

This consists of two main long forms 老架, lǎo jià 一路 yī lù, old frame or Old Frame first form and 老架二路 èr lù, lǎo jià, old frame or Old Frame second form.

老架一路 lǎo jià yī lù – Old Frame first form

This is practised slowly with large movements interspersed with releases of quick and powerful energy, known as fā jìn 发劲, which we have covered previously. As a reminder, 发劲 Fā jìn is delivered through a combination of adopting the correct body structure and alignment together with the flow of internal energy. The required execution of 发劲 fā jìn requires the rooting of the body, moving in a co-ordinated fashion, use of silk-reeling internal energy with the rotation of the 下丹田 xià dān tián and the opening and closing of the 胯 kuà. These are combined with relaxation to produce the speed and power of fā jìn.

Practising 老架一路 lǎo jià yī lù will teach you how to move your whole body in a co-ordinated way, whilst being aligned structurally with the mind focused and concentrated, keeping the body relaxed.

Later, as your practice matures and the body is strong enough, you will develop lower stances whilst still being able to maintain an upright stance and, at the same time, remain relaxed. As you further increase your skill, the movements become smaller and the legs become stronger.

Please note that there are variations in the English translations of these words. As stated previously some of these arcane Chinese terms do not lend themselves to a very precise translation. Hence you will find different terms for such the movement - Lazy About Tucking the Robe (see 3, below).

This can be described as Lazy About Tying the Coat, Lazy About Tying One's Coat and so on. Moreoever, some of the the forms even have different numbers of movements listed. This may be the case where a movement to the left and to the right are combined. You will find differences even within the same school with regard to both English translataions and the number of movements listed.

The form consists of 74 moves, some of which repeat. The number of sections may be categorised differently, depending on which level of detail is described. These are listed below:

First Section

1 Taiji Starting Posture – 太极起势 Tài jíqǐ shì

2 Buddha's Warrior Pounds Mortar – 金刚捣碓 jīn gang dǎo duì

3 Lazy About Tucking the Robe – 懒扎衣 lǎn zā yī

4 Six Sealing and Four Closing – 六封四闭 liù fēng sì bì

5 Single Whip – 单鞭 dān biān

6 Buddha's Warrior Pounds Mortar – 金刚捣碓 jīn gang dǎo duì

7 White Goose Spreads Wings – 白鹅亮翅 bái é liàng chì

8 Diagonal Posture – 斜行 xié xíng

9 Brush the Knee – 搂膝 lǒu xī

10 Twisted Steps – 拗步 ào bù

11 Diagonal Step – 斜行 xié xíng

12 Brush Knee – 搂膝 lǒu xī

13 Twisted Steps – 拗步 ào bù[81]

14 Covering Hand Punch – 掩手肱拳 yǎn shǒu gōng quán

15 Buddha's Warrior Pounds Mortar – 金刚捣碓 jīn gang dǎo duì

Second Section

16 Leaning Body Punch – 撇身捶 piē shēn chuí

---

[81] This can be translated as either Walking Obliquely or Diagonal Walking

17 Black Dragon Emerges From Water – 青 龙 出 水 qīng lóng chū shuǐ

18 Push with Both Hands – 双 推 手 shuāng tuī shǒu

19 Fist Under the Elbow – 肘 底 看 拳 zhǒu dǐ kān quán

20 Roll the Forearm Backward and Step Back – 倒 卷 肱 dǎo juǎn gōng

21 White Goose Spreads Wings – 白 鹅 亮 翅 bái é liàng chì

22 Diagonal Step – 斜 行 xié xíng

23 Flash the Back – 闪 通 背 shǎn tōng bèi[82]

24 Covering Hand Punch – 掩 手 肱 拳 yǎn shǒu gōng quán

25 Six Sealing and Four Closing – 六 封 四 闭 liù fēng sì bì

26 Single Whip – 单 鞭 dān biān

Third Section

27 Wave Hands – 云 手 yún shǒu

28 High Pat on Horse – 高 探 马 gāo tàn mǎ

29 Right Toe Kick – 右 擦 脚 yòu cā jiǎo

30 Left Toe Kick – 左 擦 脚 zuǒ cā jiǎo

31 Left Stump with thre Heel – 左 蹬 一 跟 zuǒ dèng yī gēn

32 Wade Forward with Twisted Steps – 前 趟 拗 步 qián tang ào bù

33 Punch The Ground – 击 地 捶 jī dì chuí

34 Double Flying Kick – 踢 二 起 tī èr qǐ

35 Protecting the Heart Punch – Hu Xin Quan - 护 心 拳 hù xīn quán

36 Tornado Kick – 旋 风 脚 xuán fēng jiǎo

37 Right Stump with the Heel – 右 蹬 一 根 yòu dèng yī gēn

38. Covering Hand Punch – Yan Shou Gong Quan 掩 手 肱 拳 yǎn shǒu gōng quán

39 Small Frame Grappling Strike – 小 擒 打 xiǎo qín dǎ

40 Hold the Head and Push the Mountains – 抱 头 推 山 bào tóu tuī shān

41 Six Sealing and Four Closing – 六 封 四 闭 liù fēng sì bì

42 Single Whip – 单 鞭 dān biān

---

[82] This is also known as Flash through the Back

Fourth Section

43 Cover the Front – 前 招 qián zhāo

44 Cover the Back – 后 招 hòu zhāo

45 Wild Horses Part Mane – 野马分鬃 yě mǎ fēn zōng

46 Six Sealing and Four Closing – Liu Feng Si Bi - 六 封 四 闭 liù fēng sì bì

47 Single Whip – Dan Bian - 单 鞭 dān biān

Fifth Section

48 Fair Lady Works with Shuttles – 玉 女 穿 梭 yù nǚ chuān suō

49 Lazy About tucking the Robe – 懒 扎 衣 lǎn zā yī

50 Six Sealing and Four Closing – 六 封 四 闭 liù fēng sì bì

51 Single whip – 单 鞭 dān biān

Sixth Section

52 Wave Hands – 云 手 yún shǒu

53 Double Lotus Kick followed by Falling Split – 摆 腳 跌 叉 bǎi jiǎo diē chǎ

54 Golden Rooster Standing on One Leg – 金 鸡 独 立 jīn jī dú lì

55 Roll the Forearms backwards and Step Back – 倒 卷 肱 dào juǎn gōng

56 White Goose Spreads Wings – 白 鵝 亮 翅 bái é liàng chì

57 Diagonal Step – 斜 行 xié xíng

58 Flash the Back - 閃 通 背 shǎn tōng bèi

59 Covering Hand Punch – 掩 手 肱 拳 yǎn shǒu gōng quán

60 Six Sealing and Four Closing – 六 封 四 闭 liù fēng sì bì

61 Single Whip – 单 鞭 dān biān

Seventh Section

62 Wave Hands – 云 手 yún shǒu

63 High Pat on Horse – 高 探 马 gāo tàn mǎ

64 Cross Kick – 十 字 脚 shí zì jiǎo

65 Punch the Groin – 指 裆 捶 zhǐ dāng chuí

66 Ape Reaches for Fruit – 白 猿 探 果 bái yuán tàn guǒ

67 Single Whip – 单 鞭 dān biān

Eighth Section

68 Dragon Creeps on the Ground – 雀 地 龙 què dì lóng

69 Step up with Seven Star Punch – 上 步 七 星 shàng bù qī xīng

70 Step Back and Hold with Forearms – 下 步 跨 虎 xià bù kuà hǔ

71 Double Lotus Kick – 转身双 摆 莲 zhuǎn shēn shuāng bǎi lián

72 Cannon Fist – 当 头 炮 dāng tóu pào

73 Buddha's Warrior Pounds Mortar – 金 刚 捣 碓 jīn gang dǎo duì

74 Closing Posture – 收 势 shōu shì

Repeat Sequences

- Buddha's Warrior Pounds Mortar – 金 刚 捣 碓 jīn gang dǎo duì is repeated four times in moves 2, 6, 15 and 73

- Six Sealing and Four Closing – 六 封 四 闭 liù fēng sì bì is repeated six times in moves 4, 25, 41, 46, 50 and 60.

- Single Whip – 单 鞭 dān biān is repeated seven times in moves 5, 26, 42, 47, 51, 61 and 67.

- Wave Hands – 云 手 yún shǒu is repeated three times in moves 27, 52 and 62.

- Lazy About Tucking the Robe – 懒 扎 衣 lǎn zā yī is repeated twice in moves 3 and 49.

- White Goose Spreads Wings – 白 鹅 亮 翅 bái é liàng chì is repeated three times in moves 7, 21 and 56.

- Diagonal Step – 斜 行 xié xíng is repeated four times in moves 8, 11, 22 and 57

Removing the repeated movements above, there are 62 discrete individual movements to learn in the form. Some of the 74 movements are selected for the 18 form movement that we shall cover in some detail later on in the book.

老架二路 èr lù, lǎo jià, old frame or Old Frame second form

The second empty hand form, 老架二路 lǎo jià èr lù, or Old Frame second form, is also known as 炮捶 pào chuí or Cannon Fist. The form is practised once 一路 yī lù has been mastered. It is a form that incorporates faster movements and is very martial in its appearance. It incorporates advanced martial techniques and utilises the 发劲 fā jìn methods. It should only be practised once the student has attained a level at which he or she is able to release energy in a relaxed and natural manner. Forcing the energy is counterproductive and may even harm the body if practised incorrectly.

There are 41 movements, some of them the same as in 老架, lǎo jià old frame 一路 yī lù.

1. Starting Posture – Tai Ji Shi 太极起势
2. Buddha's Warrior Pounds Mortar – 金刚捣碓 jīn gang dǎo duì
3. Lazy About Tucking the Robe – 懒扎衣 lǎn zā yī
4. Six Sealings and Four Closings – 六封四闭 liù fēng sì bì
5. Single Whip – 单鞭 dān biān
6. Leap with Protecting the Heart Punch – 护心拳 hù xīn quán
7. Diagonal Step – 进步斜行 jìn bù xié xíng
8. Turn Around with Buddha's Warrior Pounds Mortar – 回头金刚捣碓 huí tóu Jīn gāng dǎo duì
9. Leaning Body Punch – 撇身拳 piē shēn quán
10. Strike the Groin – 指裆捶 zhǐ dāng chuí
11. Chop Hand – 斩手 zhǎn shǒu
12. Dance with the Sleeves – 翻身舞袖 fān shēn wǔ xiù
13. Covering Hand Punch – 掩手肱拳 yǎn shǒu gōng quán
14. Turn Around with Cross Elbow Strike – 腰拦肘 yāo lán zhǒu
15. Upper and Lower Arm Strikes – 大肱拳小肱拳 xiǎo gōng quán dà gōng quán
16. Fair Lady works with Shuttles – 玉女穿梭 yù nǚ chuān suō
17. Ride on the Dragon Backward – 倒骑龙 dào jì lóng
18. Covering Hand Punch – 掩手肱拳 yǎn shǒu gōng quán

19. Wrapping Punches – 裹鞭炮 guǒ biān pào

20. Beast Head Posture – 兽头势 shòu tóu shì

21. Chopping Posture – 披架子 pī jià zǐ

22. Covering Hand Punch – 掩手肱拳 yǎn shǒu gōng quán

23. Tame the Tiger – 伏虎 fú hǔ

24. Brush the Eyebrows – 抹眉肱 mǒ méi gōng

25. Yellow Dragon Stirs Up Water– 黄龙三搅水 Huáng long sān jiǎo shuǐ

26. Dash to the Left and Dash to the Right – 左冲右冲 zuǒ chōng yòu chōng

27. Covering Hand Punch – 掩手肱拳 yǎn shǒu gōng quán

28. Sweep Kick – 扫堂腿 sǎo táng tuǐ

29. Covering Hand Punch – 掩手肱拳 yǎn shǒu gōng quán

30. Full Cannon Fist– 全炮捶 quán pào chuí

31. Covering Hand Punch – 掩手肱拳 yǎn shǒu gōng quán

32. Pound and Split – 捣叉捣叉 dǎo chā dǎo chā

33. Left and Right Continuous Strikes – 左二肱右二肱 zuǒ èr gōng yòu èr gōng

34. Turning Around with Cannon Fist – 回头当门炮 huí tóu dāng mén pào

35. Switch Position with Cannon Fist – 变势大捉炮 biàn shì dà zhuō pào

36. Cross Elbow Strike – 腰拦肘 yāo lán zhǒu

37. Straight Elbow Strike –顺拦肘 shun lán zhǒu

38. Cannon into the Nest– 窝底炮 wō dǐ pào

39. Turn Around and Drop the Pulley Rope into the Well – 回头井拦直入 huí tóu jǐng lán zhí rù

40. Buddha's Warrior Attendant Pounds Mortar – 金刚捣碓 jīn gang dǎo duì

41. Closing Posture – 收势 shōu shì

Repeat Sequences

金刚捣碓 jīn gang dǎo duì is repeated three times so there are 41 unique movements.

新架 Xīn Jià – new frame

This style believed to have first been introduced by 陈发科 Chén Fākē during the 1950s and he is regarded as the progenitor of the routine. Later, it was taught to his son, Chen Zhaokui, and his fellow students. This is not a new form in terms of being a completely devised system; rather, it is a different way of practising the same fundamental movements. The movements are smaller than 老架, lǎo jià, old frame, and it incorporates more of the 缠丝劲 chán sī jìn. It places more emphasis on the qìn na techniques that we mentioned earlier. It also includes the existing striking techniques.

Chen Zhaokui returned to 陈家沟 Chénjiāgōu to further promote the family's 太极拳 tàijíquán and thereafter succeeded Chen ZhaoPei. He trained today's modern generation of grandmasters. These masters are known as the Four Buddha's Warriors and include Grandmaster Chen Zheng Lei. The other three are:

Grandmaster Chen Xiao Wang

Grandmaster Chen Xiao Wang was born in 1946 and commenced his training under his father, Chen Zhao Xu, who unfortunately died when he was just 14 years old, whereupon he trained under Chen Zhao Pei and later with Chen Zhao Kui, who taught in Chénjiāgōu village, following Chen Zhao Pei's death in 1972.

Grandmaster Wang Xian

Grandmaster Wang Xian was born in Xian in 1945, and was a student of both Chen Zhao Pei and Chen Zhao Kui. He was head of Chénjiāgōu village during the persecutions of the Cultural Revolution. He was awarded the title of one of 13 Grandmasters of 太极拳 tàijíquán in China (Da Shi) in 1996.

Grandmaster Zhu Tian Cai

Grand Master Zhu Tian Cai was born in 1945 and was a student of both Chen Zhao Pei and Chen Zhao Kui. He was awarded the title of one of 13 Grandmasters of 太极拳 tàijíquán in China in 1996.

The purpose of 新架 Xīn Jià is to emphasise the silk reeling movements in order to help beginners to learn the internal principles in form.

In 陳家溝 Chénjiāgōu, 新架 Xīn Jià is traditionally learned only after 老架, lǎo jià, old frame. Like 老架, 老架, lǎo jià, old frame, 新架 Xīn Jià consists of two routines, 一路 yī lù and er lu – cannon fist. It consists of 72 movements.

The New Frame First Routine consists of 83 postures, 新架一路 Xīn Jià yī lù

1. Preparing Form – 太极起势 Tài jíqǐ shì
2. Buddha's Warrior Pounds Mortar – 金 刚 捣 碓 jīn gang dǎo duì
3. Lazy About Tucking the Robe – 懒 扎 衣 lǎn zā yī
4. Six Sealing And Four Closing – 六 封 四 闭 liù fēng sì bì
5. Single Whip – 单 鞭 dān biān
6. Second Buddha's Warrior Pounds Mortar –金 刚 捣 碓 jīn gang dǎo duì
7. White Crane Spreads Wings –白 鹅 亮 翅 bái é liàng chì
8. Twisted Diagonal Step – 斜 行 xié xíng
9. First Closing - 初收 chū shōu
10. Twisted Wade Forward – 前膛坳步 qián táng ào bù
11. Second Twisted Diagonal Step – 斜行坳步 xié xíng ào bù
12. Second Closing – 再收 zài shōu
13. Twisted Wade Forward– 前膛坳步 qián táng ào bù
14. Covering Hand Punch – 掩手肱拳 yǎn shǒu gōng quán
15. Third Buddha's Warrior Pounds Mortar – 金刚捣碓 jīn gang dǎo duì
16. Leaning Punch – 撇身捶 piē shēn chuí
17. Black Dragon Emerges from Water – 青龙出水 Qīng Lóng chū shuǐ
18. Push With Both Hands – 双推手 shuāng tuī shǒu
19. Switch the Palms Three Times – San Huan Zhang 三换掌 sān huàn zhǎng
20. Fist Under the Elbow – 肘底看捶 zhǒu dǐ kàn chuí
21. Roll the Forearm Backwards – Dao Juan Gong 倒卷肱 dào juǎn gōng

22. Press the Elbow on a Backward Step – 退步压肘 tuì bù yā zhǒu

23. Middle Winding – 中盘 zhōng pán

24. White Crane Spreads Wings – 白鹤亮翅 bái é liàng chì

25. Twisted Diagonal Step – 斜行坳步 xié xíng ào bù

26. Flash the Back – 闪通背 shǎn tōng bèi

27. Covering Hand Punch – 掩手肱拳 yǎn shǒu gong quán

28. Large Frame Six Sealing and Four Closing – 六封四闭 liù fēng sì bì

39. Single Whip – 单鞭 dān biān

30. Wave Hands – 云手 yún shǒu

31. High Pat on Horse – 高探马 gāo tàn mǎ

32. Right Toe Kick – 右擦脚 yòu cā jiǎo

33. Left Toe Kick – 左擦脚 zuǒ cā jiǎo

34. Turn around and Left Stomp Kick – 左蹬一根 zuǒ dèng yī gēn

35. Wade Forward on Twisted Steps – 前膛坳步 qián táng ào bù

36. Punching the Ground – 击地捶 jī dì chuí

37. Double Flying Kick – 踢二起 tī èr qǐ

38. Protecting Heart Punch – 兽头势 shòu tóu shì

39. Tornado Kick – 旋风脚 xuàn fēng jiǎo

40. Right Stomping Kick – 右蹬一根 yòu dèng yī gēn

41. Covering Hand Punch – 掩手肱拳 yǎn shǒu gōng quán

42. Small Posture Grappling Strike – 小擒打 xiǎo qín dǎ

43. Hold the Head and Push the Mountain – 抱头推山 bào tóu tuī shān

44. Switch the Palms Three Times – 三换掌 sān huàn zhǎng

45. Six Sealing and Four Closing – 六封四闭 liù fēng sì bì

46. Single Whip – 单鞭 dān biān

47. Cover the Front – 前招 qián zhāo

48. Cover the Back – 后招 hòu zhāo

49. Wild Horse's Part Mane – 野马分鬃 yě mǎ fēn zōng

50. Large Posture Six Sealing and Four Closing – 六封四闭 liù fēng sì bì

51. Single Whip – 单鞭 dān biān

52. Double Stomping – 双震脚 shuāng zhèn jiǎo

53. Fair Lady Works at Shuttles – 玉女穿梭 yù nǚ chuān suō

54. Lazy About Tucking the Robe – 懒扎衣 lǎn zā yī

55. Six Sealing and Four Closing – 六封四闭 liù fēng sì bì

56. Single Whip – 单鞭 dān biān

57. Wave Hands – Yun Shou 云手 yún shǒu

58. Double Lotus Kick – 双摆脚 shuāng bǎi jiǎo

59. Falling Split – Die Cha 跌岔 diē chà

60. Golden Rooster Stands on One Leg – 金鸡独立 jīn jī dú lì

61. Roll the forearm backwards – 倒卷肱 dào juǎn gōng

62. Press the Elbow on a Backward Step – 退步压肘 tuì bù yā zhǒu

63. Middle Winding – 中盘 zhōng pán

64. White Crane Spreads Wings – 白鹤亮翅 bái é liàng chì

65. Twisted Diagonal Step – 斜行 xié xíng

66. Fan the Back – 闪通背 shǎn tōng bèi

67. Covering Hand Punch – 掩手肱拳 yǎn shǒu gōng quán

68. Large Posture Six Sealing and Four Closing – 六封四闭 liù fēng sì bì

69. Single Whip – 单鞭 dān biān

70. Wave Hands – 云手 yún shǒu

71. High Pat on Horse – 高探马 gāo tàn mǎ

72. Cross Single Lotus Kick – Shi Zi Dan Bai Jiao 十字单摆脚 shí zì dān bǎi jiǎo

73. Punch to the Groin – 指裆捶 zhǐ dāng chuí

74. White Ape Reaches for Fruits – 白猿探果 bái yuán tàn guǒ

75. Small Posture Six Sealing and Four Closing – 六封四闭 liù fēng sì bì

76. Single Whip – 单鞭 dān biān

77. Cover the Ground with Silk – 铺地锦 pū dì jǐn

78. Step Up with Seven Star Punch – 上步七星 shàng bù Qī xīng

79. Step Back and Mount the Tiger – 退步跨虎 tuì bù kuà hǔ

80. Turn around with Double Lotus Kick – 转身双摆莲 zhuǎn shēn shuāng bǎi lián

81. Cannon Fist Right in the Face – 当头炮 dāng tóu pào

82. Buddha's Warrior Pounds Mortar – 金刚捣碓 jīn gang dǎo duì

83. Closing Posture – 收势 shōu shì

The New Frame Second Routine 新架二路 Xīn Jià Er Lu

As with lǎo jià 老架 the second routine of Xīn Jià 新架 is the 炮捶 pào chuí, cannon fist form.

1. Preparation Posture – 太极起势 Tài jíqǐ shì

2. Buddha's Warrior Pounds Mortar – 金刚捣碓 jīn gang dǎo duì

3. Lazy about Tucking the Robe – 懒扎衣 lǎn zā yī

4. Six Sealing and Four Closing – 六封四闭 liù fēng sì bì

5. Single Whip – 单鞭 dān biān

6. Horizontal Elbow – 搬拦肘 bān lán zhǒu

7. Protecting the Heart Punch – 护心捶 hù xīn chuí

8. Twisted Diagonal Step – 坳步斜行 ào bù xié xíng

9. Twist the Waist and Punch with Elbow Press – 煞腰压肘拳 shā yāo yā

10. Pulley Rope Drops in the Well – 井缆直入 jǐng lǎn zhí rù

11. Wind Sweeps Plum Blossoms – 风扫梅花 fēng sǎo méi huā

12. Buddha's Warrior Pounds Mortar – 金刚捣碓 jīn gang dǎo duì

13. Leaning Body Punch – 撇身捶 piē shēn chuí

14. Chopping Hand – 斩手 zhǎn shǒu

15. Fancy Waves with the Sleeves – 翻花舞袖 fān huā wǔ xiù

16. Covering Hand Punch – 掩手肱拳 yǎn shǒu gōng quán

17. Leap with Horizontal Elbow – 飞步腰拦肘 fēi bù yāo lán zhǒu

18. Wave Hands (First Three) 云手（前三） yún shǒu (qián sān)

19. Wave Hands (Second Three) 云手（后三） yún shǒu (hòu sān)

20. High Pat on Horse – 高探马 gāo tàn mǎ

21. Continuous Bombardment – 连珠炮 lián zhū pào

22. Ride Backward on a Deer – 倒骑麟 dào qí lín

23. White Snake Spits its Tongue – 白蛇吐信 bái shé tǔ xìn

24. Stir Up Turbulence Under the Sea – 海底 翻花 hǎi dǐ fān huā

25. Covering Hand Punch – 掩手肱拳 yǎn shǒu gōng quán

26. Left Side Wrap and Punch – 左裹鞭炮 zuǒ guǒ biān pào

27. Right Side Wrap and Punch – 右裹鞭炮 yòu guǒ biān pào

28. Beasts Head Pose – 兽头势 shòu tóu shì

29. Chopping Posture – 披架子 pī jià zi

30. Fancy Waves with Sleeves – 翻花舞袖 fān huā wǔ xiù

31. Covering Hand Punch – 掩手肱拳 yǎn shǒu gōng quán

32. Tame the Tiger – Fu Hu 伏虎 fú hǔ

33. Brush the Eyebrows – 抹眉红 mǒ méi hóng

34. Right Side Yellow Dragon Stirs Water 右黄龙三搅水 yòu Huáng lóng sān jiǎo shuǐ

35. Left Side Yellow Dragon Stirs Water 左黄龙三搅水 zuǒ Huáng lóng sān jiǎo shuǐ

36. Left Heel Kick – 左蹬一根 zuǒ dèng yī gēn

37. Right Heel Kick – 右蹬一根 yòu dèng yī gēn

38. Stir Up Turbulence Under the Sea – 海底 翻花 hǎi dǐ fān huā

39. Covering Hand Punch – 掩手肱拳 yǎn shǒu gōng quán

40. Sweeping Kick – 扫膛腿 sǎo tang tuǐ

41. Covering Hand Punch – 掩手肱拳 yǎn shǒu gōng quán

42. Double Punch to the Left – 左冲 zuǒ chōng

43. Double Punch to the Right – 右冲 yòu chōng

44. Backward Thrust – 倒插 dào chā

45. Stir Up Turbulence Under the Sea – 海底 翻花 hǎi dǐ fān huā

46. Covering Hand Punch – 掩手肱拳 yǎn shǒu gōng quán

47. Fight for Control with the Forearm – 夺二肱 duó èr gōng

48. Continuous Punches – 连环炮 lián huán pào

49. Fair Lady Works with Shuttles – 玉女穿梭 yù nǚ chuān suō

50. Turn around with Double Punch – 回头当门炮 huí tóu dāng mén pào

51. Fair Lady Works with Shuttles – 玉女穿梭 yù nǚ chuān suō

52. Turn around with Double Punch – 回头当门炮 huí tóu dāng mén pào

53. Horizontal Blocking Elbow – 腰拦肘 yāo lán zhǒu

54. Straight Elbow – 顺拦肘 shùn lán zhǒu

55. Piercing the Heart Elbow – 穿心肘 chuān xīn zhǒu

56. Cannon into the Nest – 窝里炮 wō lǐ pào

57. Pulley Rope Drops in the Well – 井缆直入 jǐng lǎn zhí rù

58. Wind Sweeps Plum Blossoms – 风扫梅花 fēng sǎo méi huā

59. Buddha's Warrior Pounds Mortar – 金刚捣碓 jīn gang dǎo

60. Closing Posture – 收势 shōu shì

Modern Chen forms

Forms have been adopted for competition to fit within the overall framework of 武术 Wǔshù competitions. A 56-movement Chen Competition form was developed by the Chinese National 武术 Wǔshù Association, based upon 新架 xīn jià new frame routines.

There is also a 48/42 Combined Competition form that was developed by the Chinese Sports Committee from Chen and three other traditional styles between 1976 and 1989.

Modern Grandmasters have developed short forms for students that take less time to learn and perform. This also helps in workshops, when time is limited.

The short 18 form that this book focuses on was developed by Grand Master Chen Zheng Lei some 15 years ago.

Weapon forms

Chen style 太极拳 tàijíquán has several weapons forms:

- Broadsword 大刀 dà dāo 13-posture form
- Halberd 戟 jǐ 30-posture form
- Spear 枪 qiāng solo and partner forms
- Staff 杖 zhàng 3, 8, and 13 posture forms
- Straight Sword 49 posture 剑 jiàn form
- Double weapons forms based on sword 剑 jiàn and spear 枪 qiāng and 大刀 dà dāo

As we have noted, 站桩 zhàn zhuāng is a fundamental art of the practice, as is 缠丝劲 chán sī jìn which we will cover soon.

That concludes the overview of the types of forms that you will encounter if you fully immerse yourself in learning this martial art. It may seem daunting, but please remember that these are all inter-related and, hence, once you have mastered the basics, moving through the later stages can be achieved more easily.

Regard your practice like driving a car. The car can be considered as akin to the body and must be strong, with a firm chassis, and be able to change speed with regard to conditions on the road. This requires the alignment of the spine, relaxation and a strong and flexible 胯 kuà and 下丹田 xià dān tián. It must be economical with a tank that contains sufficient fuel so the 下丹田 xià dān tián stores sufficient 气 qi for it to flow through the body. Once the car is efficient, fast and powerful, the driver, which is the mind, can navigate anywhere and in all road conditions until car and driver become one.

A beginner with say less than five years' experience may become confused when they see in real life or on YouTube many different variations of the forms. It is important to understand this difference from several perspectives.

First of all, just because you see someone with a Chinese name in a tàijí suit, it does not mean they have any meaningful skill.

Additionally, even if the description says they are a student of a bona fide teacher, who is recognised in China as having a high level of skill, it may be because that person may have only done a few workshops with a good teacher and learned the rest from videos and books.

During the 1980s and 1990s, there was lots of so-called 'masters' appearing in the West. Another misleading factor is Duan grading. When it was first introduced, it was fairly tight but with time these grades were being handed out fairly easily. As the Chinese government took power away from the traditional families that held the skills of Taiji, and in an effort to popularise Chinese culture, they created standard forms like the Beijing 24 movement form and others.

As their goal was to spread Chinese culture worldwide through sports, they created the Duan system and offered higher Duan grading for teachers who published books, especially in foreign languages.

Some people therefore ended up with higher grading than genuinely skilled people. Also, if they graded more recently, the difference in skill would be even higher.

The true benchmark for skill should always be teachers and practitioners who are recognized by those in the know in China as having a high level skill. One needs to do one's research!

At the next level, when you compare bona fide teachers' forms, these have obvious and more nuanced differences. The reasons are mainly twofold. One is that each person's body is different and so their energy flow is stronger doing the form in one way more than another. They also may be showing externally different variations of the huge number of applications of the form.

Internally, however, the principles of movement will be the same, albeit that this is not something many people will notice until they have practised for some time. The other reason is they may have studied with the same teacher but at a different moment in that teacher's development.

One has just to look at Chen Zhaopei's students Chen Xiao Wang, Chen Zheng Lei Wang Xian and Zhu Tian Cai. They all move differently but had the same teacher and at the same time! But they are each individual and, hence, there are nuances in their expression of the form. So, do not be robotic and ape things as though you are a clone.

One's movements should not be exactly like one's teacher at a higher level. But in the beginning, once you find the right teacher for you, you should follow their movements until you begin to understand taiji's inner essence.

It is healthy to try different teachers in the beginning, but later, if you have lots of teachers without good understanding, your movements will be confused and it will take you longer to understand your own internal Jin and qi.

Because of the diverse range of ways to do the same form and the lack of judges who understand the differences, the competition forms were created based on traditional forms and shortened as a way to standardise routines. It sets a clear model for people to compete so judges can evaluate things like balance, co-ordination, flexibility and power. However, these forms are not suitable to increase gong fu. In the competition form, there is a format, or a way of performing, that cannot be changed. Therefore, it provides a common ground for judging.

For the sake of competition, the forms are set with no deviation, and so they are dead in a way.

The main martial arts training form is 老架一路 lǎo jià yī lù and it should be studied with a skilled teacher and practised seriously. You can study in a group but you are advised to practise alone to enter an undistracted state that will help you to deeply co-ordinate internally and externally. We will cover this in Volume Two of this series.

If you are a beginner and need the support of a practice group to remember the movements and discipline yourself, this is fine, but you should with time be able to train alone.

Of course, it would be a mistake to think that were you to be following your internal energy flow then you would be doing the movements differently to your teacher, and hence incorrectly. When you are connected, you will not need anyone to tell you because you will intuitively and experientially know. The test to ascertain whether you are correct is to spend some time pushing hands with a Master.

Now to study the form in physical detail.

Chapter 6 – Warm-Up Exercises

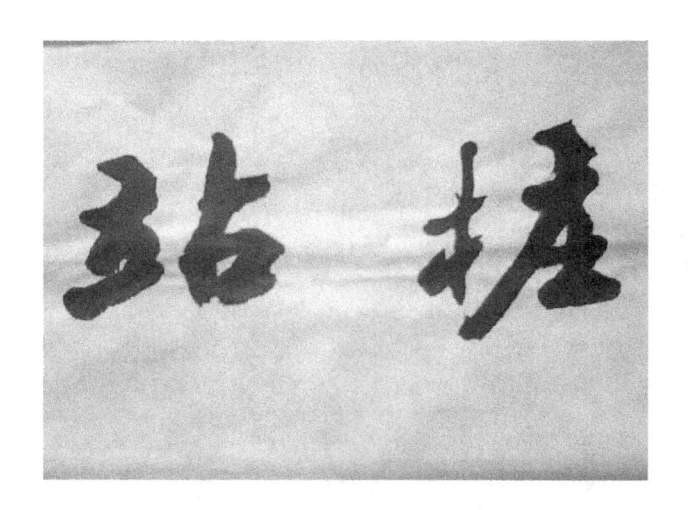

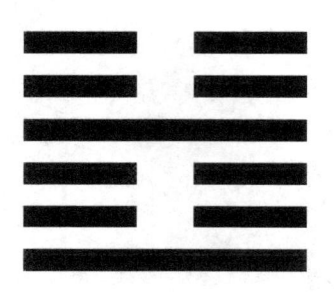

站樁

☳ Zhèn

道生一. 一生二.二生三. 三生萬物.

萬物負陰而抱陽. 沖氣以為和.

　人之所惡唯孤寡不穀, 而王公以為稱.

故物或損之而益.或益之而損.

人之所教, 我亦教之.

強梁者不得其死.吾將以為教父.

The Dao produced One; One produced Two; Two produced Three; Three produced All things. All things leave behind them the Obscurity out of which they have come, and go forward to embrace the Brightness into which they have emerged, while they are harmonized by the Breath of Vacancy. What men dislike is to be orphans, to have little virtue, to be as carriages without naves; and yet these are the designations which kings and princes use for themselves. So it is that some things are increased by being diminished, and others are diminished by being increased. What other men thus teach, I also teach. The violent and strong do not die their natural death. I will make this the basis of my teaching.

Translated by James Legge, 1891, Chapter 42

The purpose of the warm-up exercises is to loosen up all of the major joints so that the body is ready to undertake the exercises. It also ensures that you are relaxed and flexible and that the mind is calm and unperturbed. The exercises should be performed slowly and in a relaxed fashion. This is a good way to properly prepare for the practice. In a class setting, it provides a way of getting the class integrated and relaxed and ready – aside from those who perpetually turn up late half way through!

All images are to be read from left to right, top to bottom as you work through the form.

1. **Loosening the wrists**

Overall exercise

Stand with your legs at more than shoulder-width apart and with the head and neck raised but remaining relaxed. Bring both hands together and intertwine the fingers. Rotate your wrists clockwise and then counter-clockwise. Ensure that you remain relaxed and that your shoulders do not hunch or stiffen. Isolate the movement to the wrists. Let the wrists move the hands whilst the rest of the body remains relaxed. Aim for eight repetitions in either direction. Breathe in and out with the movements but do not force your breathing. Do not use any specialised type of breathing other than using the abdomen. Just take long, slow and relaxed breaths. When you are exercising alone, you have the artistic licence to undertake as many repetitions as you like.

Constantly using mobile phones or laptops to type will strain your wrists, so loosening them up in this way will help to address this modern-day problem.

Detailed exercise

Figure 6-1

Remain balanced and rooted. The shoulders are relaxed and are neither hunched nor raised up. The hands are together and the fingers are intertwined at chin height, at about six inches in front of the body. The feet are around shoulder width or more apart.

Figure 6-2

With the hands clasped together and the right hand over the left, circulate the wrists up and down and from side to side. Do not let the shoulders hunch up.

Figure 6-3

With the hands still clasped together, leading with the left hand over the right continue to circulate the wrists up and down and from side to side. Remember to remain relaxed and pay attention to your posture. Avoid the temptation to keep looking down at the floor.

2. Stretching to all sides

Overall exercise

This warm-up consists of stretching in four directions: stretching forwards, stretching up, stretching to either side, and stretching down. Please be careful to avoid stretching too far and ensure that you do not strain yourself. It is your body and so you know what the limitations are. This does not mean avoiding a stretch but doing it safely. For each stretch, loosen and extend from the feet and through to the hands. The whole body stretches, not just the upper torso. Remaining as you were standing, with your legs at more than shoulder width apart and the head and neck raised but staying relaxed with both hands together and the fingers intertwined.

Detailed exercise

2.1. Stretching Forwards

This stretch is forwards with the arms, with the body kept relaxed, the hips open and the legs slightly bent.

Figure 6-4

Remain in the relaxed and balanced posture and lace the fingers together. Stretch out from the shoulders, leading with the hands and wrists.

Figure 6-5

Stretch out fully. Avoid the temptation to raise the shoulders. Relax and stretch out all of the way through to the fingers.

Stretching Up

Detailed exercise

This stretch is upwards in direction with the arms. The body remains relaxed, the hands clasped together, the hips open and the legs slightly bent.

Figure 6-6

Remain in the relaxed and balanced posture with the fingers still intertwined. Stretch out from the shoulders upwards, leading with the hands and wrists. Stretch the whole body from the feet upwards. Keep looking forwards and remain relaxed.

Figure 6-7

Stretch out fully and follow the hands with the eyes, raising your head. Avoid the temptation to raise the shoulders. Relax throughout. This is a qigong exercise in its own right in some traditions and is very beneficial for your health.

2.2. Stretching down

Slowly bend from the waist and straighten the legs with the hands still clasped together, and very slowly bend forwards with the hands leading and the head following. Do not stretch further than you feel is comfortable. These are not gymnastic exercises. In fact, they are the opposite, designed for relaxing and slowly lengthening the tendons.

Detailed exercise

Figure 6-8

Stretch down slowly and follow the hands with the eyes, lowering your head. Ensure that your feet are solidly placed and that you feel balanced. Slowly lower from the waist and only stretch as far as is comfortable. Do not jolt. Perform the movement slowly, ensuring that you do not place too much strain on the lower back.

Figure 6-9

Continue to stretch down as far as you can from the waist and follow the hands with your eyes, lowering your head further. Ensure that your feet are planted solidly and that you feel balanced. Try to touch the ground with your hands. Stretch as far as you can but do not force this as you may injure your back. Also, remember that in class you are not in a competition. Some people are more flexible in some directions and positions than others. This exercise is for you and nobody else, so pay attention to yourself. This stretching should be performed slowly and in a relaxed manner. Stretch down and slightly upward a few times and then raise yourself slowly back upright to the standing position.

295

2.4. Stretching to the side

This is a stretch to both the left and the right, commencing with the former. Keep your feet firmly planted on the floor and stretch to both sides with your hands intertwined. Stretch as far as is comfortable without straining. Of course, there is no cardinal rule as to which side to stretch to first. In class, follow the lead of your teacher. On your own, follow your own preference.

Detailed exercise

Figure 6-10

Keep your feet planted on the floor and with the fingers intertwined, stretch to the left in a sideways direction. Stretch as far as you feel you can comfortably and remain relaxed. The stretching is through the hands from the waist. If you are not used to performing this exercise, you will notice some stiffness, but each time you will be able to stretch a little bit further. It is a particularly effective antidote to sitting down at a desk all day feeling as though you are being treated as a battery hen.

Figure 6-11

With your feet remaining planted on the floor, and with the fingers intertwined, stretch to the right in a sideways direction. Stretch as far as you feel you can comfortably and remain relaxed.

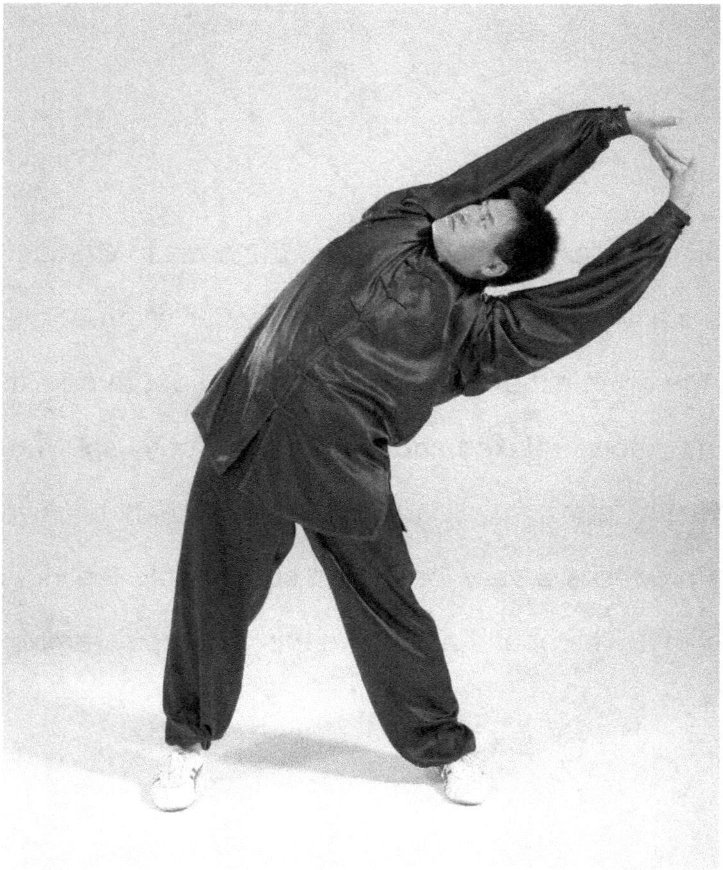

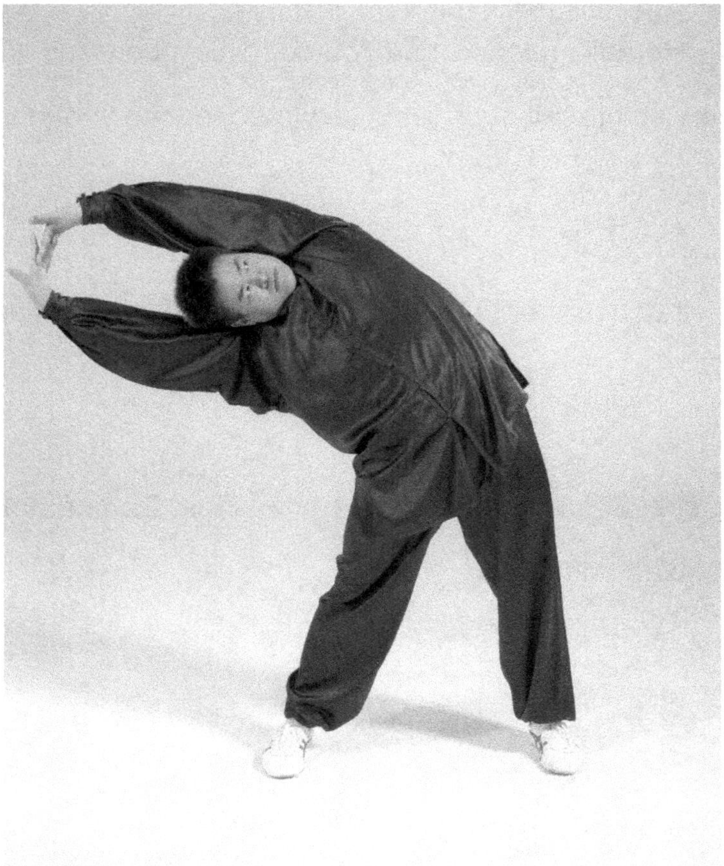

3. Opening the Chest

Overall exercise

This warm-up exercise will open up the chest by using the arms to stretch it. Remain as you were in a position where you are standing with your legs at shoulder width apart, with the head and neck raised but relaxed. Drop the arms either side of your body about three fists distance from the torso and with the shoulders and elbows relaxed and the fingers slightly splayed and pointing to the floor. Using your elbows, circle the arms towards the body and then backwards until you achieve a comfortable rhythm. This produces a very loose and comfortable feeling. Aim for eight repetitions, or more if you like the feel of it. The open hands lead this. The body remains relaxed and rooted.

Detailed exercise

Figure 6-12

Clench the hands into a fist but without applying any tension. Stretch back the shoulder blades, leading from the elbows. Remain relaxed and rooted. This opens up the shoulder blades. The shoulders often can be very stiff for all of us living as we do too much of a sedentary lifestyle.

Figure 6-13

Leading with the elbows, stretch further to the back and sides. Remain relaxed and rooted.

Figures 6-14 and 6-15

Leading with the elbows, stretch further to the back and sides. Do not strain or stretch further than is comfortable. Remain relaxed and rooted.

299

Figure 6-16

Stretch out fully to the sides and open the hands from the fist position by loosening the fingers and let the arms stretch fully to the sides. The palms are open and facing upwards.

Figure 6-17

Form the hands back into a fist and move back towards each other towards the centre position. Soon you will get into a nice rhythm, repeating the movements in a cycle.

Figure 6-18

Continue to move each hand towards the centre so that the fists cross over. Remain relaxed and move smoothly. Remember to remain rooted.

Figure 6-19

Continue to move each hand so that each fist is under the shoulder of the opposite side. This is the furthest stretch inward before repeating the opening-out sequences.

Figure 6-20

Repeat the movement outwards with the fists leading the elbows and chest, and opening out as previously.

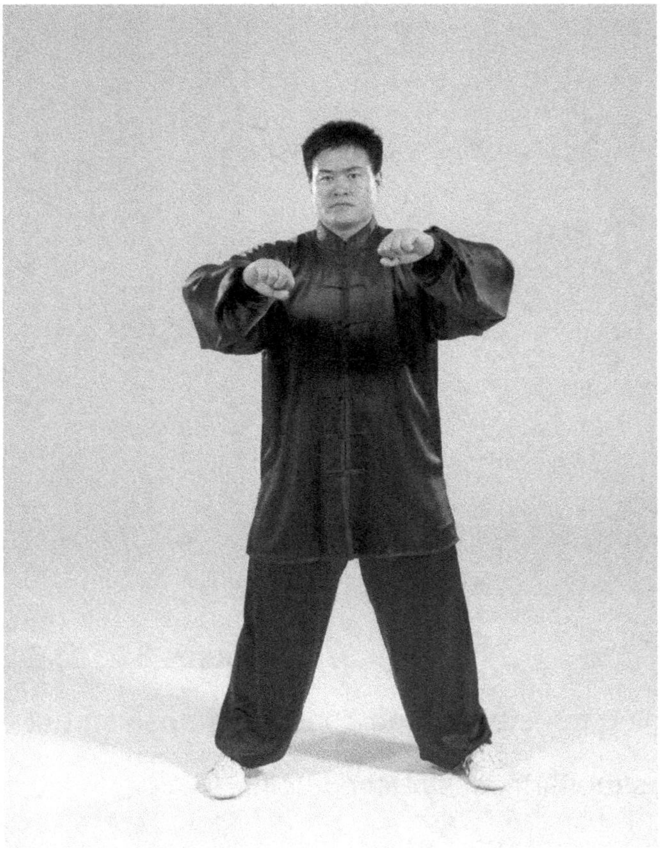

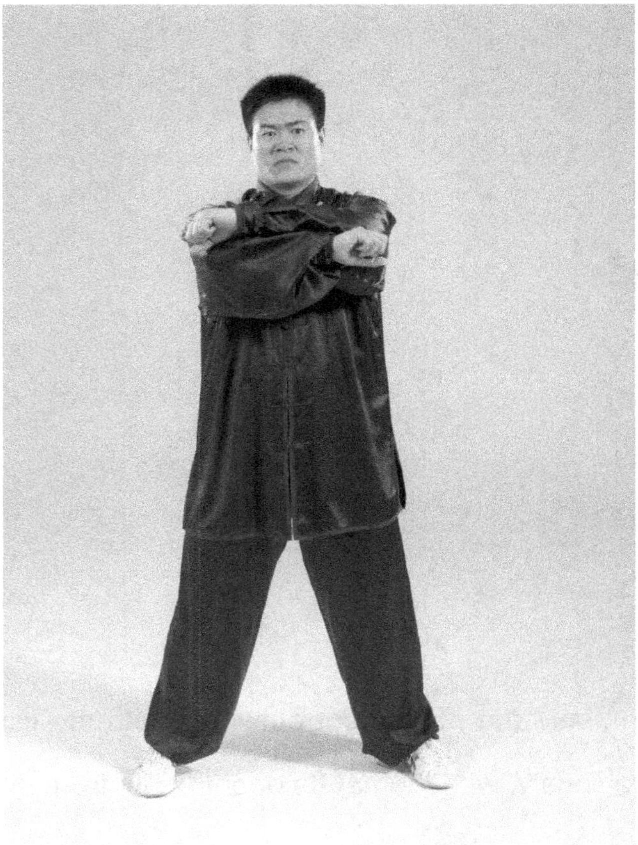

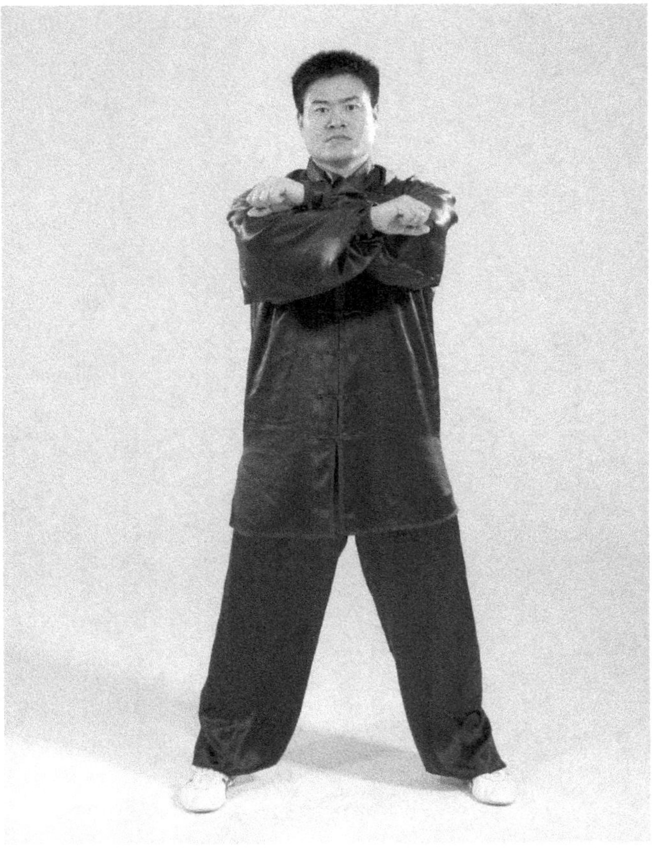

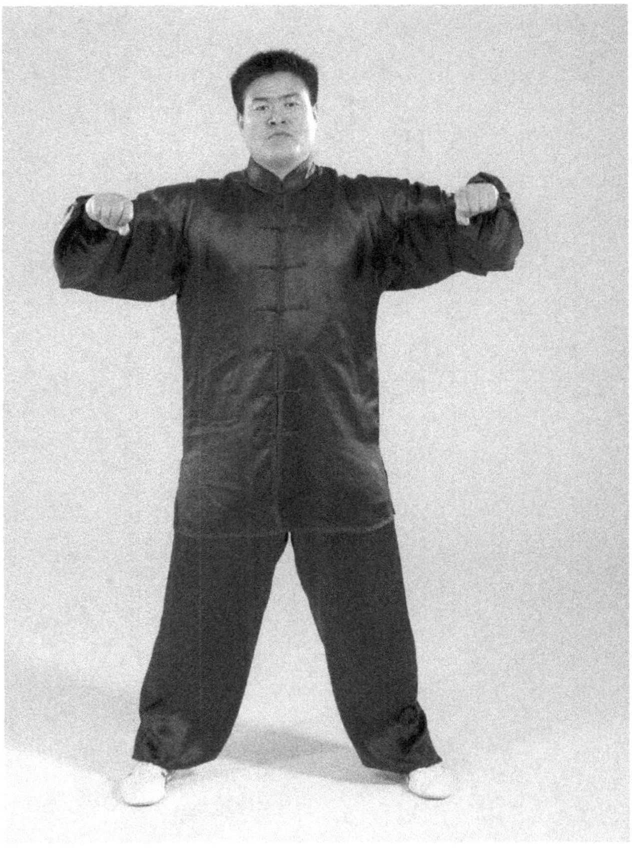

Figures 6-21 to 6-24

Continue the movement outwards with the fists leading the elbows and chest, and opening out as previously, with the hands opening out from the fist as the furthest stretched point.

303

Loosening the Arms

Overall exercise

Remain relaxed with a rooted posture and breathe naturally. This warm-up exercise will loosen the joints in the arms. The left arm and then the right arm are exercised– although it does not matter if you prefer to exercise in the opposite sequence.

Detailed exercise

Figure 6-25

Both hands are loosely formed into a fist. The right arm remains relaxed by the side and the left leads with the elbow and stretches back away from the body slightly. The right arm is bent, not straight. And the fist moves over the head.

Figure 6-26

Move both arms forwards and backwards together stretching out. Remain relaxed and feel the stretch at the top of the shoulders and neck.

Figure 6-27

Continue to move both arms forwards and backwards together stretching out.

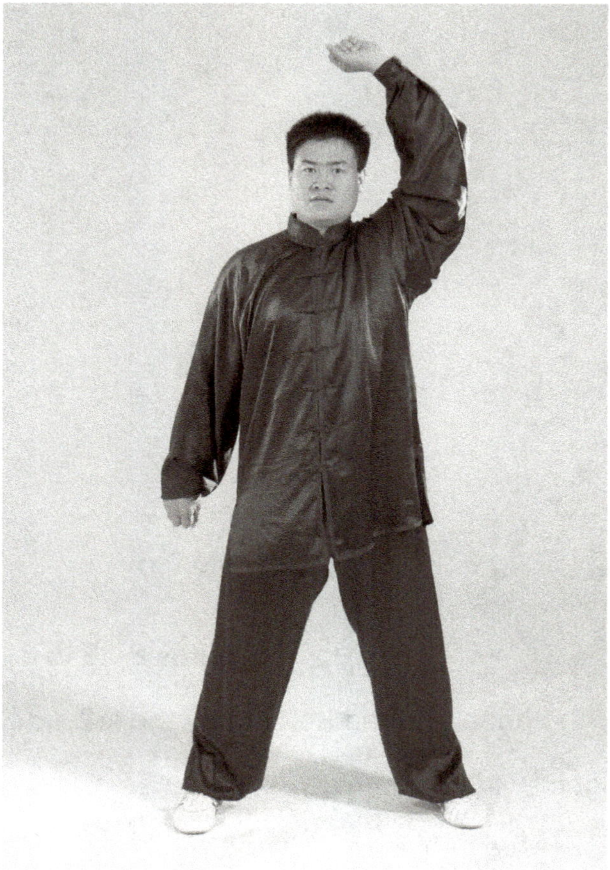

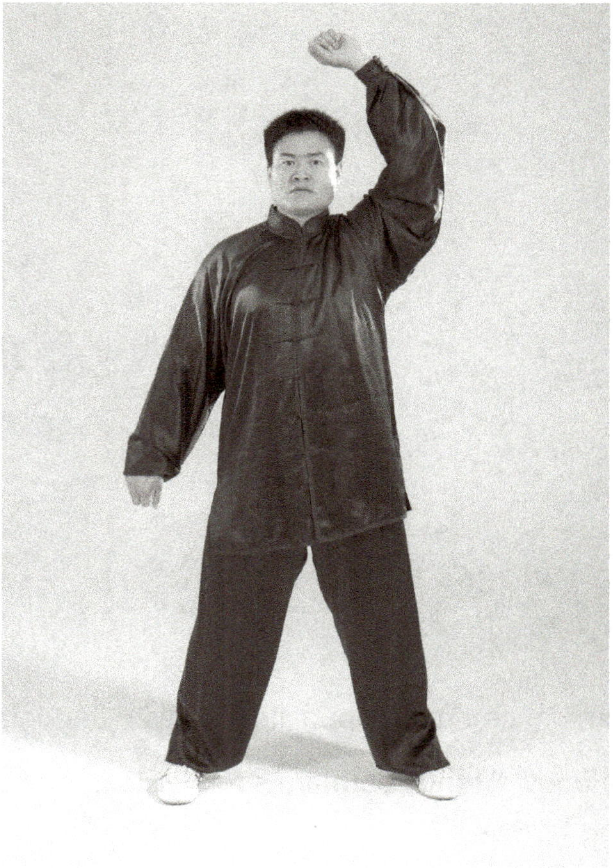

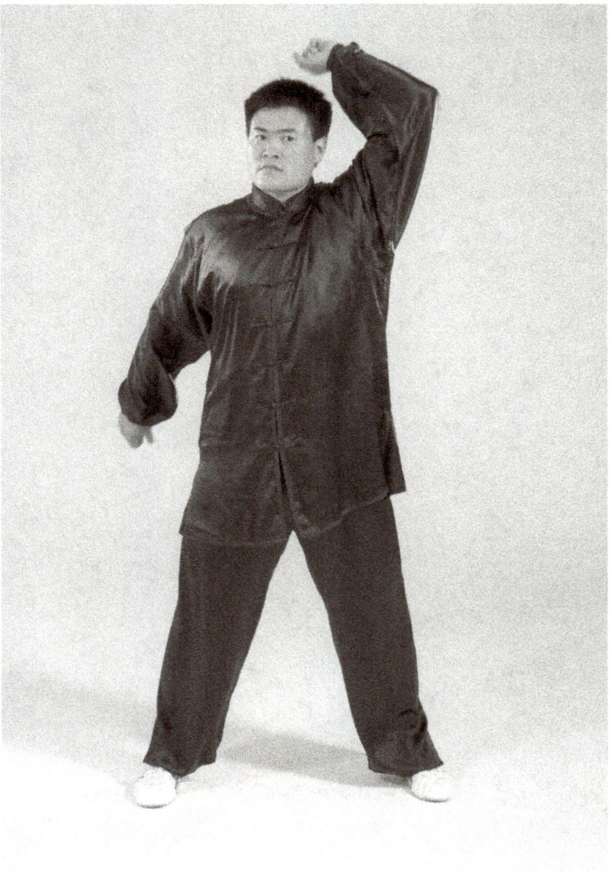

Figures 6-28 to 6-31

Commence on the opposite side with the right arm over the head and the left arm by the left side of the body. Stretch both arms forwards and backwards together in unison.

307

Loosening the Shoulders

Overall exercise

This warm-up exercise is to engender a loosening up of the shoulders by rotating in both clockwise and anti-clockwise directions, leading with the elbows. As always, remain rooted and relaxed. Both hands point in to the clavicle. The shoulders act as pivots for the movement.

Detailed exercise

Figures 6-32 to 6-41

Place each hand under the corresponding clavicle, with the elbows pointing to the side and with the shoulders relaxed. After practising for a short period of time, you will learn to listen to your body and hence will find it easy to discern your level of relaxation. Roll the shoulders forward with the hands and elbows moving forwards. The shoulders are relaxed and not hunched and there is a space between the arms and the armpits. The rolling motion is repeated forwards and backwards so that the joints in the shoulders are rotated and opened up. Remember to isolate the movement by ensuring that the body remains rooted and relaxed.

309

310

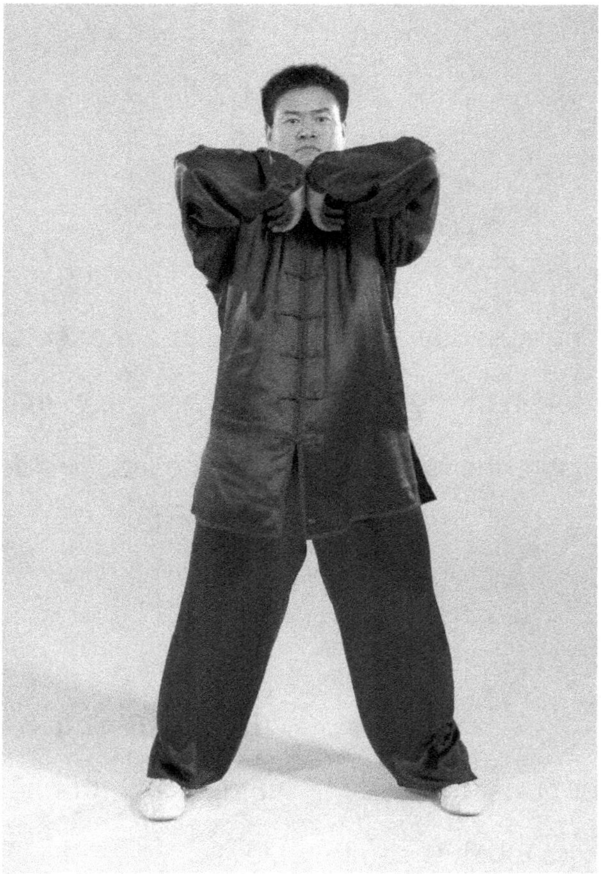

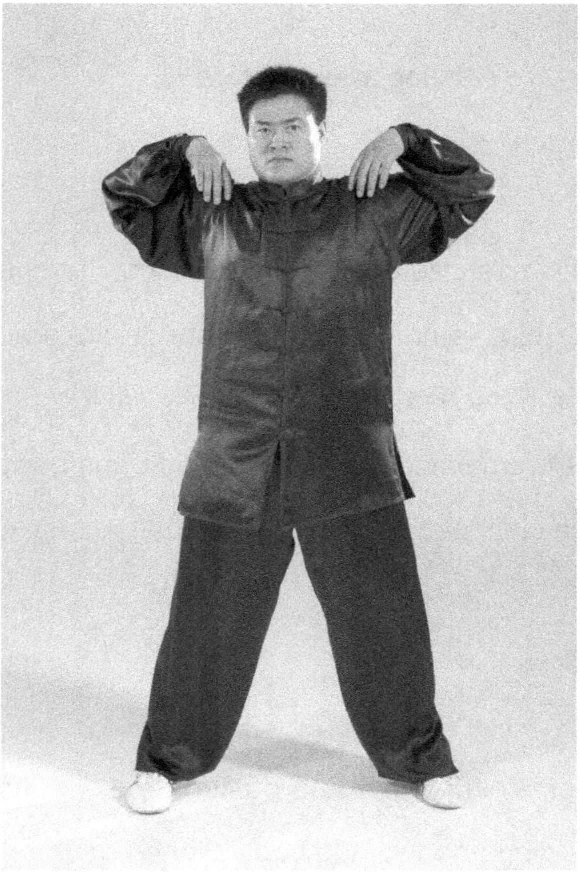

5. Twisting to either side

Overall exercise

This exercise is performed in order to open up the hips and both sides of the body. Opening up of the hips is crucial for the correct practice of the form. You may feel that your hips are indeed stiff and not loose at all. This is natural if you have not undertaken this type of exercise before. With the feet remaining in a fixed position, the body is stretched first to the right and then to the left. The fists are clenched but not tightly, but in a relaxed fashion.

Detailed exercise

Figures 6-42 to 6-44

Remaining firm in the posture and with hands intertwined, twist the body to the right with the head leading and with the body aligned and upright. Only twist as far as is comfortable.

Figure 6-45 to 6-47

Remaining firm in the posture and with the hands intertwined, twist the body to the left with the head leading and the body aligned and upright. Only twist as far as is comfortable.

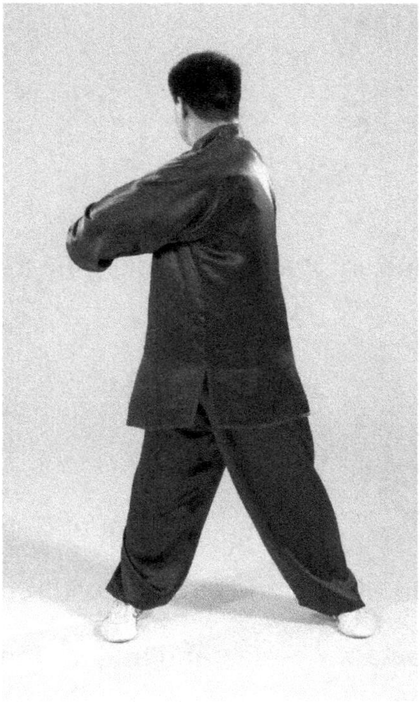
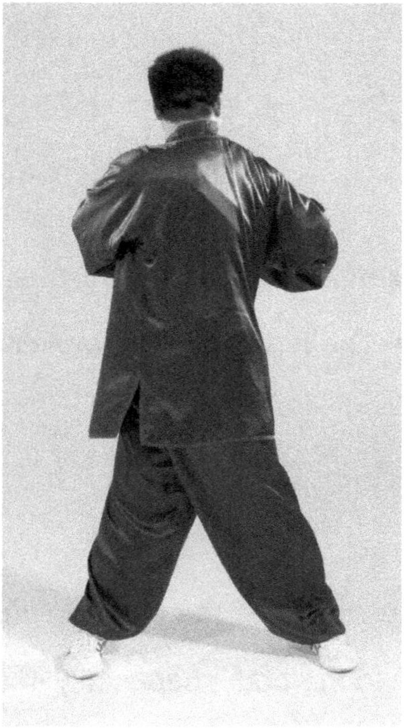
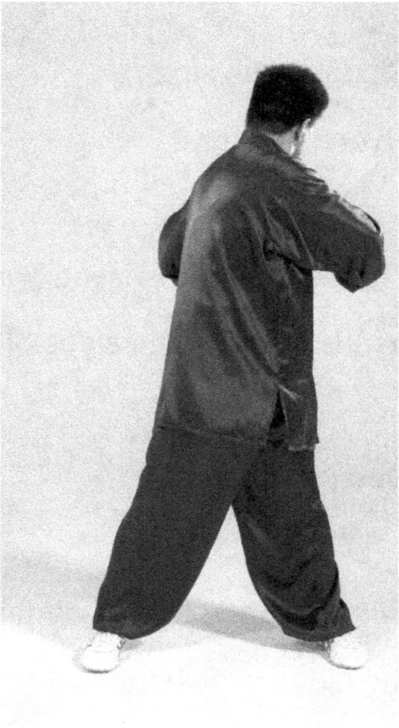
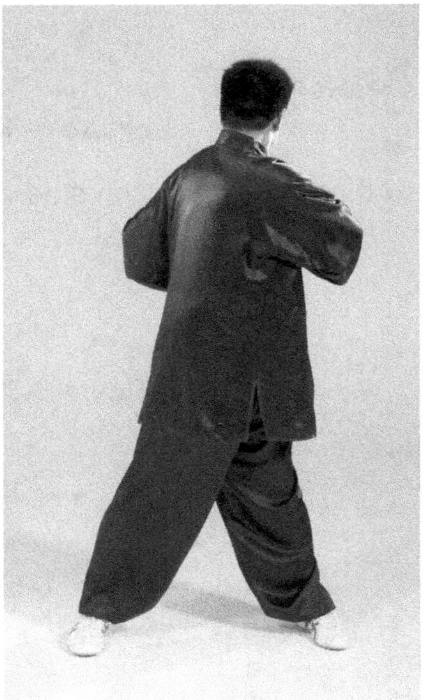
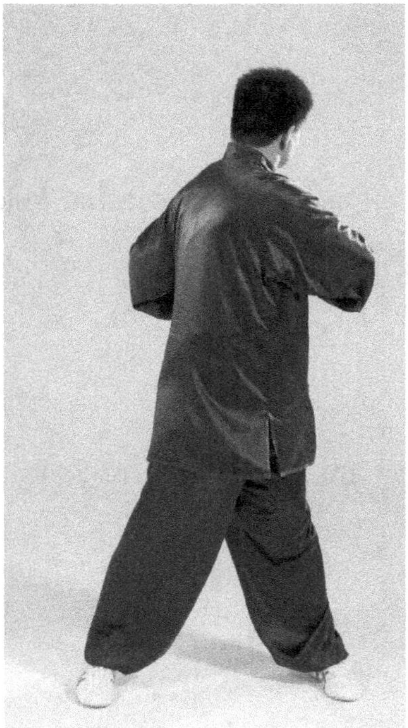
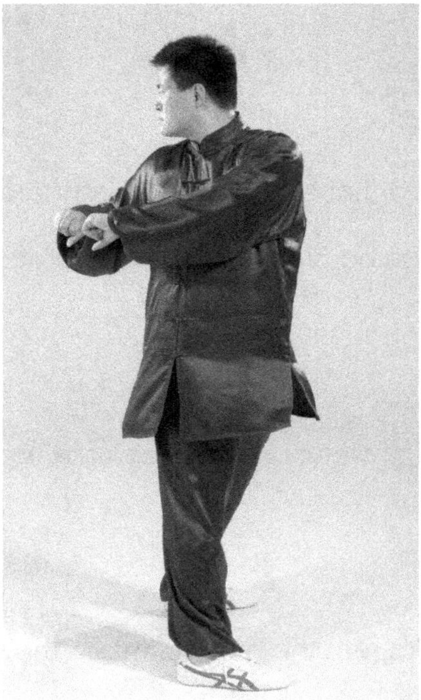

Slapping the sides

Overall exercise

This is another twisting warm-up exercise similar to the previous one in Section 6, but with the hands slapping the sides of the body. The hips drive the movement.

Detailed exercise

Figure 6-48

With the arms loose and falling and held out by the sides at a distance of around twice the width of the hips, resolve to create no tension in the body, especially in the shoulders. Relax and ensure that you are rooted. Concentrate on the movement: ignore the thoughts that may arise about the day's events or anything you spot in the room.

Figure 6-49

Twist the body to the left with the arms following so that they slap each side of the body as the twist of the hips generates their movement. The arms should feel as though they are as light as a feather.

Figure 6-50

Continue to further twist the body to the left with the arms following so that they slap each side of the body as the twist generates their movement.

Figure 6-51

Centre yourself in the starting position and prepare to move to the right.

Figures 6-52 and 6-53

Repeat the movement on the right-hand side.

315

6. Circling the Hips

Overall exercise

Overall exercise

This warm-up exercise consists of moving the hips in a pivotal fashion – in clockwise and then counter-clockwise directions so as to open the joints. As stated before, the hips need to be loose to perform the movements. The hips will tend to be stiff if you are sitting at a desk all day.

Detailed exercise

Figures 6-54 and 6-55

With the hands placed on the small of the back and with the legs firmly rooted, circle the hips backwards, bending the knees. Circle to the right and then to the left. The circles are small to begin with, but get bigger as the warm-up continues.

Figure 6-56

Turn the body in a circular motion, leading from the hips. Circle forwards slightly.

Figures 6-57 and 6-58

Continue to the right and left. The circles become even bigger.

Figures 6-58 to 6-62

Continue to the right and left and backwards and forwards. The circles become even bigger. Keep the hips relaxed and ensure that the hands remain covering the small of the back. Do not strain yourself: revolve the hips only as far as is comfortable. Maintain the rooted position of the feet.

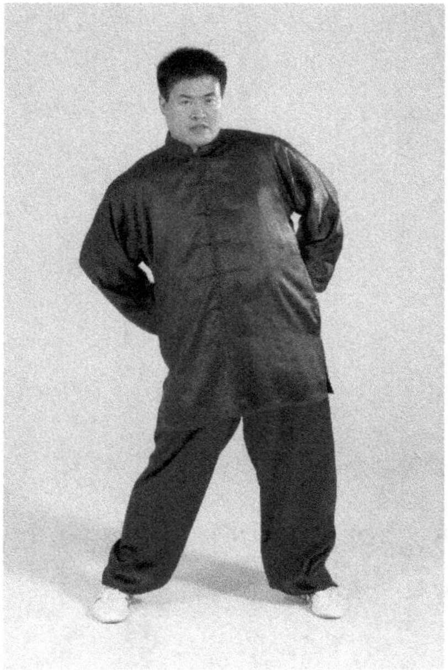
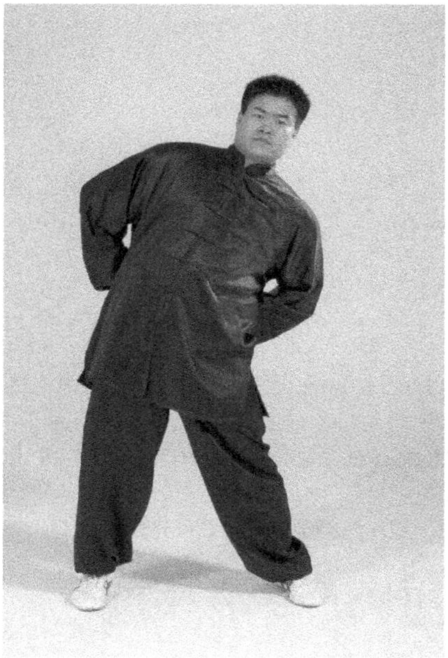
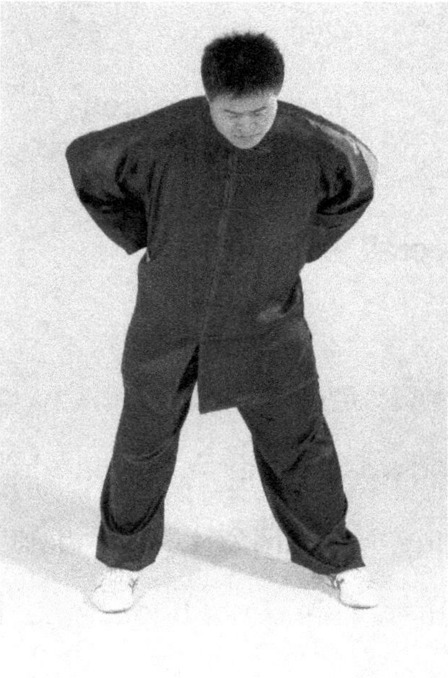
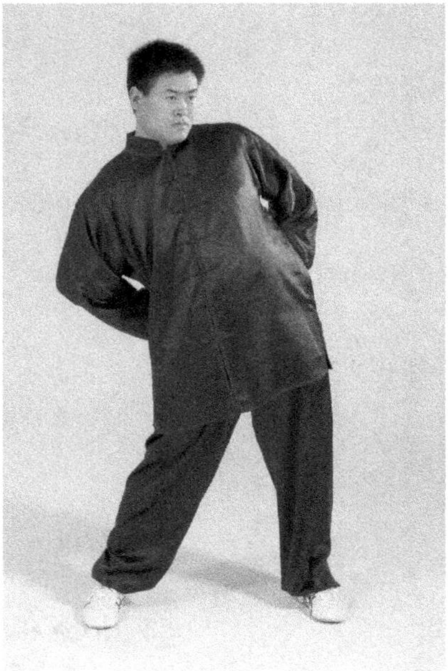
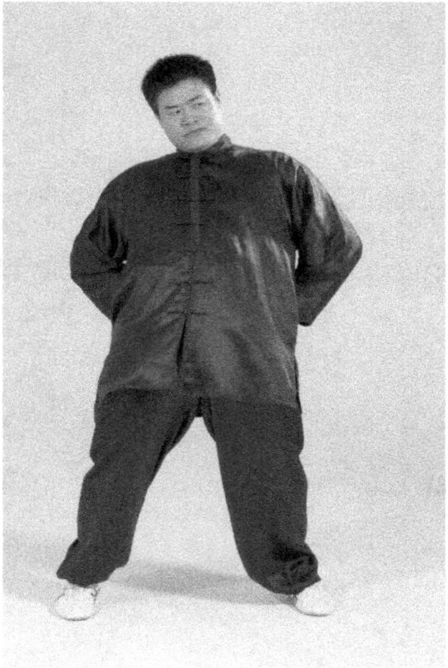
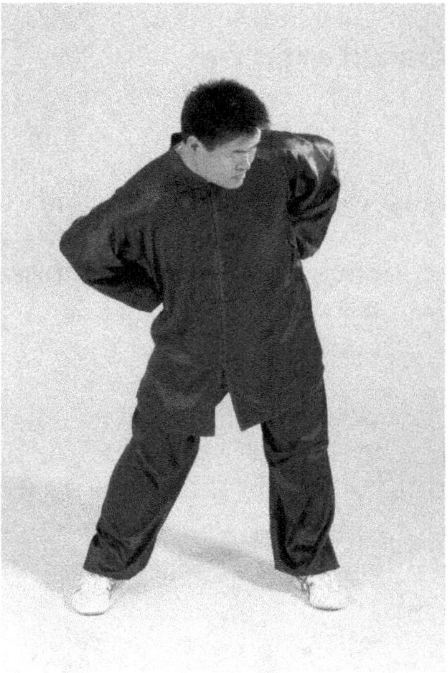
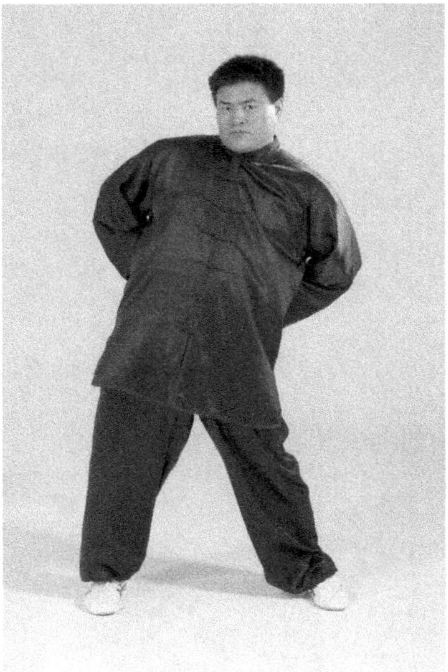
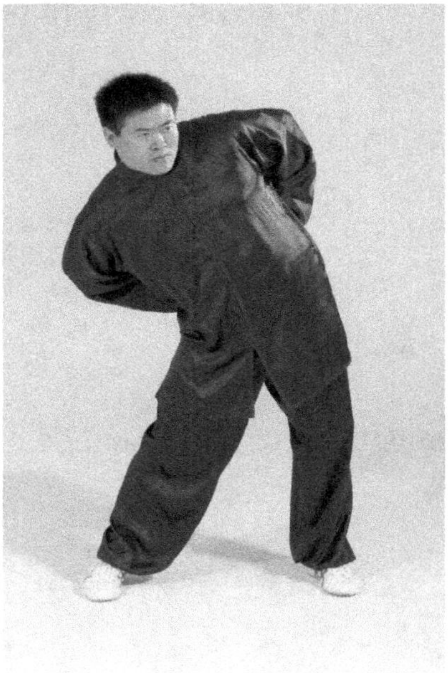

7. Circling the knees

Overall exercise

This warm-up exercise is to loosen the knee joints and consists of both a clockwise and a counter-clockwise rotation. Be careful if you have had knee problems in the past. As always, listen to your body. You are not in competition with anyone and there are no awards given out for making yourself suffer. This is not boot camp. Relax, relax relax.

Detailed exercise

Figure 6-63

Place the feet together with the knees touching. Bend the knees and place each hand over each respective knee with the fingers pointing down. Commence to circle the knees to the left.

Figure 6-64

Continue circling the knees to the left.

Figure 6-65

Circle the knees to the right back towards the centre.

Figure 6-66

Continue circling the knees to the right.

Figures 6-67 to 6-77

Continue to circle the knees back to the left and continue left to right and back again.

319

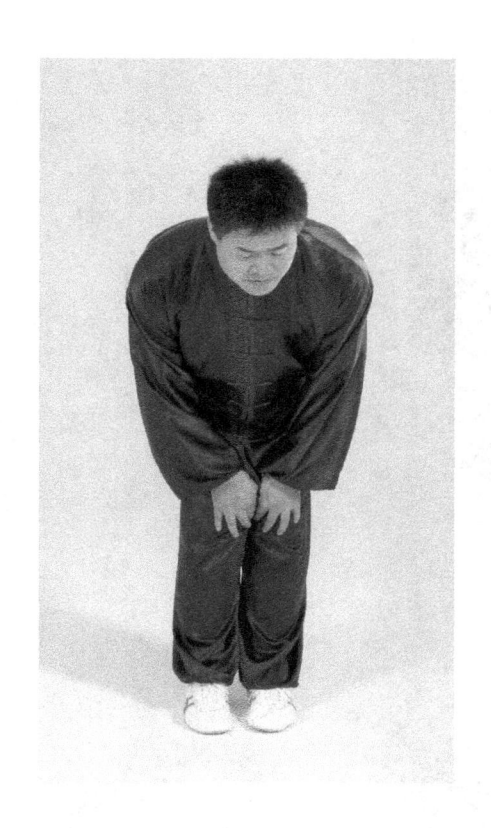

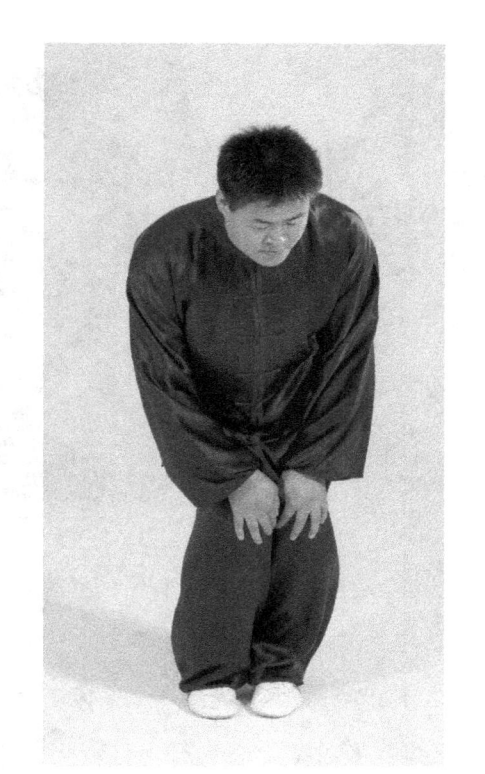

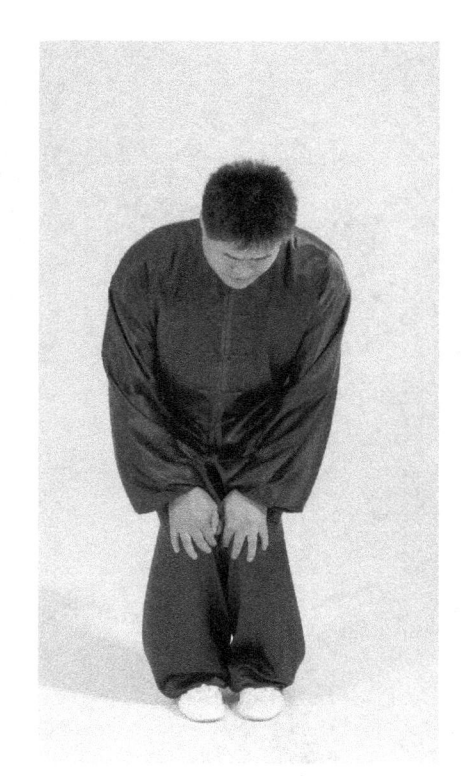

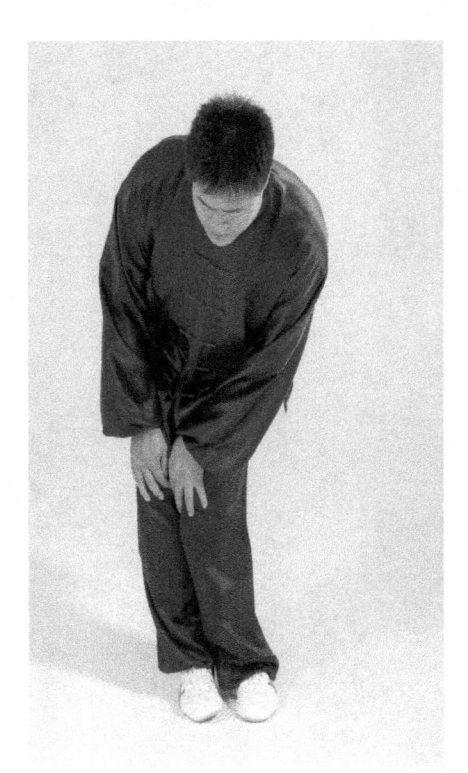

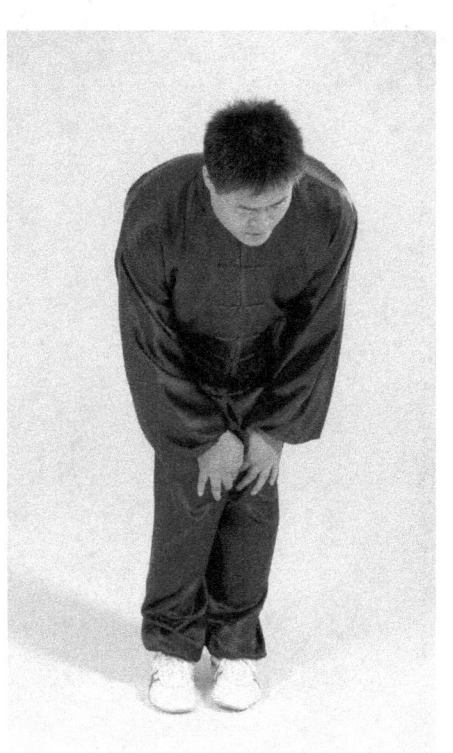

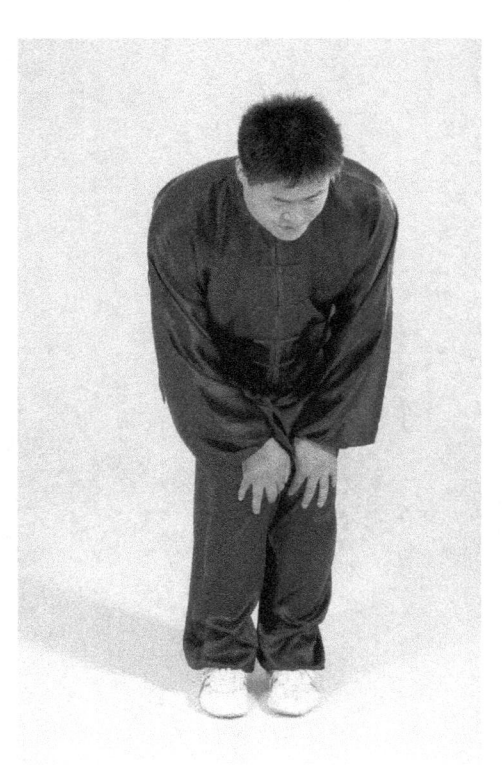

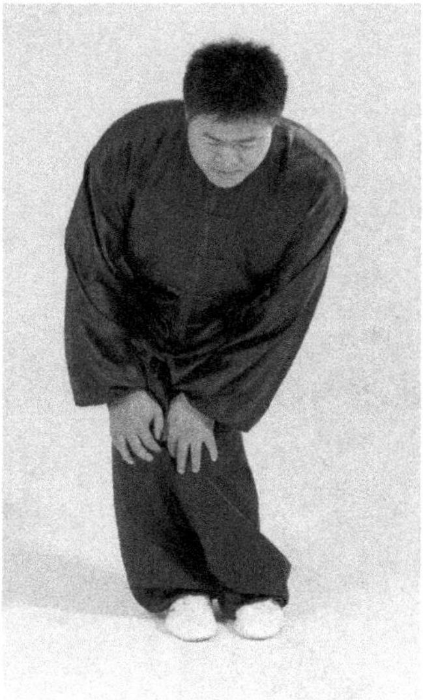
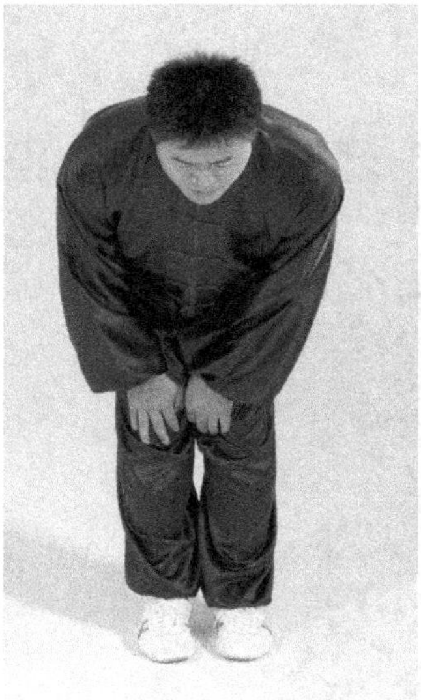
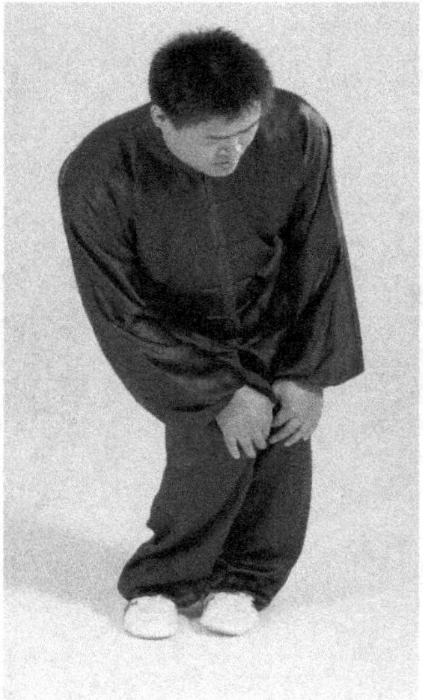
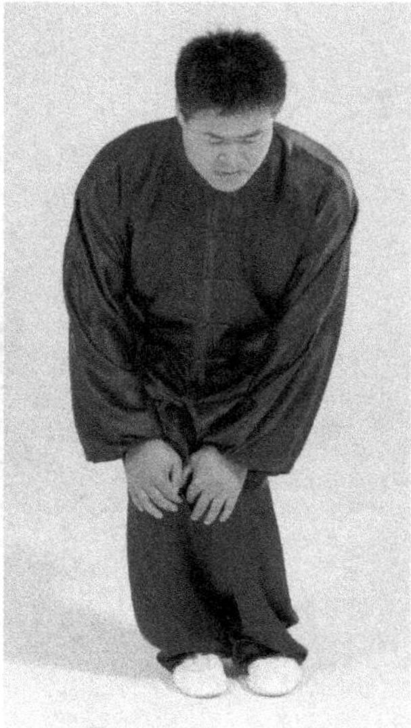
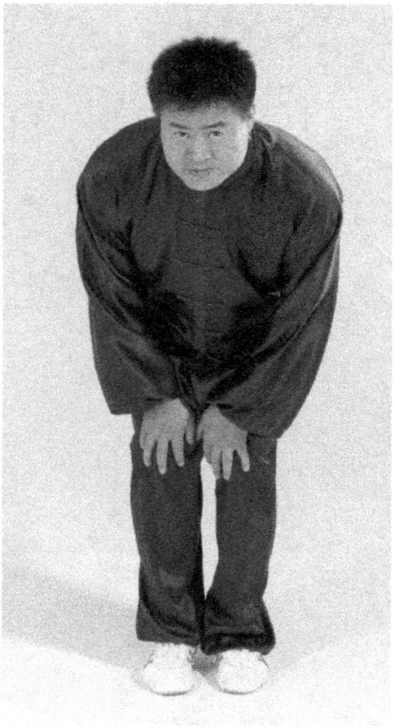
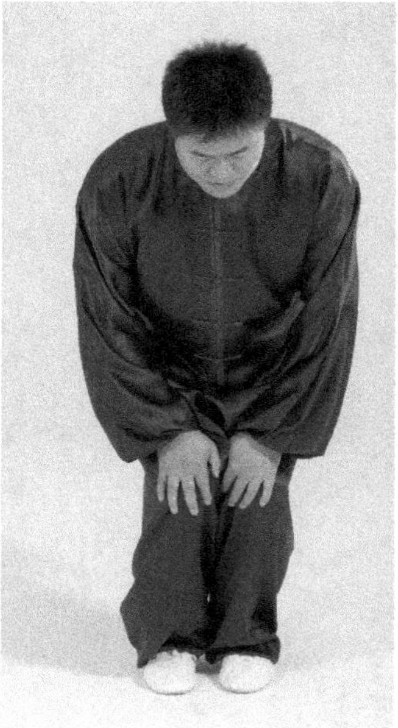
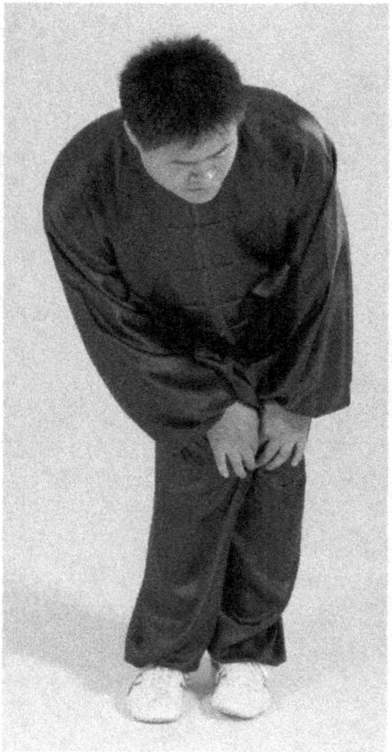
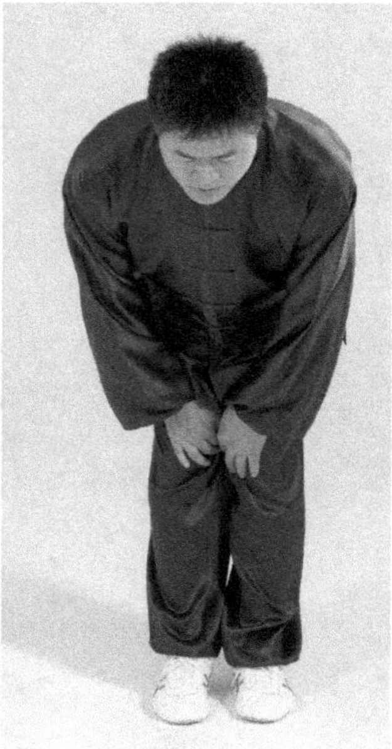
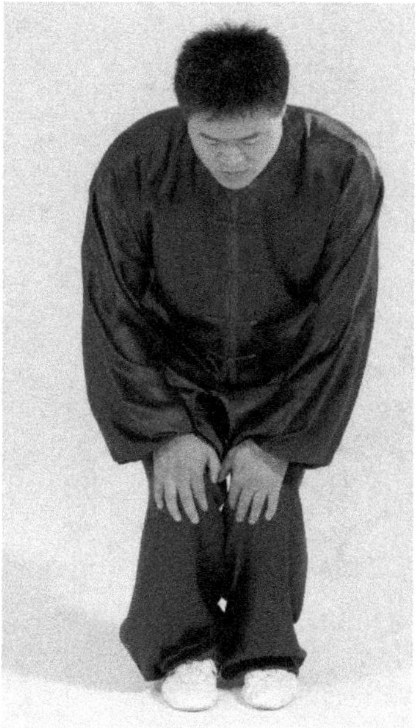

Circling the ankles

Overall exercise

Stand with your legs at shoulder width apart and with your head and neck raised but relaxed. Imagine that something is gently pulling your head upwards and stretching your spine. Place your hands on either side of the waist about a fist width away from your torso and with the fingers pointing downwards towards the floor. Shift your weight onto the right leg and raise the left foot so that the heel is raised and just the toes remain on the floor.

Rotate the ankle clockwise and then counter-clockwise eight times. The feet are just a few inches apart. Now bring the left foot down, shift the weight onto the left leg and rotate the right ankle clockwise and counter-clockwise eight times. Place the toes of the left foot in line with the heel of the right foot to ensure full rotation of the ankle. Having the rotating foot too far forward tends to only work the flexion and extension of the ankle. Remain balanced and ensure that you do not fall over!

Detailed exercise

Figures 6-78 and 6-79

Raise the right heel off the ground so that the toes are pointing downwards. The hands rest gently just over the hips. Start to move the ankle backwards in a circular motion.

Figures 6-80 and 6-81

Continue to move the ankle backwards in a clockwise direction from the tips of the toes through the ankle. Remain relaxed and balanced.

Figures 6-82 to 6-84

Continue to move the ankle backwards in a clockwise direction and then forwards in the opposite (counter-clockwise) direction from the tips of the toes through the ankle. Make the circles as big as you can. Remain relaxed and balanced.

Figures 6-85 to 6-93

Rotate the left ankle. Move the ankle backwards in a clockwise direction and then forwards in the opposite (counter-clockwise) direction from the tips of the toes through the ankle. The circular movements of the ankles should get wider. Remain relaxed and balanced.

323

Stretching the legs and 胯 kuà
Overall exercise

This exercise stretches the 胯 kuà and the legs and also opens up the groin and hips. The stretch is performed on either side of the body, starting with the left and then the right. Or, of course, the opposite sequence (rtight, then left) can be used. As always, the sequence of left or right is unimportant. This stretch is popular in all martial arts systems and is known as a bow stance (as in bow and arrow. It is primarily a hip flexor stretch. The hand pressing on the pelvis helps to guide the tilt of the pelvis in the front of the rear hip.

Detailed exercise

Figure 6-94

Turn to the left with the feet at a distance of around double shoulder width apart. With the right heel placed flat on the floor, bend the left knee and sink down. Place the left hand over the knee. Place the right hand on the right hip. Only sink as far as is comfortable.

Figures 6-95 to 6-99

Move the right rear hip down and then up again so that there is a constant movement sinking up and down and slowly stretch.

Figures 6-100 and 6-101

Repeat on the right-hand side. Turn to the right with the feet placed around double shoulder width apart. With the left heel placed down, bend the right knee and sink down. Place the right hand over the knee. Place the left hand on the left hip. Only sink as far as is comfortable.

325

Stretch down and open the hips

Overall exercise

This is quite a difficult exercise for beginners to undertake, especially if the hips are tight. And it is normal for the hips to be tight for new students so do not worry if it takes some time to get low down. Remember always that you are not in competition with either yourself or anyone else. So there are two points to make: the first is that it is worth the effort to practise slowly and patiently and the second is to warn you to be careful. So that if you are not able to get down low, do not force it. The stretch opens up the hips and 胯 kuà and is performed on either side – on the left and on the right.

This is known as crouching stance. It is a leg adductor stretch for the Adductor Magnus, brevis, longus, gracilis and pectineus . This is excellent for preventing groin strains for those who play sports or if one accidentally slips on ice etc. initially one can hold ones knees then when more flexible one can hold ones feet. Try to point both feet in the same direction once you are used to the exercise.

Figure 6-102

Place the feet apart at around double shoulder width. The feet are on the same line of direction. The weight is placed on one side of the body and the other leg is stretched out, with the foot at a 45 degree angle. Lower the right hip and sink the knees as low as you can. Place the right hand on the right knee and sink down with the left leg straightening. Place the left hand over the left knee.

Figures 6-103 to 6-104

Sink up and down slightly so that there is a dynamic stretch. Do not bounce vigorously but maintain a slow movement up and down. The left leg is straight, and thus the stretch is from the hips and groin and you sink the body down.

Figure 6-105 and 6-106

Place the feet apart at around double shoulder width. The feet are on the same line of direction.

Lower the left hip and sink the knees as low as you can. Place the left hand on the left knee and sink down with the right leg straightening. Place the right hand over the right knee.

Sink up and down slightly so that there is a dynamic stretch. Do not bounce vigorously but maintain a slow movement up and down.

Now you have performed a nice stretching exercise that is of benefit in its own right regardless of whether you then go on to practise the form. Keep practising and your body will become flexible. You can perform these stretches with limited room available, for example, if you are indoors. If you perform the warm-ups immediately in advance of practising the form, you will be nice and relaxed and loose and ready.

Chapter 7 – 缠丝劲 Chán sī jìn

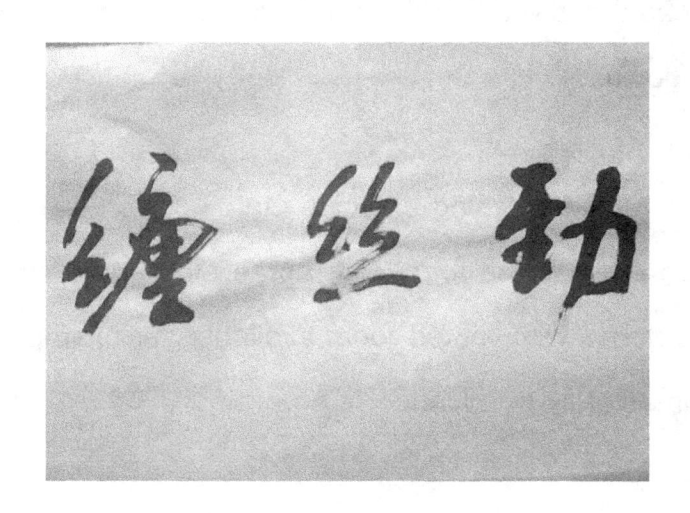

缠丝劲

震 zhèn

曲則全

枉則直

窪則盈

弊則新

少則得

多則惑

是以聖人抱一為天下式

不自見故明

不自是故彰

不自伐

故有功不自矜故長

夫唯不爭, 故天下莫能與之爭

古之所謂曲則全者, 豈虛言哉

誠全而歸之

Be humble, and you will remain entire.

Be bent, and you will remain straight.

Be vacant, and you will remain full.

Be worn, and you will remain new.

He who has little will receive.

He who has much will be embarrassed.

Therefore the Sage keeps to One and becomes the standard for the world

He does not display himself; therefore he shines.

He does not approve himself; therefore he is noted.

He does not praise himself; therefore he has merit.

He does not glory in himself; therefore he excels.

And because he does not compete; therefore no one in the world can compete with him.

The ancient saying 'Be humble and you will remain entire'

Can this be regarded as mere empty words?

Indeed he shall return home entire.

– Translated by Ch'u Ta-Kao, 1904, Chapter 22

缠丝劲 chán sī jìn, or Silk Reeling Energy, is a fundamental component in the practice of Chen style taijiquan. The term is derived from the observation of the twisting and spiralling movements of the silkworm larva. This becomes a metaphor for the movements of wrapping itself in its cocoon and reeling the silk. And so this term is used to describe the internal movements within the taijiquan form.

It is embedded within every movement. Without it, there is no taijiquan and certainly no martial arts application. Hitherto, it was not separately taught in traditional circles. As is equally the case with 站桩 Zhàn Zhuāng, it is only in recent times that has it been separated out as a form of practice. Before, it was merely part of the training as practised within the forms.

It creates great power through the development and release of internal force through 发劲 fā jìn, using the unified force of the whole body. 发劲 Fā jìn, is a very important concept in Chen style taijiquan. It is the power expressed that makes it a martial art. 劲 Jìn, which can be roughly translated as power, is not the same as 精 jīng, which we explained earlier. It is an expression of how the body moves the energy from the axis of the structure and spirals qi from the feet and through the 下丹田, Xià Dān tián to the rest of the body to discharge energy.

缠丝劲 chán sī jìn energy release starts from the 下丹田 Xià Dāntián where the hips and legs combine in unison with the waist and the chest. All of the body moves together seamlessly as one. Hence, this causes the whole body to move as though it were one indivisible unit. Special attention must be paid to the shoulders. This part of the body is the one most likely to be exhibiting tension due to the modern lifestyle, which often involves stress, use of computers, etc., thereby inhibiting the natural flow of energy. The shoulders should remain relaxed so as not to move independently of the rest of the body. The practice of Zhan Zhuang that we covered in Chapter Five provides the fundamentals of correct body alignments and concentration, together with relaxation, so that there is a buildup of 气 qì in the 下丹田 Xià Dāntián and elimination of stiffness. This relaxed state and body alignment can then allow the energy to be expressed throughout the body by the movements of 缠丝劲 chán sī jìn.

When we practise the routine, 缠丝劲 chán sī jìn can help our form get smoother and rounder, and so the movements are continuous and unbroken. Silk Reeling energy is nothing less than the expression of the movement of yin to yang and yang to yin in a continuous circle, in harmony with how the universe expresses itself.

In the early stages, it is important that we practise the routine very gently. This will allow us to get the rhythm of the movement and allow the qi to be built up and then latterly expressed. At this stage, the movements of the routine activate our 气 qì, that is a by-product that is present mainly in the movements themselves. Later, as you become more advanced, the 气 qì becomes stronger, and this drives the movements, hence it becomes the progenitor of the movements. Hence, if we maintain the correct body alignment and move in the spiralling and circular manner required, it becomes much easier for the 气 qì to direct the movements through an inner movement. This process is that of moving from a more external approach to the movements to that of an inner one.

Ultimately, the practice becomes that of an internal energy movement, with the meridians opened and an effortless control of the external through the internal.

Figure 7-1 and 7-2 below are from Chen Xin's 'Illustrated Explanations of Chen Family Taijiquan'.

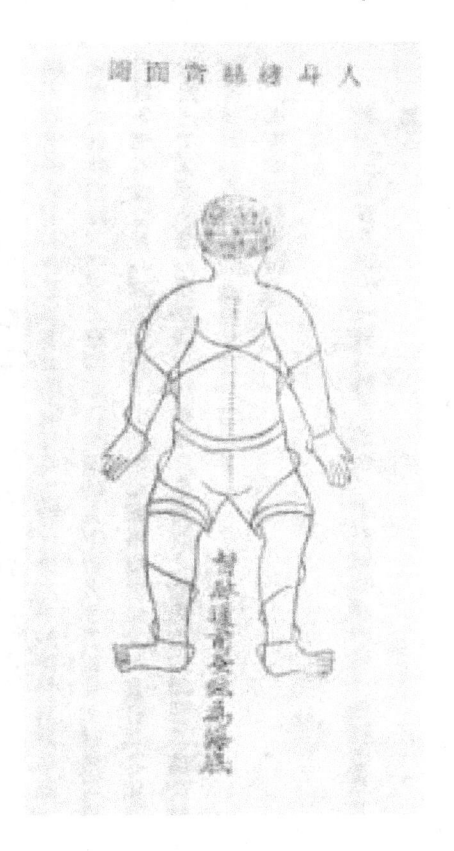

Figure 7-1

太極拳纏絲精圖

吾讀諸子而悟太
極圖,而打太極拳
須明體圓
明打太極猶須
絲者運中氣精纏
之法者明門不動
明此法即不動
拳明

Figure 7-2

Figure 7-3: Grandmaster Wang Hai Jun practising

In Figure 7-2, you will observe the manifestation of the Dao as described by the two forces of Yin and Yang; as the white and black paths in the circular movement from Heaven and Earth to the Five Phases, expressed as energy movement.

Practising 缠丝劲 chán sī jìn as a separate set of exercises apart from the formal frames consists of both static and moving exercises.

Static 缠丝劲 chán sī jìn

There are eight different practices for 缠丝劲 chán sī jìn in the static form. These are:

Single Handed Front Sided Silk Reeling

The opening position requires attention to those of posture, relaxation, balance and rooting. As previously stated, these basics apply to each 太极拳 tài jí quán movement. The head is held as if being pulled up by a thread. The shoulders are relaxed and not hunched. The kuà is slightly sunk so that the knees bend a little. The body is relaxed and alert, and the mind is calm.

Step out to a distance of about two shoulder widths apart or thereabouts – at a distance that makes you feel centred and stable. You can start with either the left or right hand leading the movement. We will start with the right side, that is, with the right arm extended and placing around 60% or 70% of the weight on the right leg.

Extend the right arm stretching internally through the tendons as the right wrist is positioned at shoulder height. The palm is open but relaxed, with a slight tension to express the energy, and is facing forward. The left hand rests on the side of the waist. Sink slightly and attain a state of relaxed awareness – ready to move.

Once you are relaxed and stable and feel ready, the movement may commence. The waist moves slightly to the right before fully to the left and the weight accordingly shifts to the left leg. This moves the right hand down and across the face of the body but not beyond the centre line - at waist height in a circular and clockwise direction. The left hand remains attached to the waist but the movement of the body turns the arm slightly to the left and back in a counter-clockwise direction. The eyes follow the movement. The eyes look straight ahead, with the head and gaze following the waist direction as the hand moves down, with the peripheral vision still being able to see the arm as it passes at waist height.

Continue the movement until the right arm has moved upwards on trhe centre line of the chest, maintaining its circular and clockwise direction. This movement reaches its apex and the body is now ready to return to the original position.

Relax the right hip and start the return movement to the right until it reaches the commencing position, with the right hand at shoulder height and level with the right knee. Please note that the circular movement arises from the waist and the arm moves in synchronisation. Co-ordinate your breathing intake in conjunction with the outward and return movement by breathing in as the arm closes to the 下丹田 Xià Dāntián and the weight shifts to rhe left. Breathe out as the arm opens to the right at shoulder height. Perform the movements slowly and smoothly and, as always, refrain from the temptation to bob your body up and down or hunch the shoulders.

Pay attention to the following:

- The hips and waist must be relaxed and loose.
- The hips and legs should move in a figure-of-eight Infinity movement, as per Figure 7.4. below.
- Ensure that the knees do not extend too far; so, sink from the hips rather than just stiffly bending your knees.

- Do not over-emphasise one part of the body – the movement should be integrated as one composite whole.
- Do not use the shoulder to drive the movement – it leads from the grouding andis led by the 胯 kuà and 下丹田, Xià Dān tián
- Ensure that the shoulders remain relaxed, and do not hunch or rise up as this will restrict the flow of qi to the fingers.
- Ensure that the shoulders and elbows remain relaxed and are ready to move with the direction of the whole body
- The legs will tire as the exercises are intended to strengthen them, but do not sink too far and thereby lose the relaxation and looseness of the hips, lower back and 下丹田, Xià Dān tián
- The exercise will create torque in the legs, but this is gained through relaxation and repetition so that the legs gradually develop strength.

Figure 7-4: The movement of the body in 缠丝劲 Chán sī jìn

Right Hand Side

Image 7-5

The left hand is cupped on the left hand side of the waist. The body is sunk to a comfortable position. Do not sink too low and feel like you are straining. Between 60% and 70% of the weight is on the right leg. The toes of the right foot point out at 45 degrees. The toes of the left foot are turned in slightly. The eyes look at the right hand that is relaxed, with the index finger slightly tensed and forming a link with the thumb. Relax and compose yourself before you are ready to begin the movement.

Image 7-6

Begin to prepare for the movement to the left. Do not start the movement until you are relaxed and focused.

Image 7-7

Start to move the body to the left, guided by the grounding and with movement of the 胯 kuà and 下丹田 Xià Dāntián. The movement starts from the rooting at the feet. The hand moves in unison and guides the arm, which remains relaxed.

Image 7-8

Continue to move the body to the left, guided by the 胯 kuà and 下丹田 Xià Dāntián. The right hand passes over the knee, with the eyes following the movement. The weight is shifting to the left. The movement is seamless, as one whole, not a series of jerky, discrete, individual ones.

Image 7-9

The right hand passes over the knee, with the eyes following the movement. The weight continues to shift to the left. The right hand moves upwards as the movement is circular, with the hips moving in a circle also. The eyes continue to follow the hand movement.

337

Image 7-10

The right hand continues to move in unison with the body and guides the arm until it has reached the zenith of its movement. The body has turned to the extent of its circle and the hips have moved to the left in a semi-circle, ready to repeat the return to the right.

Image 7-11

Start to move the body to the right, guided by the 胯 kuà and 下丹田, Xià Dān tián. The hand moves in unison and guides the arm. The hands are relaxed and the palm faces outwards.

Image 7-12

Continue to move the body to the right, as always, guided by the 胯 kuà and 下丹田, Xià Dān tián. The hand moves in unison and guides the arm. The hands are relaxed and the palm faces outwards. The movement is slow and circular and takes you back towards the original starting position. This is the completion of one cycle.

Image 7-13

You are back ready to start to move again. Repeat eight cycles or more if you are not tiring. Undertake this practice in a relaxed way. So, do not continue if you are getting tense or if your mind is wandering. It is far better to complete one proper set than several more that are incorrect. The objective is not to achieve multiple repetitions like bench presses in the gym, but to practice in a co-ordinated, relaxed, aligned way, with the breath acting as the link between mind and body.

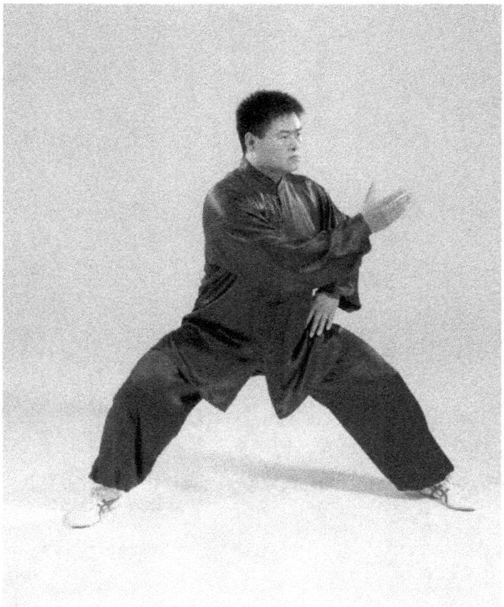

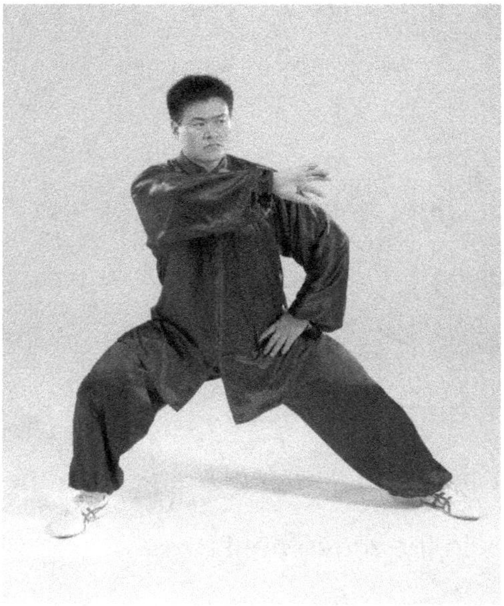

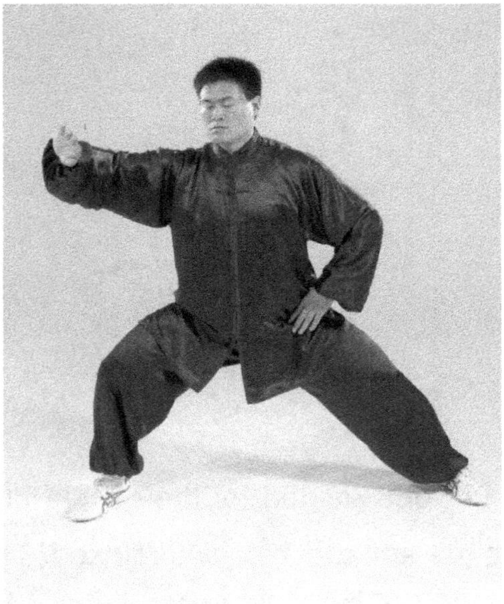

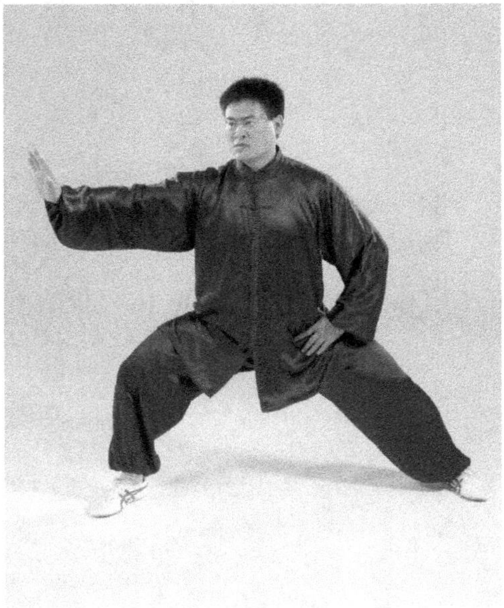

Left Hand Side

Image 7-14

The right hand is cupped on the right hand side of the waist. The body is sunk to a comfortable position. Do not sink too low so that you feel like you are straining. Between 60% and 70% of the weight is on the left leg. The toes of the left foot point out at 45 degrees. The toes of the right foot are turned in slightly. The eyes look at the left hand, which is relaxed, with the index finger slightly tensed and forming a link with the thumb. Relax and compose yourself before you are ready to begin the movement.

Image 7-15

Begin to prepare for the movement to the left.

Image 7-16

Turn very slightly to the left.

Image 7-17

Continue to move the body to the right, guided by the 胯 kuà and 下丹田, Xià Dān tián. The left hand moves in unison and guides the arm that remains relaxed. The left hand passes over the knee with the eyes following the movement. The weight is shifting to the left. The movement is one whole, not discrete, individual ones.

Image 7-18

The weight continues to shift to the right. The left hand moves upwards as the movement is circular, with the hips moving in a circle also. The eyes continue to follow the hand movement.

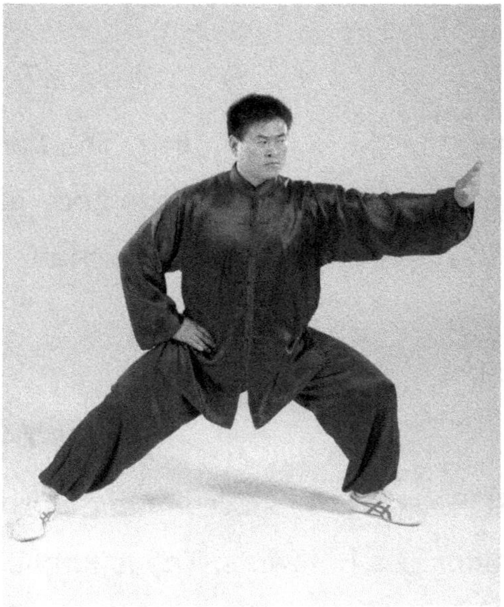
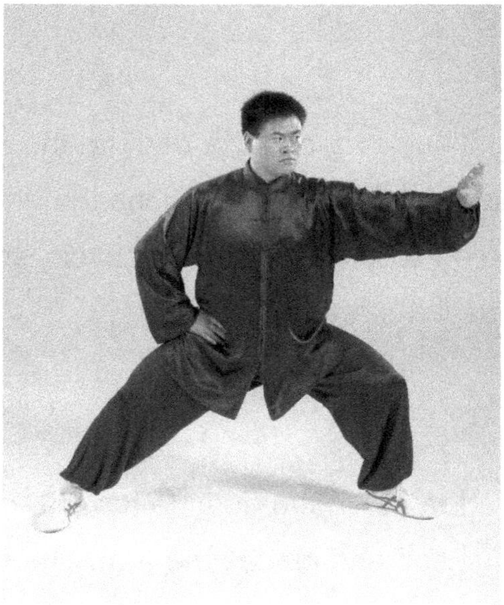
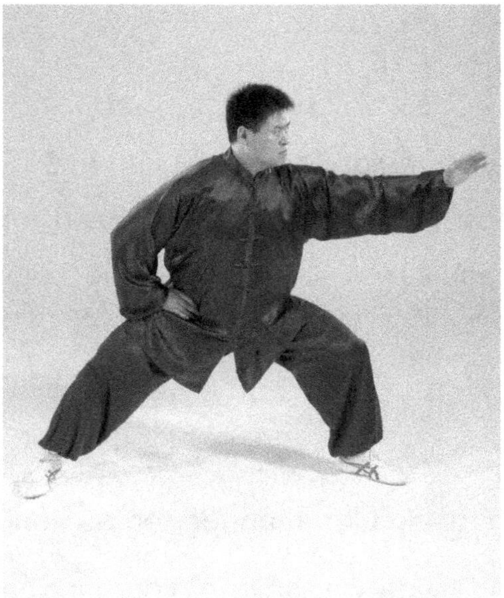
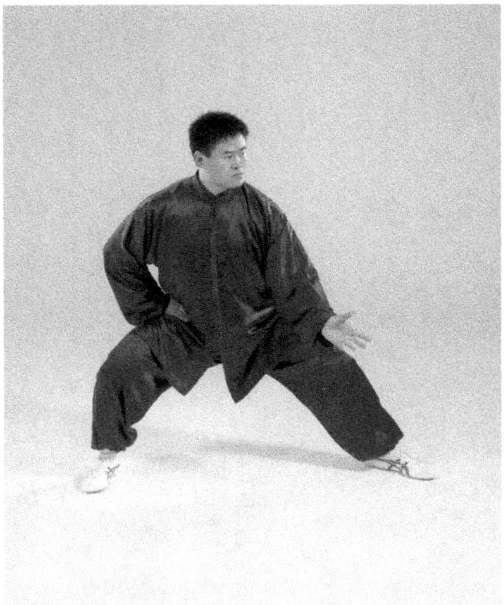
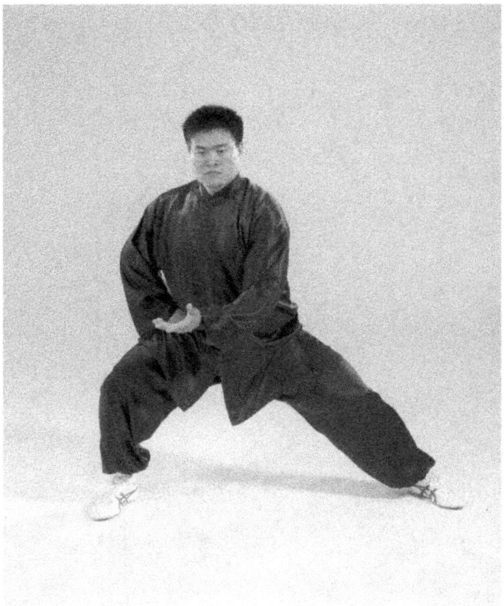

Image 7-19

Continue to move the body to the right, guided by the 胯 kuà and 下丹田, Xià Dān tián. The hand moves in unison and guides the arm until it has reached the zenith of its movement. The body has turned to the extent of its circle and the hips have moved to the right in a semi-circle, ready to repeat the return to the left.

Image 7-20

Start to move the body to the left, guided by the 胯 kuà and 下丹田, Xià Dān tián. The hand moves in unison and guides the arm. The hands are relaxed and the palm faces outwards. The left hand is facing palm outwards. Ensure that your shoulders and left elbow do not rise.

Image 7-21

Continue to move the body to the left, guided by the 胯 kuà and 下丹田, Xià Dān tián. The left hand moves in unison and guides the arm. The hands are relaxed and the palm faces outwards. The movement is slow and circular and takes you back to the original starting position. This is the completion of one cycle.

Image 7-22

Continue with the movement in a circular manner for as long as you feel comfortable. The assessment for you is how much feeling you have of your internal energy and how effortless your movements seem to be.

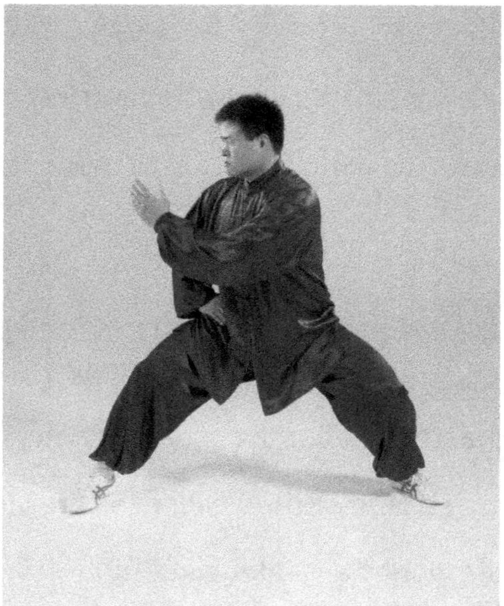

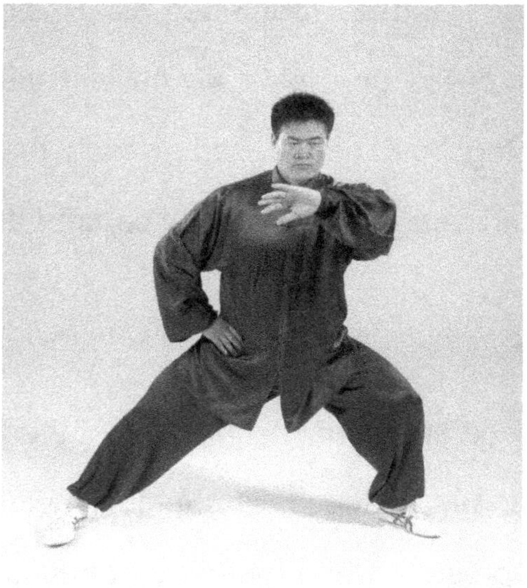

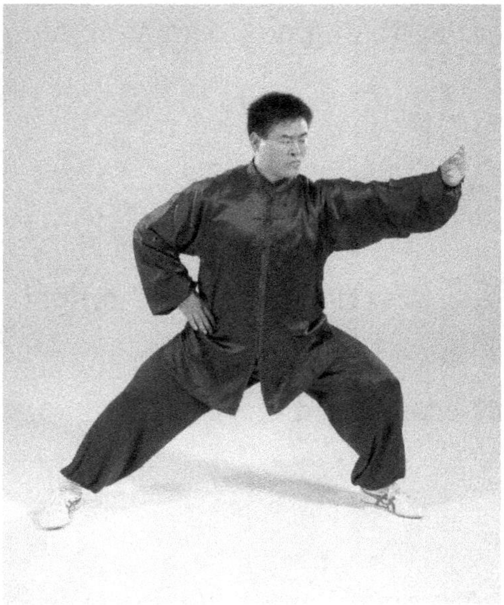

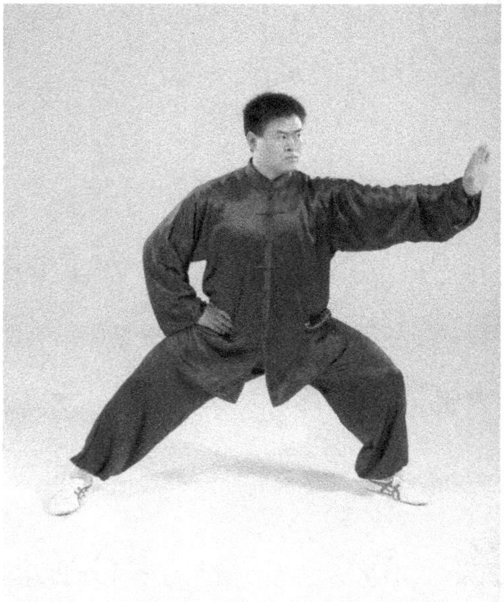

Two Handed Front Sided Silk Reeling

All of the principles we have discussed apply here. It is just that we add the extra dimension of using both hands, rather than the single one. Step out to about two shoulder widths apart or thereabouts until you feel centred and stable.

Image 7-23

Place around 60% and 70% of the weight on the left leg. The left hand is as per the previous single-handed exercise; at shoulder height with the palm open but relaxed and with a slight tension to express the energy and by direction facing forward. Now place the right hand at waist height level with the groin of the right leg. The palm is open and the fingers face forwards. Eyes face forwards. The left leg is positioned pointing diagonally with the toes pointing out at 45 degrees. The right foot is pointed out slightly.

Image 7-24

Start to turn to the right. The dan tián drives the movement, which is slow and circular. The right hand palm faces out towards the left leg. The left hand palm is open and rests in line with the left foot, which points out at 45 degrees. The 胯 kuà is open and the hips are relaxed.

Image 7-25

Using the waist, turn the body to the right, with the left hand moving downwards in a clockwise circular movement and, simultaneously, the right arm moves upwards to shoulder height.

Image 7-26

Continue to use the waist to turn the body to the right, with the left hand moving upwards in a clockwise circular movement to a position that rests in line with the right knee, with the palm pointing upwards to the right elbow. The right hand palm faces down. Simultaneously the right arm moves upwards to shoulder height. The eyes have followed the movement and remain alert and aware, looking at the right hand. This is the complete circle to the right.

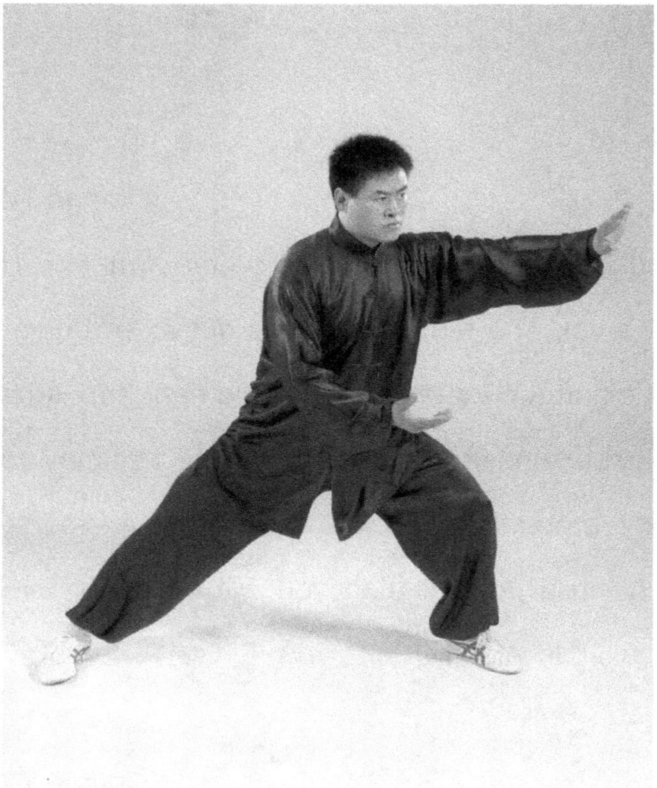

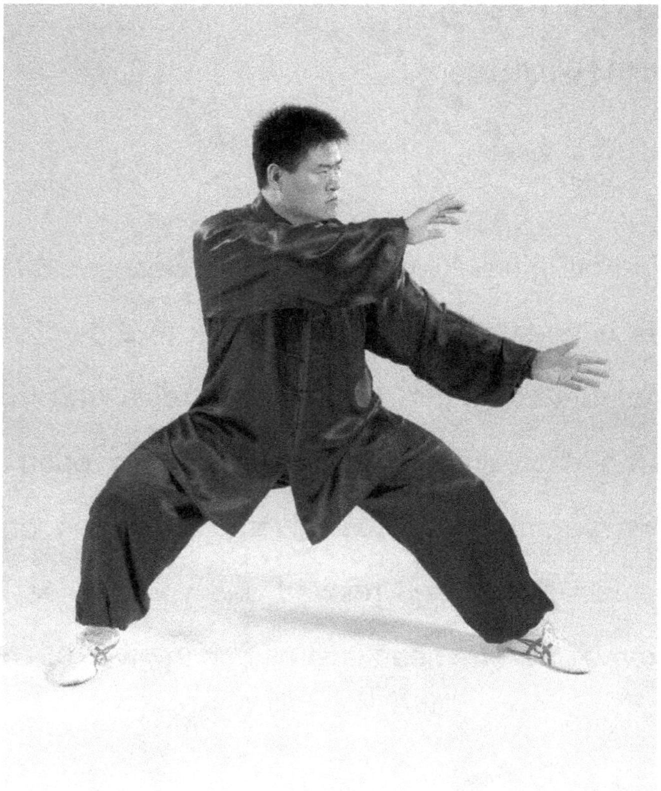

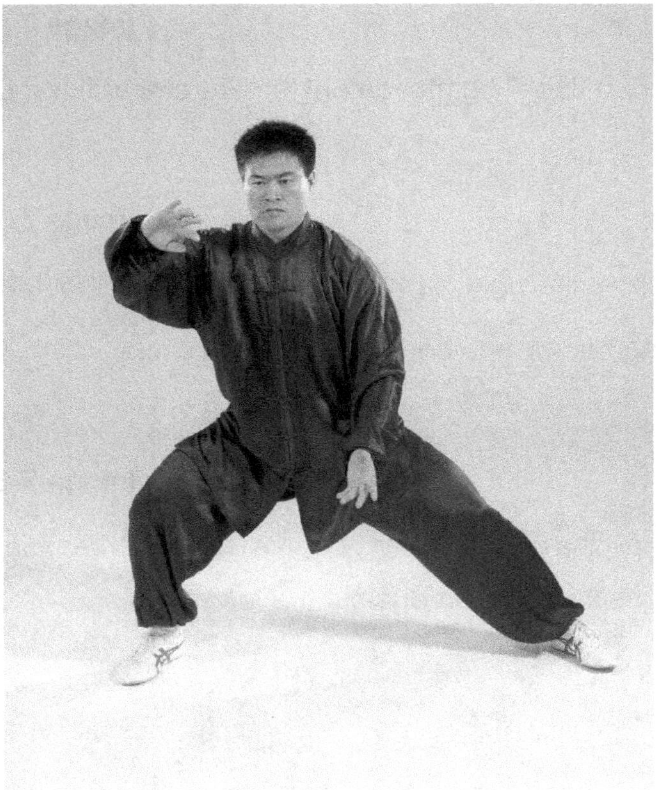

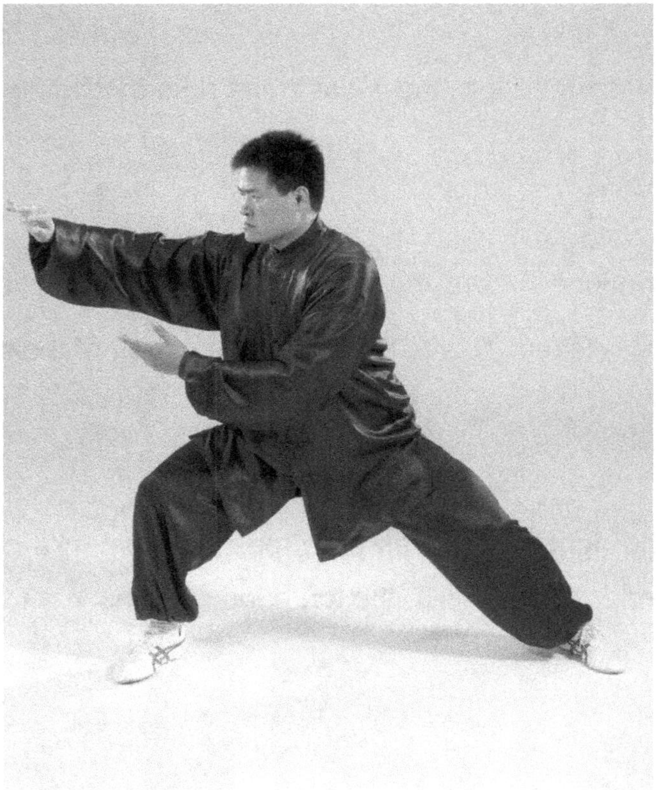

Side Silk Reeling
Right Hand Side

Image 7-27

Start off in the Right Bow Stance. Between 60% and 70% of the weight rests on the right leg. The kua is open and the body is sunk to a comfortable level. The right foot points out at an angle of around 45 degrees. The left hand rests over the left hip in a cupped position. The right arm is bent at the elbow and points out with the palm open and just extending over the right foot – this may vary from person to person slightly depending on body proportions. Relax and compose yourself. Remain rooted and relaxed and you will feel the qi build up and then you will be ready for its expression. The head is stretched up, with the chin pulled in slightly.

Image 7-28

Turn the waist to the right by circling the dan tián. This moves the arm and the fingers that lead. The torque is built from the legs and the rooted posture.

Image 7-29

Continue to turn the waist to the right by circling the dan tián. The movement is natural and flowing.

Image 7-30

Continue to turn the waist to the right. The eyes follow the right hand and the hips and shoulders are relaxed. This is the maximum extent of the turning backwards and the top of the circle.

Image 7-31

The right arm moves from top to bottom in a circle as the 下丹田, Xià Dān tián directs it back in a circle. All the while, the left hand remains in a fixed position cupped against the left hip.

347

Image 7-32

Continue to turn the waist back to the left and shift the weight from the right to the left leg. The right arm continues to move upwards in a circular motion, with the palm slightly stretched and relaxed.

Image 7-33

Continue to turn the waist back to the left and shift the weight from the right to the left leg according to the original weight distribution of between 60% and 70%. The right arm continues to move upwards in a circular motion, with the palm slightly stretched and relaxed. This is the end of one cycle.

Left Hand Side

Image 7-34

Stand in 懒扎衣 Lǎn zhā yī posture. Between 60% and 70% of the weight rests on the left leg. The 胯 kuà is open and sunk to a comfortable height. The left foot points out at around 45 degrees. The right hand rests on the right hip in a cupped position. The left arm is bent at the elbow and points, out with the palms open and just extending over the left foot – this may differ slightly depending on your body proportions, but you will soon be able to gauge this. Relax and compose yourself. Remain rooted and relaxed and you will feel the qi build up and be ready for expression. The head is stretched up, with the chin pulled in slightly.

Image 7-35

Turn the waist to the right by circling the dan tián. This moves the arm and the fingers that lead. The torque is built from the legs. Follow the left hand with the eyes. Maintain the fixed position of the right hand, cupped against the right hip.

Image 7-36

Turn the waist to the right by circling the dan tián. This moves the arm and the fingers that are leading.

Image 7-37

Continue to turn the waist to the right. The eyes follow the right hand and the hips and shoulders are relaxed. This is the maximum extent of the turning backwards and the top of the circle. The weight has moved to the left leg.

Image 7-38

Turn the waist back to the right and shift the weight from the left to the right leg. The left arm moves from top to bottom in a circle as the 下丹田 xià dān tián moves it back in a circle. All the while, the right hand remains in a fixed position cupped against the right hip.

Image 7-39

Continue to turn the waist back to the right and shift the weight from the left to the right leg. The left arm continues to move upwards in a circular motion, with the palm slightly stretched and relaxed.

Image 7-40

Continue to turn the waist back to the right and shift the weight from the left to the right leg, according to the original weight distribution of between 60% and 70%. The left arm continues to move upwards in a circular motion, with the palm slightly stretched and relaxed. This is the end of one cycle on both sides. Repeat on both sides as often as you can whilst maintaining your relaxed awareness.

Two Handed Side Silk Reeling

Image 7-41

This looks more complicated than it actually is. Stand about two shoulder widths apart. When you commence, the weight is balanced. The body is sunk to the extent that you still feel comfortable and not under any strain; otherwise you will lose the relaxed posture that is required. The toes on both feet point out at about 45 degrees. The right arm hangs down by the side of the body, with the palms of the right hand open and facing upwards, level with the right foot. The left arm is at shoulder height, with the palm of the left hand facing away from the body but vertically level with the left foot. The eyes look to the right. Relax and remain rooted. Compose yourself so you are ready to start the movement.

Image 7-42

Start to shift the weight to the left by turning the body as it shifts on the axis of the dan tián and the 胯 kuà.

Image 7-43

Continue to shift the weight to the left by turning the body. The right hand rises, driven by the momentum of the movement. The weight shifts to between 60% and 70% on the left leg.

Image 7-44

The right hand rises, driven by the momentum of the movement. The weight shifts to between 60% and 70% on the left leg. This is the extent of the movement to the left.

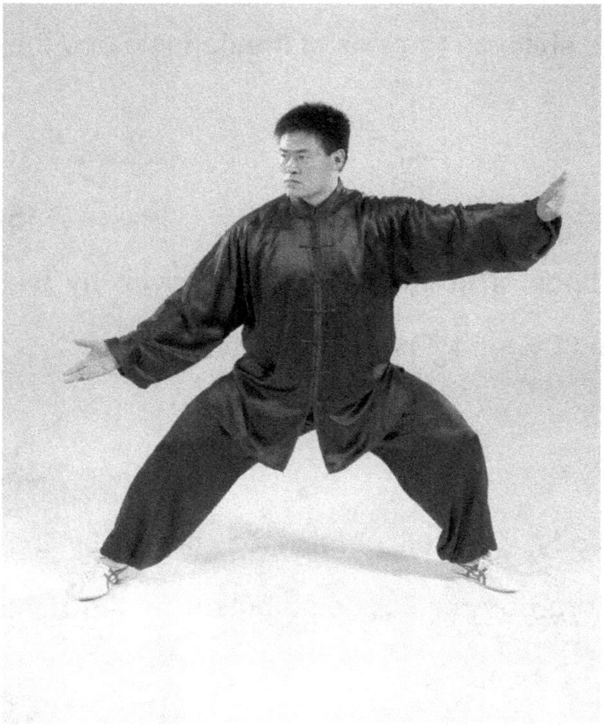

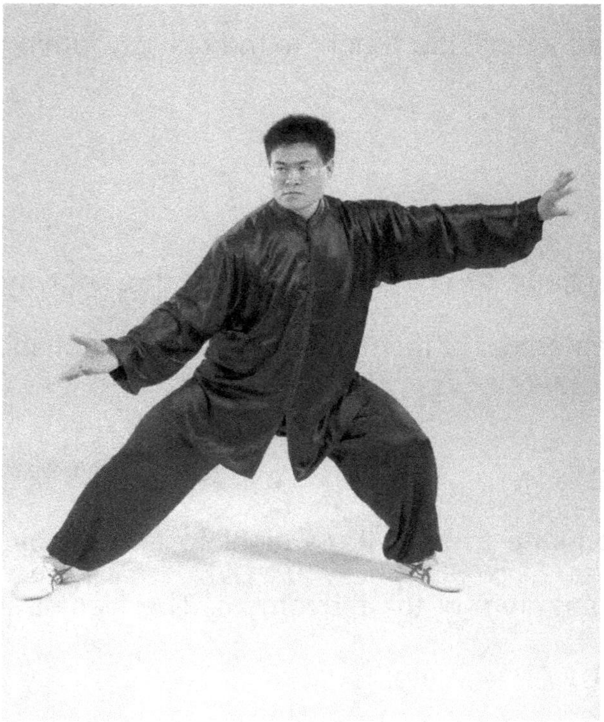

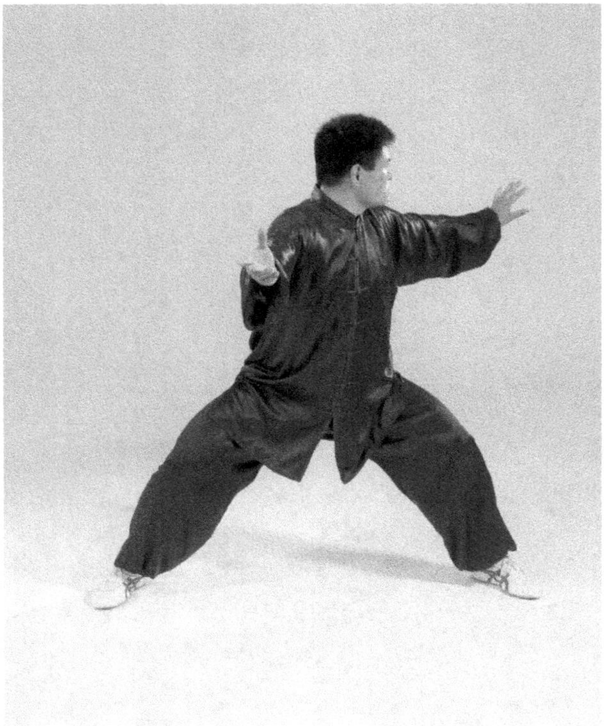

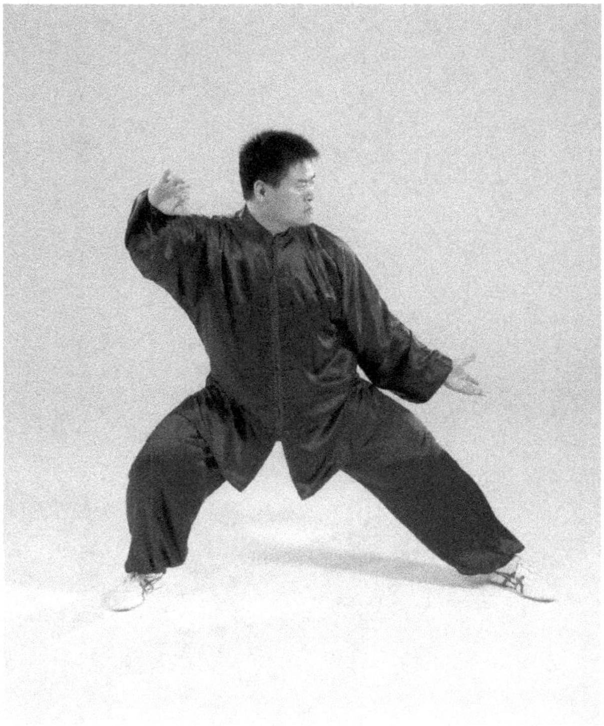

Image 7-45

Start to shift the weight to the right by turning the body as it shifts on the axis of the dan tián and the 胯 kuà.

Image 7-46

Continue to shift the weight to the right by turning the body. The left hand rises, driven by the momentum of the movement. The weight shifts to between 60% and 70% on the left leg.

Image 7-47

Continue to shift the weight to the left by turning the body. The left hand rises, driven by the momentum of the movement. The weight starts to shift to a 60% or 70% distribution on the right leg.

Image 7-48

Continue to shift the weight to the left by turning the body. The left hand rises to shoulder height, driven by the momentum of the movement. The right hand is in line with the right foot and the elbow is sunk. The weight shifts to between 60% and 70% on the right leg.

355

High Rollback Silk Reeling

Image 7-49

Starting on the right hand side the body is positioned at a 45 degree diagonal. The body is sunk to a level that you are comfortable with. Between 60% and 70% of the weight is on the right leg. The toes of the right foot point out slightly. The distance between the feet is approximately two shoulder widths apart. The shoulders are relaxed and the elbows are sunk. The right hand palm is turned out slightly and the left hand palm is slightly turned upwards. There is a very slight tension in the hands. Relax and breathe before commencing the movement. Remember to stretch the spine from the baihui point at the top of the head.

The body starts to move and turn to the right, the 下丹田, Xià dān tián directing the movement. The hands naturally follow the movement of the waist.

Image 7-50

The body continues to move to the right in a figure of eight, like the Infinity symbol introduced earlier. The movement is smooth and circular, and the hands follow the direction of movement.

Image 7-51

The movement continues to the right and slightly back, maintaining the infinity movement. The power in the movement is directed from the feet and the rooted strength in the legs. This is the zenith of the movement to the right.

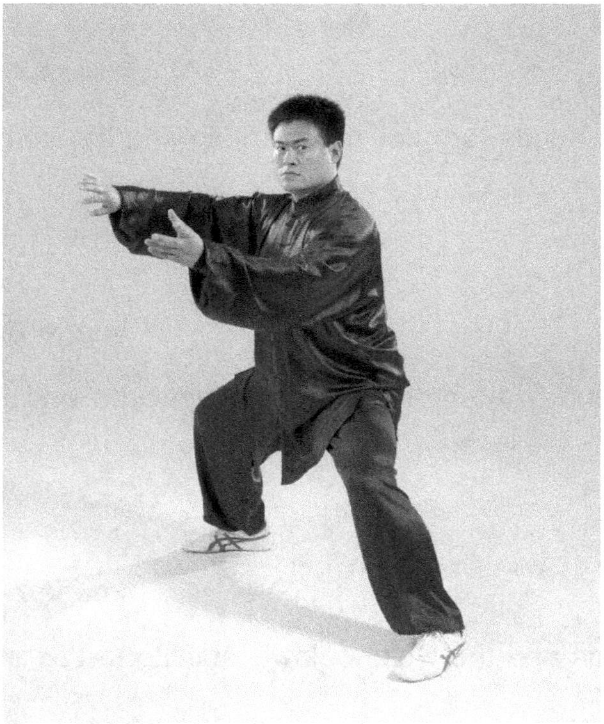

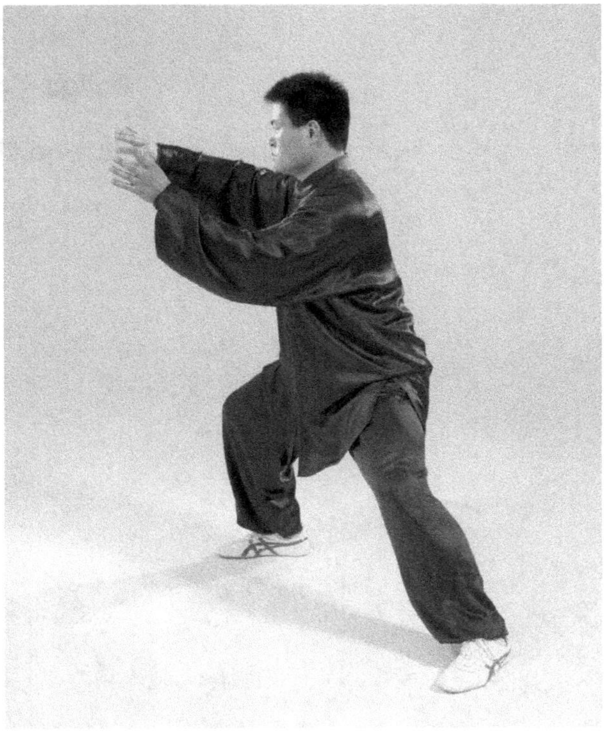

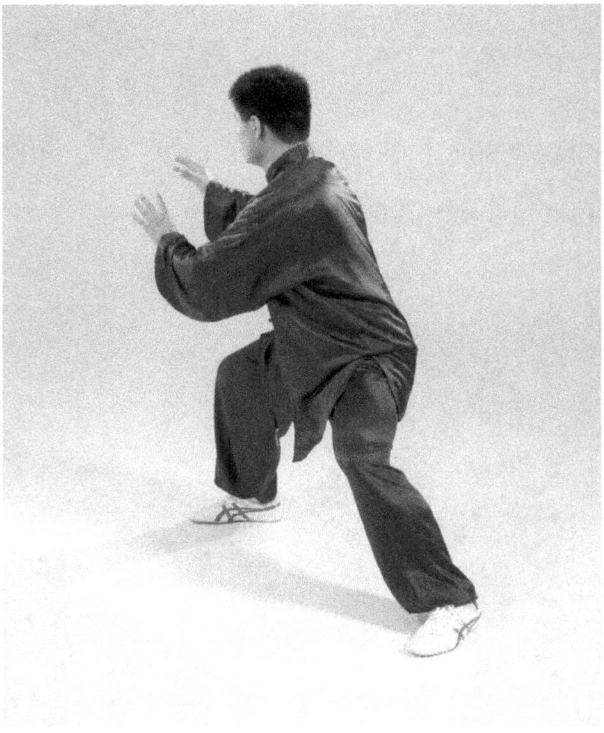

Image 7-52

The body is now ready to move back to the left as the hands turn downwards following the body's lead.

Image 7-53

Continue the movement, with the left hand leading and in co-ordination rising upwards and the weight shifting to the left leg.

Image 7-54

Continue the movement with the hands turning upwards and the weight slowly shifting to the front left leg.

Image 7-55

Continue the movement with the left hand leading and in co-ordination. The weight has shifted to between 60% and 70% on the leading left leg. Do not lean too far forward and lose your balance. This is the zenith of the movement of the left.

Image 7-56

Continue the movement, with the left hand leading and in co-ordination rising upwards and the weight shifting to the left leg.

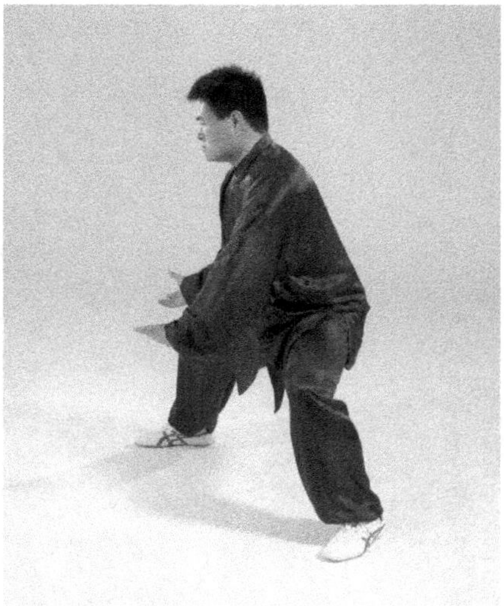

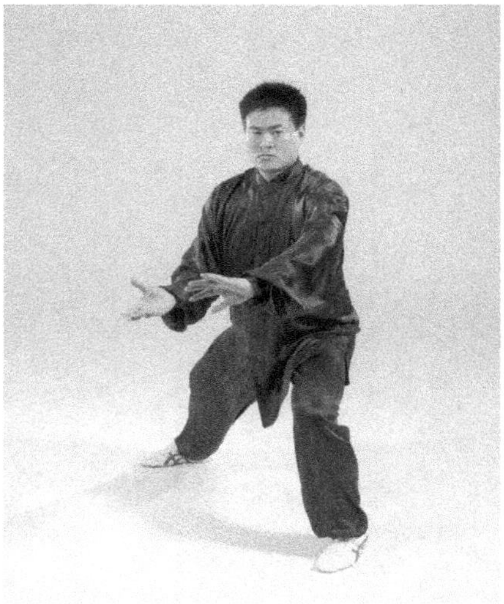

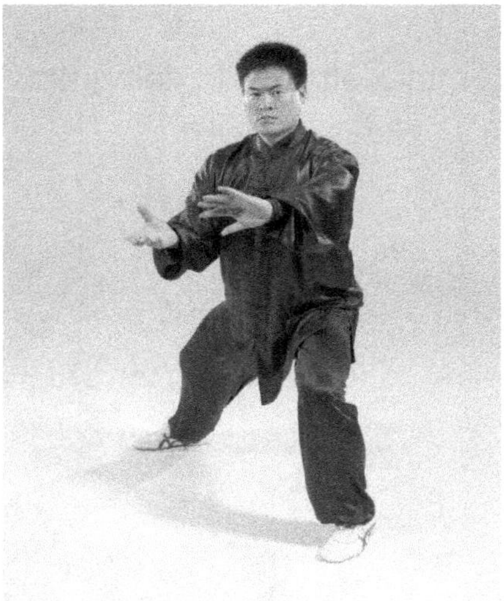

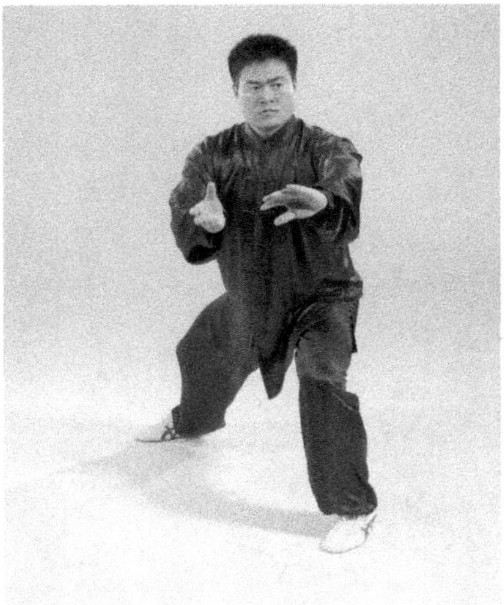

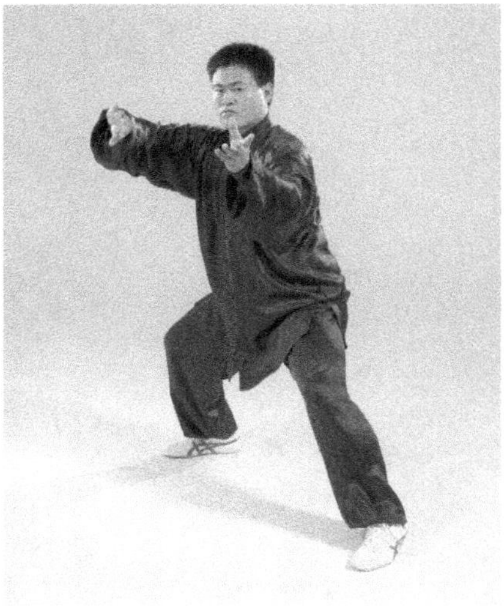

Images 7-57 to 7-64

Repeat the sequences, this time from the left side building up a rhythm of up to eight sequences if you feel that is comfortable for you.

361

Peng, Lu, Ji, An Silk Reeling

This is a great Silk Reeling exercise, incorporating the four 劲 jìn energies of péng 履 lǚ 擠 jǐ, 按 àn.

Image 7-65

This is by now a familiar position for you to start with. Commencing on the right hand side, the body is positioned at a 45 degree diagonal. The body is sunk to a level that you are comfortable with. Between 60% and 70% of the weight is on the right leg. The toes of the right foot point out slightly. The distance between the feet is approximately two shoulder widths apart. The shoulders are relaxed and the elbows are sunk. The right hand palm is turned out slightly and the left hand palm is slightly turned upwards. There is a very slight tension in the hands. Relax and breathe before commencing the movement. Remember to stretch the spine from the 百會 bǎi huì point at the top of the head.

Image 7-66

The body starts to move to the right, in the figure of eight movement – the 下丹田 Xià dān tián directing the movement. The hands naturally follow the movement of the waist.

Image 7-67

The body continues to move to the right in a figure of eight, like the Infinity symbol introduced earlier. The movement is smooth and circular and the hands follow the direction of movement. The body is now ready to move back to the left as the hands turn downwards following the body's lead.

Image 7-68

Continue the movement, with the left hand leading and in co-ordination rising upwards and the weight shifting to the left leg.

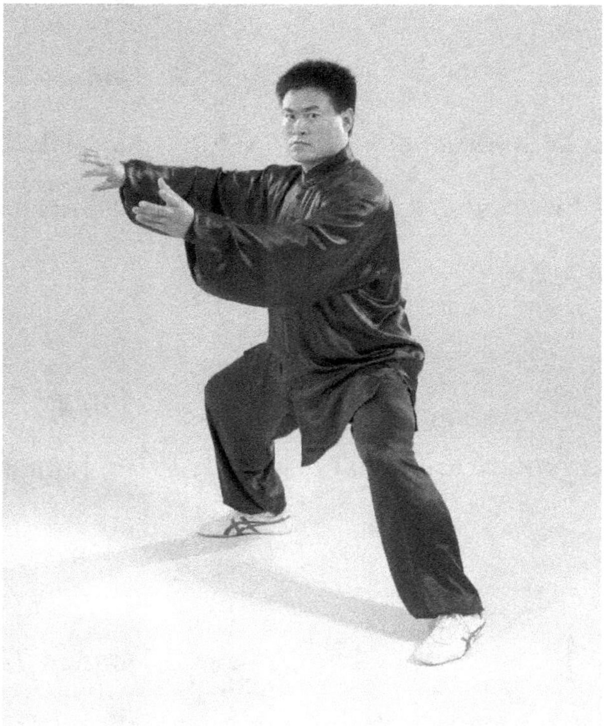

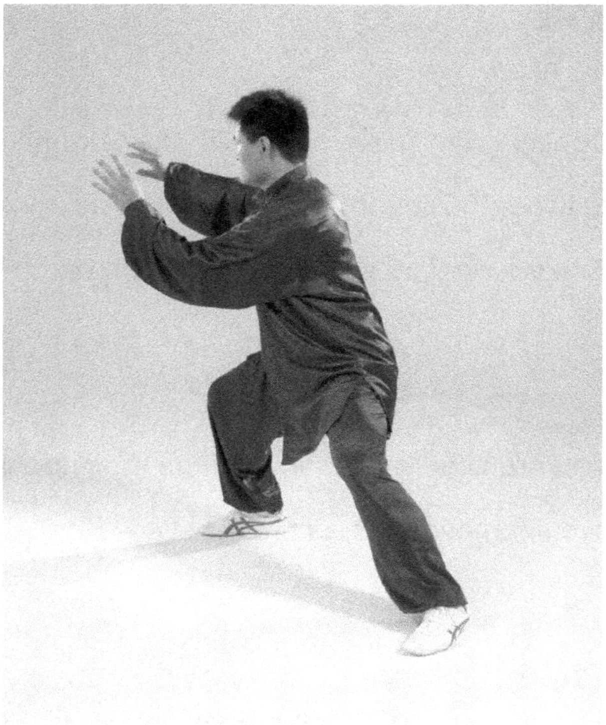

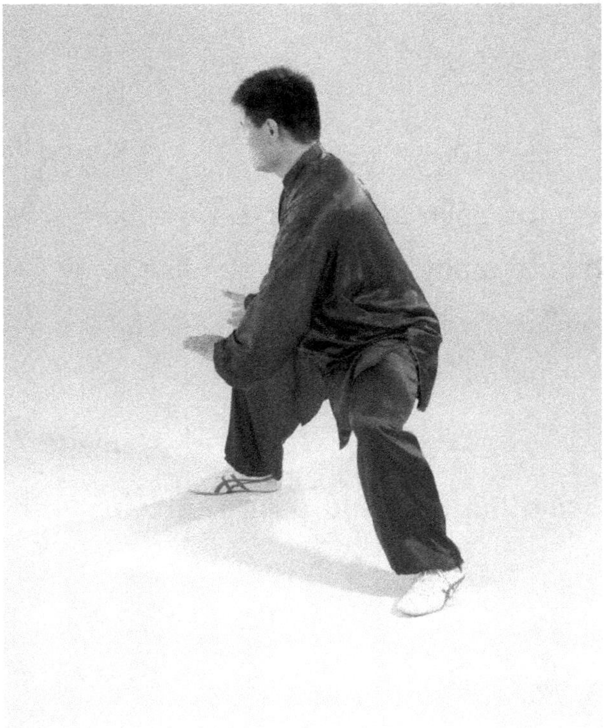

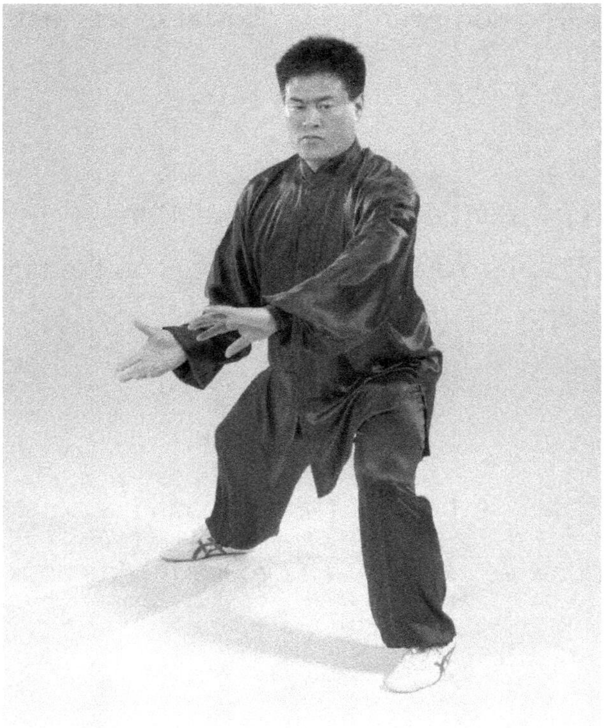

Image 7-69

Continue the movement, with the left hand leading and in co-ordination. The weight has shifted to between 60% and 70% onto the leading left leg. Do not lean your upper body too far forward and lose your balance. This is the zenith of the movement of the left.

Image 7-70

You are now at the zenith of the movement to the left and ready to move backwards to the right in a circular movement as previously.

Image 7-71

Shift the weight onto the right leg in a circular, pulling-down motion with the arms. As always, the 下丹田 Xià Dān tián creates the movement.

Image 7-72

Continue to shift the weight onto the right leg in a circular, pulling-down motion with the arms. The hands are positioned as though you were gripping the wrist and elbow of an opponent. You are still pulling downwards in a circular movement.

Image 7-73

Turn your body to the right and move the right hand with the palm towards the left wrist. Between 60% and 70% of your weight is on the right leg, and you are ready to use the stored energy in the 下丹田 Xià Dān tián to turn.

Image 7-74

Place the back of the right hand against the left wrist and the left hand palm just over the right elbow as you turn in a circular motion back to the left.

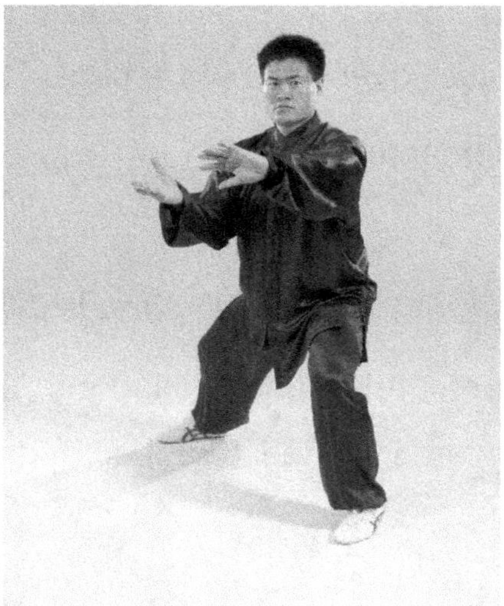
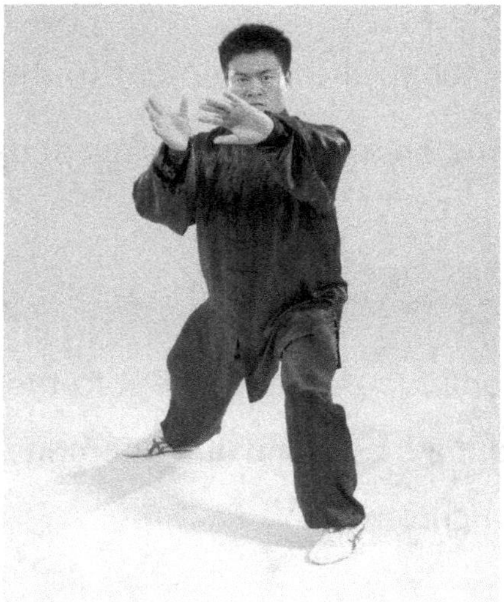
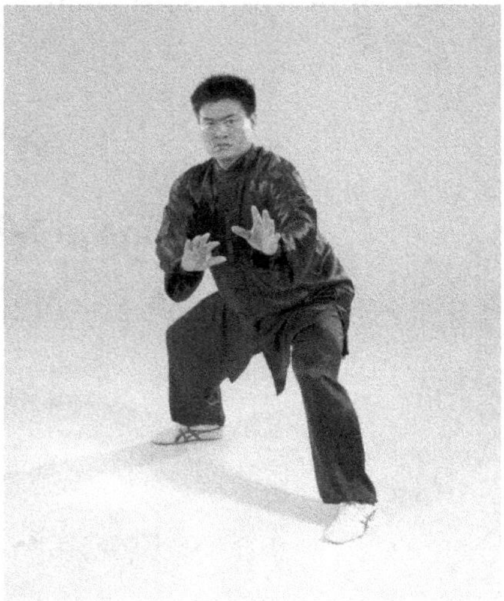
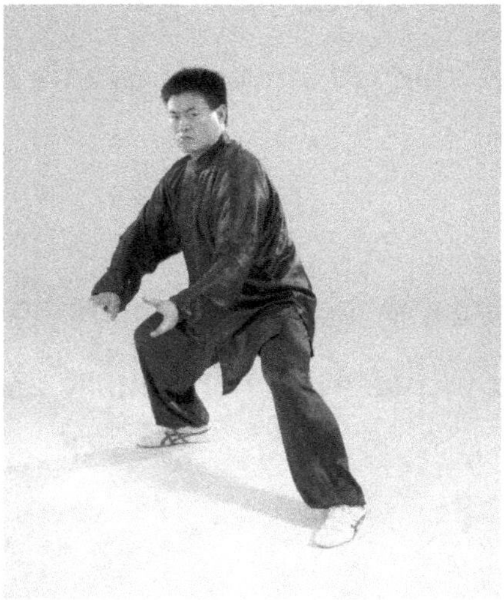
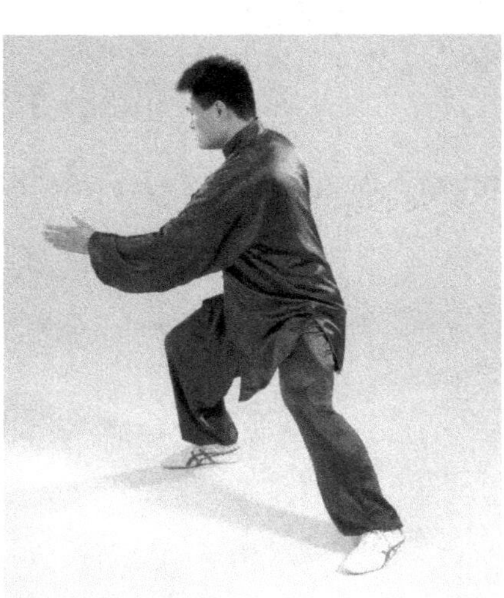
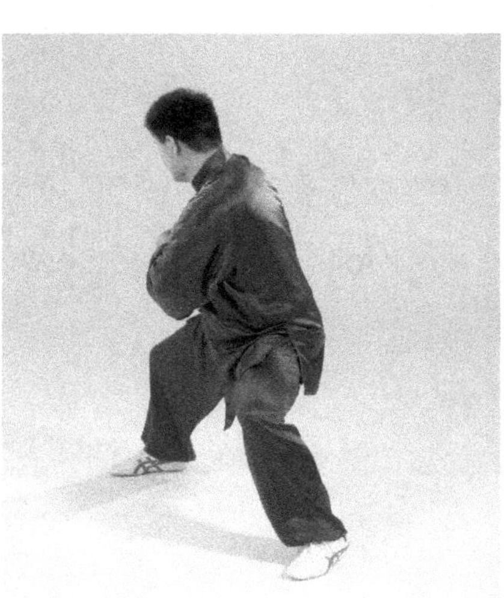

Image 7-75

Continue the movement to the left, with the left hand palm placed just over the right elbow. The arms are moving in a circular motion and moving upwards with the momentum.

Image 7-76

Continue the movement to the left, with the weight shifting to the front left leg and sinking into the right 胯 kuà and the momentum continuing to move the arms upwards, joined together and in synchrony.

Image 7-77

This is the zenith of the movement to the left and the turning of the circle. Open up both palms to get ready to move backwards in the circle back to the right.

Image 7-78

Continue the movement to the left, with the left hand palm moving to a level with the left ear about six inches away horizontally and the right hand vertically under the ear.

Image 7-79

Shifting the weight back to the right in a circular movement, move both hands down together in a pushing-downwards movement, with the palms facing the floor.

Image 7-80

The elbows are open and the hands are relaxed following the movement of the arms. As always, the eyes follow the movement, resting on the hands.

367

Chapter 8 – 18 Movement Short Form

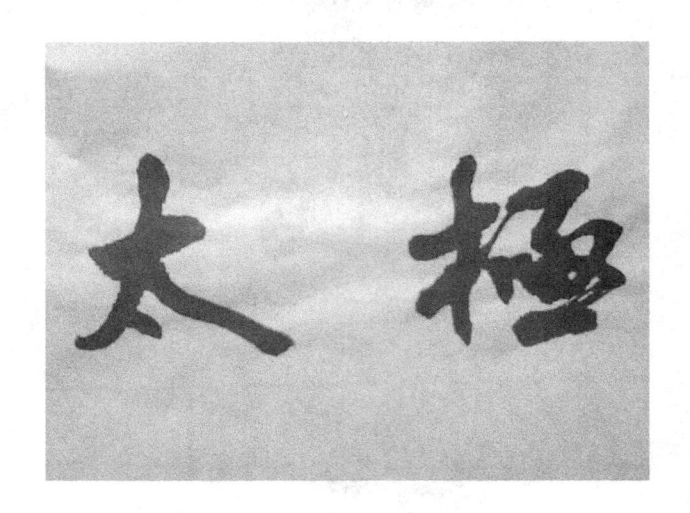

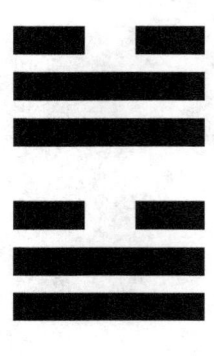

兌　duì

兵者，詭道也。

故能而示之不能，

用而示之不用，

近而示之遠，

遠而示之近

All warfare is based on deception.

Hence, when we are able to attack, we must seem unable;

When using our forces, we must appear inactive;

When we are near, we must make the enemy believe we are far away; when far away,

We must make him believe we are near

孙子兵法 – Sūnzǐ Bīngfǎ, The Art of War, Verse 18

This short form, consisting of eighteen movements, was developed by Grandmaster Chen Zheng Lei in order to provide an introductory level Chen style 太极拳 tàijíquán to beginners or for those with little experience of the form. The purpose behind the invention of this form was to provide ease of access for new students who could often be overwhelmed by the prospect of learning forms with more than 70 or 80 movements.

It incorporates the 缠丝劲 chán sī jìn and some 發勁 fā jìn movements from the 老架, lǎo jiā, old frame and 新架 xīn jia routines, although more of the former are incorporated than the latter. Hence, it is not a series of new movements, but a selection of existing movements to be learned later with the longer forms.

Each of the 18 movements of the short form has a Chinese name that may sound very esoteric when translated into English. This is true for all forms of 太极拳 tàijíquán and some of the forms are common in all styles, for example, Single Whip. A full list of the forms with their pinyin and Chinese characters is shown in Table 8-1 below.

As mentioned earlier, please note that these English translated names may differ in other literature and information. There are many variations on these translations.

Table 8-1

The 18 Movements and their correspondence with Old Frame First Form, 老架 lǎo jiā, old frame Yī lù, 74 Movements

No.	English Name	Chinese	Pinyin	Lao Jiā
1	Starting Posture	太极 起势	tài jí qǐ shì	The same
2	Buddha's Warrior Pounds the Mortar	金刚捣碓	Jīn gāng dǎo duì	2,6,15,73
3	Lazy About Tucking the Robe	懒扎衣	Lǎn zhā yī	3,49
4	Six Sealing and Four Closing	六封四闭	Liù fēng sì bì	4, 25, 41, 46, 50, 60
5	Single Whip	单鞭	Dān biān	26, 42, 47, 51, 61, 67
6	The White Goose Spreads Wings	白鹤亮翅	Bái hè liàng chì	7, 21, 56
7	Diagonal Step	斜行拗步	Xié xíng	8, 11, 22, 57
8	Grab the Knee	搂膝	Lōu xī	9, 12
9	Twisted Step	前蹚拗步	Qián táng ào bù	10, 13
10	Covering Hand Punch	掩手肱拳	Yǎn shǒu hōng quán	24, 38, 59
11	High Pat on Horse	高探马	Gāo tàn mǎ	28, 63
12	Left Heel Kick	左蹬一根	Zuǒ Dèng yī gēn	31
13	The Fair Lady Works at Shuttles	玉女穿梭	Yù nu chuān suō	48
14	Wave Hands	运手	Yùn shǒu	27, 52, 62
15	Turn around with Double Lotus Kick	转身双摆莲	Zhuǎn shēn shuāng bǎi lián	71
16	The Cannon to the Face	当头炮	Dāng tóu pào	72
17	Buddha's Warrior Pounds Mortar	金刚捣碓	Jīn gāng dǎo duì	2, 6, 15, 73
18	Closing Posture	收势	Shōu shì	74

Just to remind you - All images are to be read from left to right, top to bottom as you work through the form.

Hand Positions

Before we move to the form itself, it is useful to analyse in some detail the precise hand positions that are used in it.

The Palm

Image 8-1

The palm of the hand is open and relaxed, with a small gap created between each finger.

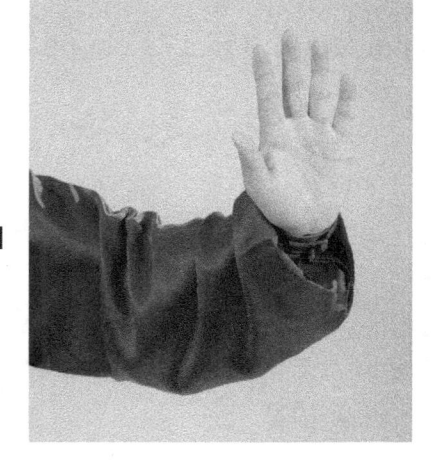

The Fist

Images 8-2 to 8-4

The fist is shown from the side. The thumb holds the middle finger just below the knuckle, crossing the index finger. The remaining fingers bend into the palm. As always, relax and do not pretend that you are gripping a pint glass of beer. Equally, sometimes it is easy to tense your fist when performing punches. Remember the maxim of relaxed power.

The Hook

Image 8-5

Relax the hand down from the wrist. Move the fingertips together around the tip of the thumb. It may be difficult to relax fully at first, especially for those – alas, now the majority of us – who spend a disproportionate amount of time typing into computers and mobile phones.

Now let us delve into the movements.

The 18 Form Movements

1. Starting Posture of 太极 Tài jí – 预备式 Yù bèi shì

Please remember that all images below are to be read from left to right, top to bottom as you work through the form.

Overall Movement

This starts off with stepping out to move into the stance in the 无极 wú jí posture. We have described 无极 wú jí previously as the Unmanifest, the primordial state of stillness before movement begins, and the Manifest unfolds through the enfoldment and interaction of yin and yang. This is the representation of the 无极 wú jí state in our 太极拳 tàijíquán preparation.

无极 wú jí exists before the differentiation begins between movement and quiescence. It is located in the space-time between kun 坤, or pure Yin, and fu 復, the return of the Yang. In the first part of the Dao de Jing, Chapter 42, it is stated:

> 道生
>
> 一生
>
> 二生
>
> 三生万物
>
> Out of Dao, One manifests;
>
> Out of One, Two;
>
> Out of Two, Three;
>
> Out of Three, the created universe

The body, therefore, is held in a state of relaxation and stillness, ready for the initial movement. This is the posture that we have described previously, where you get ready for some Zhan Zhuāng practice. The body is erect, relaxed and upright; the head is pulling up slightly from the 百会 – bǎi huì point to lengthen the body that is erect and aligned. The bottom of the pelvis is slightly tucked in. The chin is slightly tucked down towards the chest, but not exaggeratedly so. The tongue is placed gently at the roof of the mouth behind the teeth.

The shoulders are relaxed and not hunched up. This becomes difficult for those who use computers a lot as tension builds up and the shoulder blades can become tight. The hands rest naturally alongside of the body, with each palm facing inwards towards the torso and resting by the thigh, with the fingers relaxed. Each arm is held slightly out so that a fist could be placed between the top of the inside arm and the armpit.

The feet are kept together. Do not move until you feel yourself perfectly poised and your mind settled. Relax and breathe naturally and let the mind and thus the 气 qì settle. Slowly and gently let the weight shift to the right leg. Lift the left foot and step to the left, placing the toes down first. The toes should be spread slightly apart. Keep the 湧泉 – yǒng quán, the Kidney Point One and source of yin energy from the earth, hollow. The distance moved should be such that the stance as measured between the feet is at shoulder width. This shoulder-width distance is measured by reference to the points of the outside of the shoulders so it may appear to be wider than you imagine. Look forward.

Ensure that you do not commit your weight to the left leg all at once. The movement is akin to mimicking how a cat would place its paw into water to test whether to fully commit itself. The movement therefore allows for you to withdraw to the original position before committing yourself if needed. This is a principle that can be utilised in a combat situation.

Firstly, the toes are placed on the floor and then this is slowly followed by the rest of the foot. This is a cautious and smooth movement of gradually shifting the weight. As we have stated, but to emphasise again, the width of the feet when stepped apart equals the distance between the outside of the shoulders and thus may perhaps appear as though your feet are actually wider apart than they ought to be. The feet should be pointing out at a slight angle. It is important that you feel balanced and floored. This feeling is a sensory one, and practice will enable you to intuitively move to a balanced posture.

The position reached is now that of the 无极 wú jí stance described above. The body is relaxed and the correct posture allows the yin qi energy to move from the lower 丹田 dān tián, through to the 湧泉 yǒng quán acupuncture point – Kidney One, also somewhat vividly described as the Bubbling Springs on the sole of the foot – and to sink the energy down to the earth.

Raise the hands slowly to shoulder height, hands facing down whilst at the same time lowering the body by sinking down from the hips. This causes the knees to bend slightly. Adopting this movement will enable the body to remain upright and thus straight. This avoids losing the natural flow of 气 qi.

Inhale slowly and evenly and allow the arms to rise to shoulder level as though someone is pulling the wrists upwards with a string. Keep looking forward. The shoulders should not hunch up as you effect this movement. The elbows should be bent slightly as the arms rise up. The palms face down. Relax the chest, back, shoulders and ribcage to allow the 气 qì to sink to the abdomen. From the abdomen, the 气 qì will sink to the 涌泉 yǒng quán along the inner thigh. The 气 qi will naturally rise from the outside of the legs through the 督脉 dū mài to the shoulders and elbows.

Lower the hands down whilst keeping the body upright and try to avoid the common mistake of the buttocks sticking out as you do so.

Individual breakdown of the movements

Image 8-6

Stand upright, with the spine stretched from the 百会 – bǎi huì point at the top and centre of your head. Relax and prepare yourself. The feet are placed together and the hands rest effortlessly by the sides, with the palms pointing inwards against the tops of the legs. Breathe slowly. Press the tongue to the top of the palate. Prepare to move by slowly starting to shift the weight on to the right, as the left leg gets ready to move out. Maintain your balance and concentration. This is a slow and deliberate movement not a quick shift of weight where you may lose balance. Slightly and gently lift the left foot slowly. Please note that your body does not lean over to the right. You remain balanced. You are looking forward.

Image 8-7

Maintain your balance on the right leg and lift the left leg up from the hip that is relaxed until the thigh is level with the floor. The left foot is pointed slightly down. At any time you can stop the movement or return back to where you started. Hence, the movement is slow and controlled and you retain your balance.

Image 8-8

Remember that this movement is performed very slowly. You must maintain your balance and act as though you were mimicking that of a cat stepping with one paw into water – ever ready to move backwards if danger lies ahead.

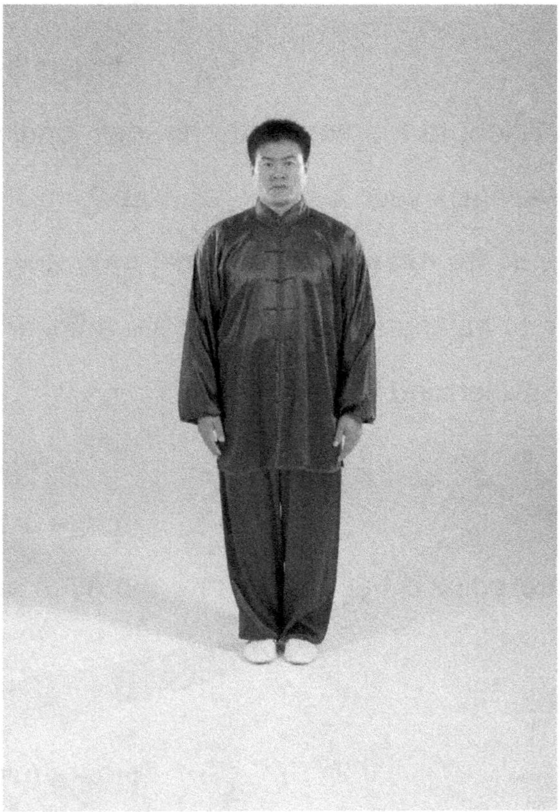

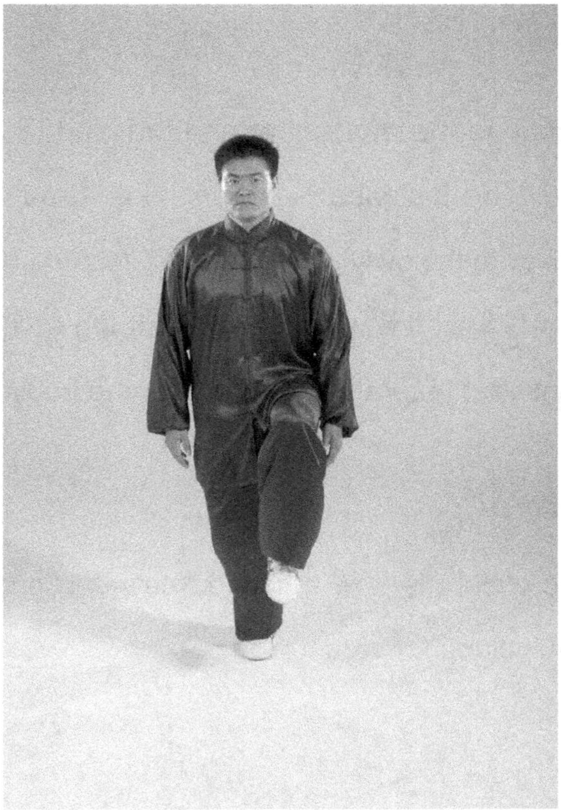

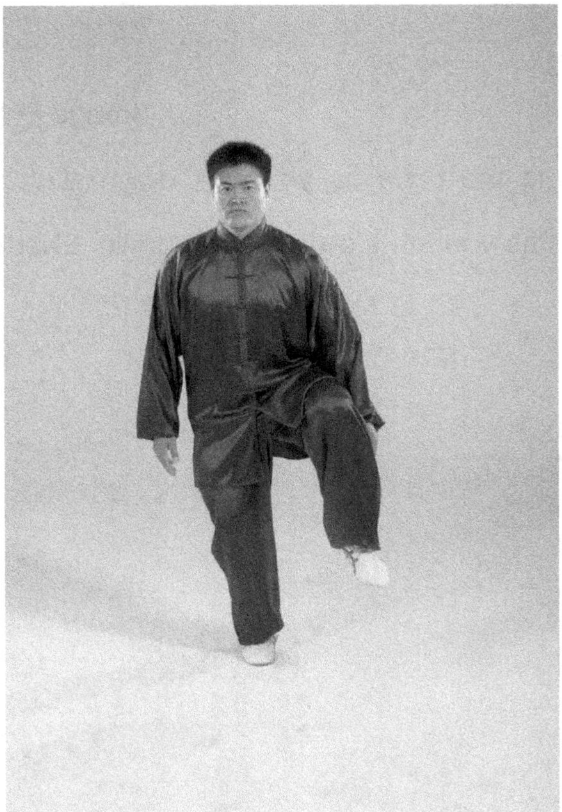

Image 8-9

Continue the movement and proceed to slowly drop the left foot to the floor with the toes landing first. The distance is around one shoulder width apart. Maintain your balance and perform the movement slowly. The eyes are looking forward and the arms are relaxed and rest naturally down. Slowly shift the weight to the left leg so that the ratio starts to balance at a 50% to 50% ratio. The arms rest naturally at the side of the body and the eyes are still looking forward.

Image 8-10

The weight is now balanced between both legs, the feet are pointed out slightly and the eyes are still looking forward.

Image 8-11

Relax both knees and hips and remain balanced and ensure that your posture is aligned and that your spine is stretched. Sink the body very slightly as you prepare to move. The hands are relaxed, with the fingers pointing to the floor and held in line with the feet.

Image 8-12

Slowly raise the arms, with the palms of each hand facing the body as you sink down slightly, evenly balanced. Relax the shoulders and lower the elbows as you are slowly sinking down. Ensure that the shoulders remain relaxed and avoid hunching up.

379

Images 8-13 to 8-15

Continue to slowly raise the arms, with the palms of each hand moving to a position facing the floor, down to the abdomen. The body is relaxed and upright and retains the feeling of being pulled up from the crown of the head, the 百会 - bǎi huì point. The arms rise to shoulder level whilst ensuring that they are both relaxed and not hunched up. The eyes face straight ahead.

Image 8-16

The body now sinks down, with the arms following and with the palms pressing down. This movement is performed by the sinking of the hips and the knees naturally bending as though you are about to sit on a chair that is behind you but that you cannot see, whilst at the same time maintaining a straight and upright posture. The sinking is from the 胯 kuà – that is relaxed and not by leaning from the waist.

Images 8-17 to 8-19

Continue to sink downwards, with the elbows lowering until the hands rest with palms facing down just above the hips. Relax and maintain your balance, posture and rooting.

381

2. Buddha's Warrior Pounds the Mortar – 金刚捣碓 – Jīn gāng dǎo duì

Overall Movement

This next movement involves a turning of the body slightly to the left. Following on from the last position, first relax and then sink the hips. Then turn the waist and hips together in unison so that the body both moves smoothly and continues to remain grounded. Let the 气 qì descend to the 涌泉 yǒng quán as the body turns left.

Slowly shift your weight on to the right leg and at the same time, turn both palms of your hands upwards and forwards in a clockwise direction and raise the arms. The palm of the left hand is facing outwards and is at eye level.

Check that the left hand is in line with the left knee. The palm of the right hand is now facing upwards and is positioned in the centre of the chest. The fingers are pointing upwards and are relaxed and open. This position held by the hand is as though you are balancing a ball within the palm.

Relax the left hip as you slowly shift the body weight onto the left leg and turn 90 degrees. Turn the hands over in an anti-clockwise direction. The left palm moves to a position of facing diagonally upwards. The right palm moves to a position that faces outwards.

Turn the body to the right at an angle of 90 degrees and at the same time, gently lift the toes and pivot open the right foot. The weight shifts from the right to the left leg. The eyes should follow the body, looking left. This is another of the fundamental principles, and one that often is overlooked. Often students are concentrating on the movements so much that the eyes remain static. The 气 qì spirals up the right leg to the 下丹田, xià dān tián, down the left leg and from the waist to the arm and hands to produce a pushing strength.

Inhale and move the weight to the right leg and lift the left foot a few inches from the floor, with the toes pointing down and alongside the heel of the right foot. Lower the body and slightly turn right.

Do not bend the waist and maintain the body upright. There will be a tendency to stick out your buttocks, but maintaining your posture is important. Raise the hands upwards and look forwards and to your left. Remain strong in your 下丹田, xià dān tián.

Maintaining the weight on the right leg, slide out the left and maintain your gaze in that direction. The 气 qì moves to the hands from the 下丹田, xià dān tián. Now, move the weight from the right to the left leg. Open the left foot and move it forward and place it down.

Move the left palm forwards and circle it back towards the inside of the right forearm. Move the right palm upwards and circular whilst moving the right foot to step forward in conjunction. The right foot passes the inside of the left foot and the toes press to the ground. Keep the weight on the left leg.

Cross the left palm out and lower it in front of the abdomen facing up. The right palm then changes into a fist. Lower it into the left facing up palm. Lift the right fist to shoulder height with a relaxed right hip. Ensure that this movement is both co-ordinated and relaxed.

Drop the right hip in a relaxed way to the ground in co-ordination with the right fist that meets the left pam.

Individual breakdown of the movements

Image 8-20

Turn the body slightly to the left, with the arms following and the palms of both hands turning to the side. The movement is from the 下丹田, xià dān tián and is accompanied by relaxed hips.

Image 8-21

Continue turning to the left led by the 下丹田, xià dān tián. The distance between the arms remains fixed and the palms face outwards. Keep the upper body relaxed and loose. Try not to stiffen up during this movement.

Image 8-22 to 8-23

Keep on turning to the left. Remain relaxed and remember that the arms follow the body. Do not hunch the shoulders. This is the fullest extent of the move to the left and you are ready to turn back to the right.

Image 8-24

Raise the toes of the right foot slowly from the floor as you turn to the right.

Image 8-25

Keep on turning to the right. Remain relaxed and once again remember that the arms follow the body and the hands maintain their distance between each other. The turning is made on the right heel. Maintain your concentration to maintain balance.

Image 8-26

Keep on turning to the right and shift the weight to the right leg that is pivoting on the heel of the right foot. Sink down slightly and, as always, remain relaxed. Do not hunch the shoulders. Concentrate on the movement, with the eyes alert and following the movement.

385

Image 8-27

Place more weight on the right leg as the toes lower to the floor. You are balanced and ready to step out back to the left.

Image 8-28

Lift the leg with the knee at thigh level and the left foot positioned so that the toes are facing down towards the floor at a 45-degree angle. All the weight is on the right leg. You need to sink your weight and ensure that your left hip remains relaxed and does not stiffen up, as may well be the case until you become more familiar with the movement. Make sure that the upper body remains straight and does not bend. Be careful to maintain your balance and not wobble. Practice will ensure that the correct posture is achieved.

Images 8-29 and 8-30

Step out to the left by sliding the left heel along the floor and keeping the weight on the right leg, with the knee slight facing outwards. The hands are still the same distance apart.

Images 8-31 to 8-33

Now turn the body slightly to the right to create a circular movement back to the left. The eyes follow the hands. The left foot remains positioned, with the heel on the floor with most of the weight on the right leg.

Image 8-34

Place the left foot flat to the floor and turn the hands downwards in a co-ordinated circular motion over the right knee as you make ready to move back to the left.

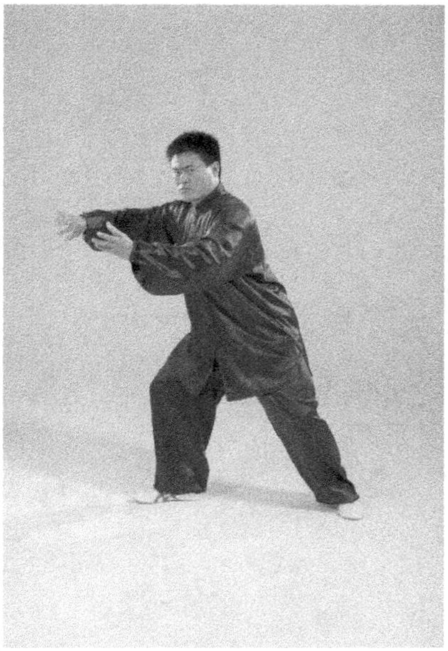
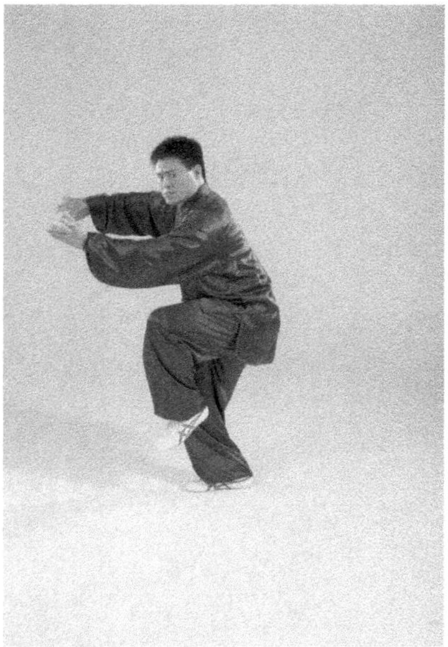
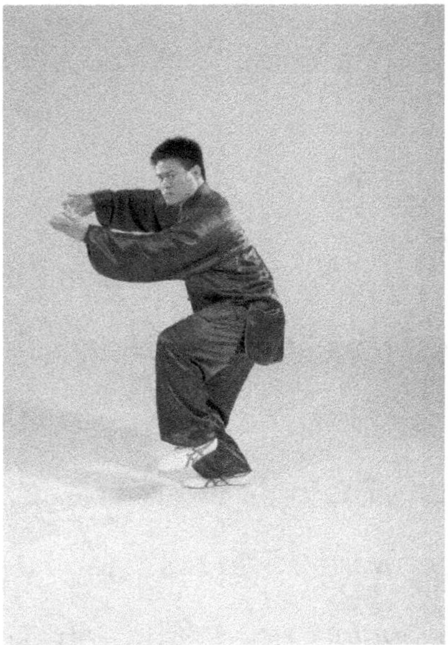
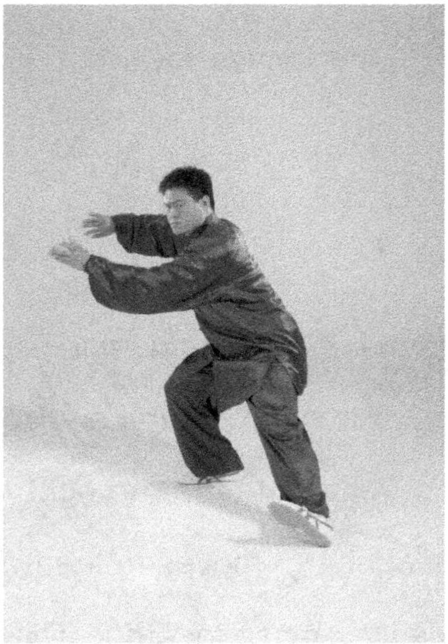
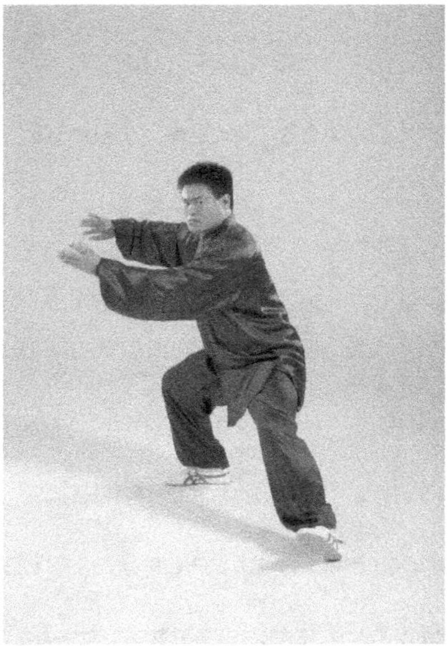
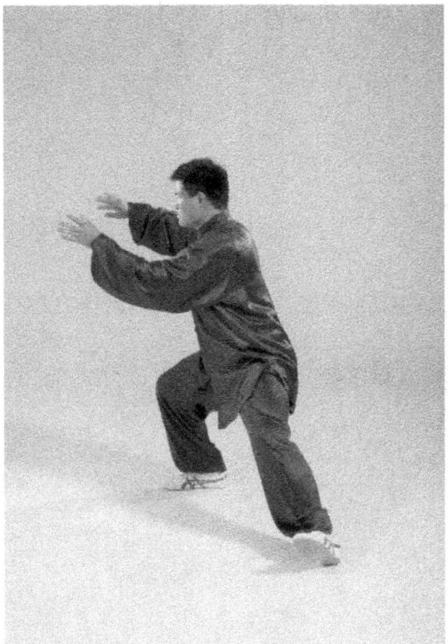
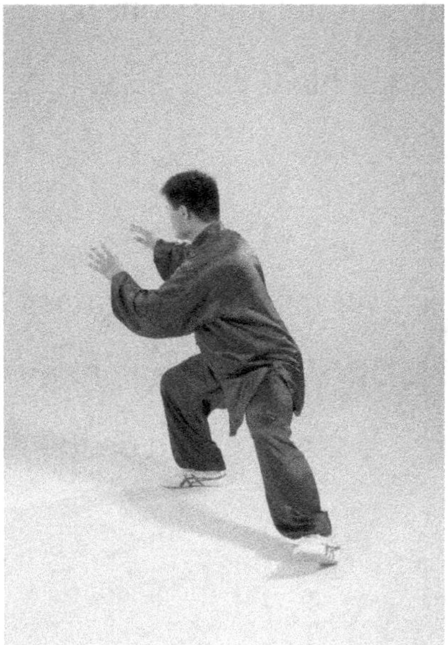
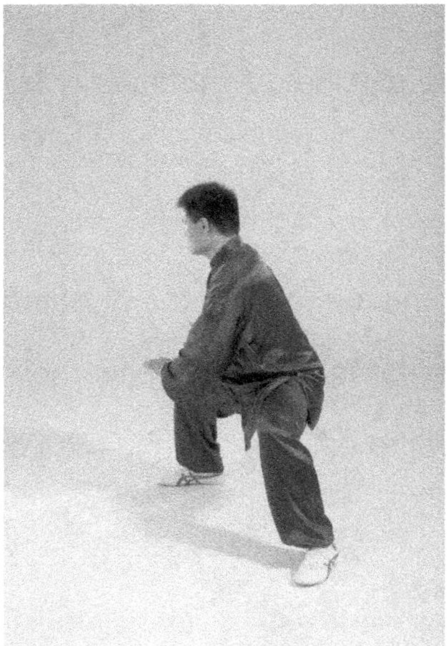

Image 8-35

Shift the weight to the front left leg as you turn the body from the back to the front, with the arms following, keeping the same distance between the hands. The left hand is positioned with the palm pointing to the floor, with the side facing forwards just over the abdomen and with the elbow of the left arm in line with the left knee. The right hand is open, with the palm facing forwards over the right foot.

Image 8-36

Shift the weight to the front left leg as you turn from the back to the front. The upper body turns at a ratio of around 45 degrees. The left hand moves towards the centre of the chest, with the elbow bent and as always relaxed.

Image 8-37

This movement looks very busy in terms of motion and this especially seems so whilst explaining it in the form of a photograph and text. It is quite natural and relatively simple. The action is that of moving forward from the back to the front and from the right to the left. The right leg is sweeping forward in a circular movement. The right arm moves to the front and upwards. The right hand and right foot that is sliding forward on the toes move together. There is little weight on the right leg. The left leg bears the weight. The left knee is naturally bent but does not extend over the toes as you are sinking down naturally by relaxing the hips.

Images 8-38 and 8-39

Continue to slide the right foot in a circular movement from left to right, with the hands co-ordinating as the left hand presses down next to the wrist of the right hand. 60 to 70 per cent of the weight is on the left leg, and the toes of the right foot are resting on the floor.

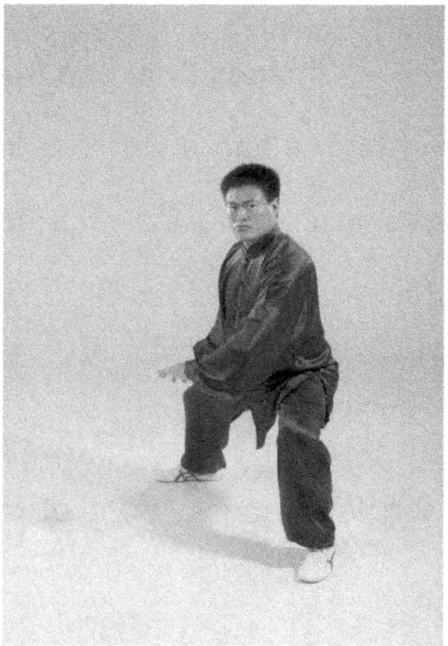

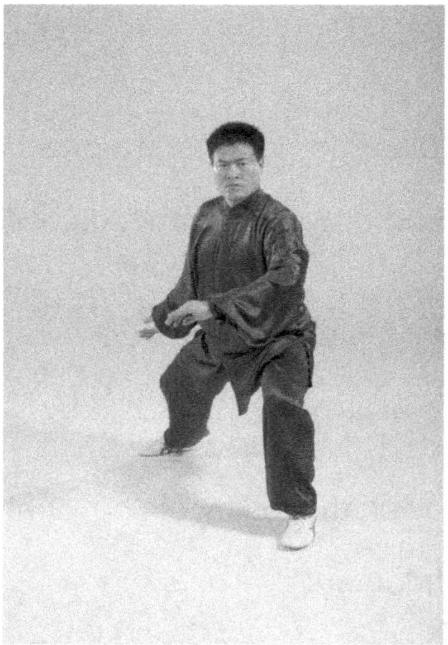

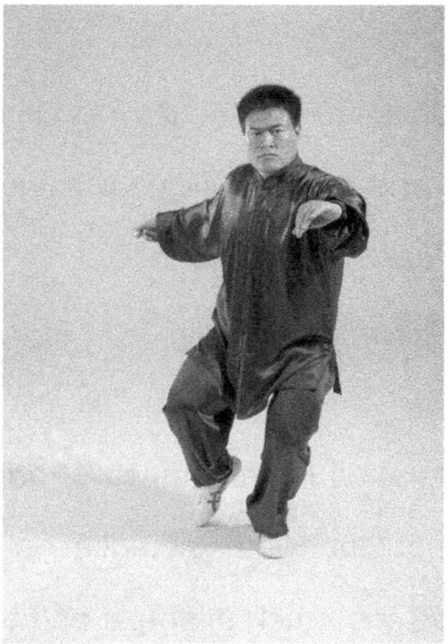

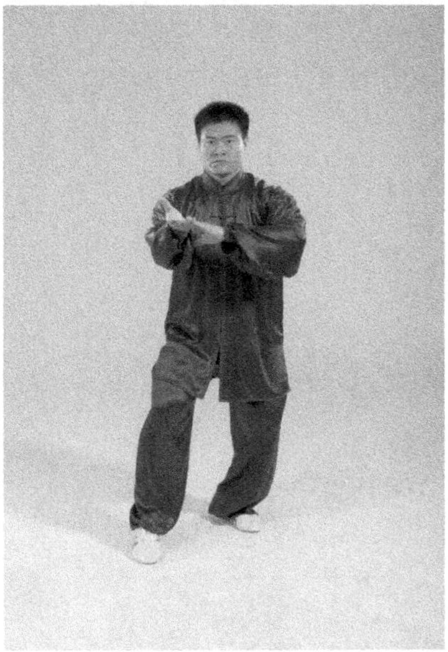

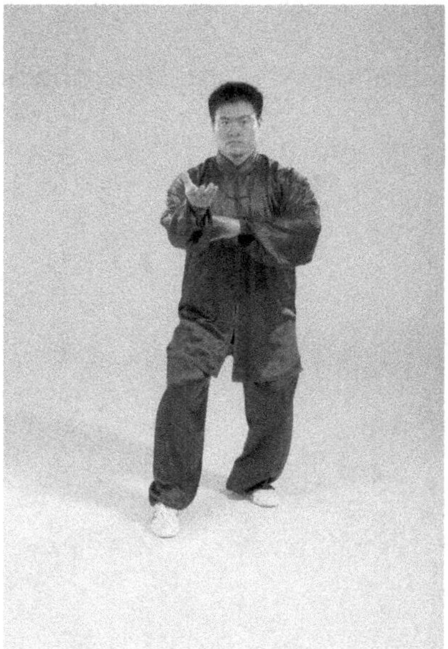

Images 8-40 and 8-41

Raise the right hand upwards to just above the centre of the chest and form it into a fist. The left hand moves downwards, with the palm facing upwards resting at a level just above the 丹田 dān tián.

Image 8-42

Lower the palm of the right hand down towards the open palm of the left hand in front of the abdomen.

Image 8-43

Raise the right fist and right leg simultaneously. The right fist is level and to the side of the eyes and placed just in front. Remain relaxed and do not lose your balance. The right toes point to the floor. Ensure that the shoulders remain relaxed and do not rise.

Image 8-44

Stamp the right foot down quickly in one fluid motion whilst bringing the right fist into the left hand palm. The movement creates a shoulder width distance between the feet. The eyes are look forward. There is a temptation to use too much stiffness and to lose the relaxed state. The key is co-ordination and keeping the hips relaxed. The power, as always, comes from the co-ordination, relaxation and internal power. Try not to scare the downstairs neighbours if you are living in an upper apartment.

391

3. Lazy about Tucking the Robe – 懒扎衣 Lǎn zhā yī

Overall Movement

Start this movement by turning the body slightly to the left. The 下丹田, xià dān tián is sunk before the waist turns. At the same time, transfer your weight on to the right leg. The right hand changes from a fist into a palm. The palm then moves slightly to the left a fraction and continues to move in an upward direction and to the right on a slow circle. This movement is undertaken in co-ordination with the feet and waist. It is one co-ordinated whole body movement. The movement of the waist energises the right arm.

At the same time, the right arm rotates in a counter-clockwise motion as the left arm moves in the opposite, clockwise direction, and the left palm presses down in a circular motion across the front of the body. Then bring the right arm down to a position in front of the body to the left of the chest, with the palm facing upwards. The left palm moves to the left and then rotates upwards in a counter-clockwise direction. It crosses with the right arm moving to the front of the chest.

Continue moving to the left and shifting the weight to the right and push the right hand upwards in a circular motion. Turn right and circle the right hand outwards to the right so that it finishes at eye level with the fingers pointing up. The left hand moves downwards in front of the abdomen and rests on the left hip with the thumb pointing to the back and the fingers to the front.

This movement consists of the upper body leading in an opponent, and the legs control those of the attacker with the left hand over the right arm, positioned to defend an attack to the face.

Individual breakdown of the movement

Image 8-45

Turn the body slightly to the left and start to separate the hands. The movement comes from the waist and the body is slightly sunk.

Image 8-46

Continue to turn to the left and move the right hand upwards in a circular movement, with the left hand following but retaining its position under the right.

Image 8-47 to 8-48

Both arms cross over. The right hand palm faces down and is level with the chin. The shoulders and elbows are relaxed. The left hand palm also faces to the floor and is open and over the right knee. Continue to move both arms outwards from their positions, moving to the left and right as they separate to the sides. Maintain most of the weight on the right leg.

Images 8-49

Shifting your weight to the left, continue to spread the arms out in a circular and wide motion. The right arm is positioned so that the right hand is slightly higher than the shoulder and the left hand is level with the abdomen.

Image 8-50

Balance yourself, ready to lift your right leg whilst maintaining a shoulder width distance in your stance. The left arm is raised to shoulder height with the palm open and the right arm is lowered down slightly, also with open palms facing the floor.

Image 8-51

Relax your right hip and raise your right leg and lower your right arm slightly, turning over the palm so that it faces forwards and is at a height level with the right knee. Both movements are performed together. The left arm rises so that the left hand palm faces to the side at the height of the left ear. The toes of the right foot point to the floor. Maintain the relaxation in the hips and sink down slightly.

Image 8-52

Sink your body a little by relaxing the left hip and get yourself ready to step out with the right leg to the right-hand side.

47

48

49

50

51

52

Image 8-53

Step out to the right by placing the heel of the right foot about two shoulder widths apart from the left. The arms come to meet each other as you do so, with the palms facing each other and about 12 inches apart.

Image 8-54

Both arms cross over. The palms of the right hand points upwards. The left palm rests on top of the right forearm with the fingers slightly raised.

Images 8-55 to 8-56

Turn to the left and start to lower the right foot to the floor by gently pushing the toes down to the floor.

Image 8-57

The body continues from the waist, with the body firmly rooted to the floor to turn to the left as the left hand rests under the bicep of the right arm, with the palm facing upwards. The right hand faces the body. Most of the weight is moved on to the right leg.

Image 8-58

Turn the right arm over in a circular movement as the body turns to the left. Avoid the mistake of either raising or restricting the shoulders. The left arm drops down, with the fingers of the left hand pointing to the ribcage. The right arm protects the front of the body, with the elbow facing out and aligned in front of the centre of the chest. The weight shifts to the left.

Image 8-59

The right arm continues to move in a horizontal circle to the right and across the body, with the palm facing downwards. The left hand moves downwards to the abdomen, with the palm facing upwards. The movement is slow, circular and relaxed.

r   l   r   l

53   54

55   56

57   58

397

Image 8-60

The right arm continues to move in a horizontal circle to the right, with the palm facing outwards. The left hand moves from the abdomen to the left hip, where it will rest in a cupping position.

Image 8-61

The right arm moves out to a placement just over the right foot as the left hand fixes itself into a cupped position over the left hip as 60 to 70 per cent of the weight now rests on the right leg. Relax and let the body sink down.

Image 8-62

Sink down further as you move the right arm slowly and slightly to a position, with the right hand palm facing out. Relax and remain focussed.

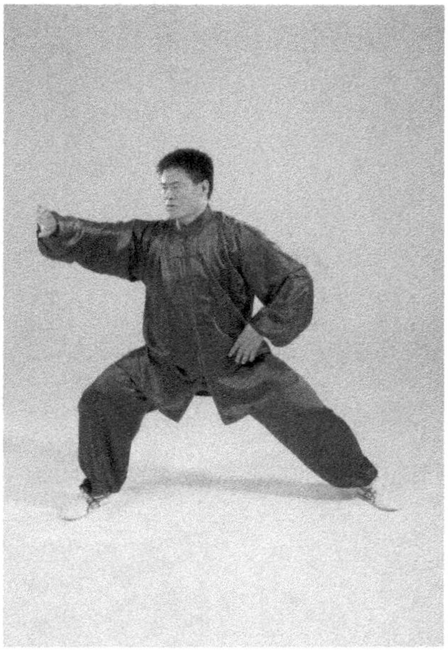

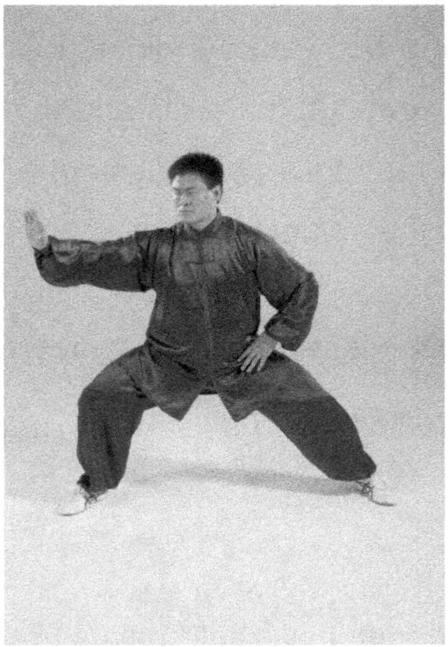

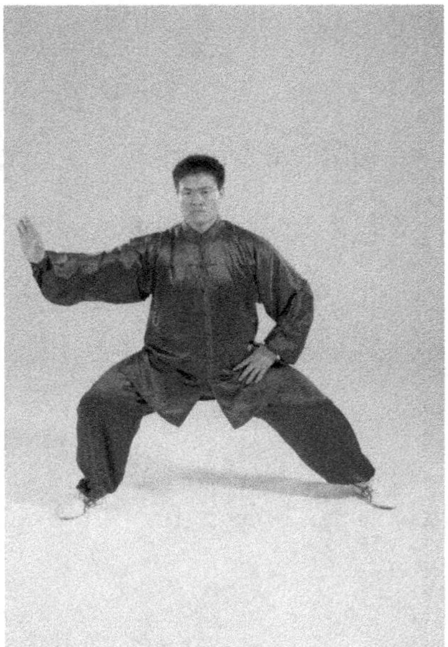

4. Six Sealing and Four Closing – 六封四闭 Liù fēng sì bì

Overall Movement

Please remember that as we explain each movement by breaking it down into individual parts, it is nevertheless all one continuous movement. So, from the last section of Lazy about Tucking the Robe, we next turn just slightly to the right whilst at the same time simultaneously shifting the weight on to the right leg. The left hand moves in a circular direction to move towards the right hand.

Both hands sink as the weight shifts. Make sure that your shoulders do not move up involuntarily as you do so. As always, the turn of the body is co-ordinated with the turning of the hands, and the lower 丹田 dān tián moves the 气 qì through the arms and then sinks. The position of the hands in its martial application is to best adhere to the opponent's arms, and, hence, this controls not just his -or her!) arms but also the body and the balance as well.

Turn right and shift the body weight slightly. Lift and move the left palm in an upwards circular motion towards the right hand and move the right hand slightly to meet it. Look at the tip of the middle finger of the right hand.

Turn and shift your weight to the left and move both palms to the left with the left hand moving in a circular closing motion and the right hand making an opening circular movement. Look to the front.

Continue to move left and shift the weight with right hand opening and circular and the left hand closing motion. Keep shifting your weight to the right and change the left hand into an opening movement and the right hand to a closing one. Lift the hands upwards towards each other and in front of the left shoulder. Look to the right.

Turn slightly to the right and sink the body. Press the palms down. Move the left foot towards the right so that it is about 8 inches away.

Individual breakdown of the movement

Image 8-63 to 8-64

From the last position, prepare to move to the right in a circular motion, guided by the waist. You are relaxed and the body is sunk and straight. Turn the left hand upwards and over in a circular motion as the right hand also moves to the right. Both hands move together as the weight shifts to the right leg.

Image 8-65

Move the left hand towards the right as you shift your weight to the right leg. The circular movement of the left hand is directed from the elbow, which remains relaxed.

Image 8-66

Continue to move to the right, shifting your weight, sinking slightly and remaining relaxed, with both palms facing outwards over the right knee. This is the extent of the movement to the right.

Image 8-67

Pull down the arms together in a circular movement over and across the right knee and as you sink. As always, as you will by now be sick of hearing, the shoulders are relaxed, the waist drives the movement and the eyes follow the hands.

Image 8-68

Shift your weight to the left, with both hands pulling down together and over the right knee and to the left across and under the abdomen.

Image 8-69

Continue to move to the left shifting your weight, with the palm of the right hand moving to face upwards and the left palm remaining downwards. Both arms move together, with the shifting of the weight driven by the 丹田 dān tián. The hands maintain their relative distance between each other.

Image 8-70

Continue to move to the right, shifting your weight to a 60% or 70% ratio to the left as the arms move together over the left knee. This is the extent of the move to the left. This is a very smooth and relaxed movement and is performed by having relaxed hips and rooted legs.

403

Image 8-71

Raise both arms up together to the level of the chin, with the left hand level with the eyes and with the right hand a few inches from the chin. The palm of the left hand faces out and that of the right hand, the floor.

Image 8-72

Move both hands over in a circle so that the right hand palm is level with the left shoulder and facing down, and, with the left hand level and six inches or so from the left ear, the palm is open and facing forward and aligned with the left foot. The hands are just a few inches apart.

Image 8-73

Lower your body by sinking from the hips and prepare to sweep down to the right, with the hands pushing down. The weight shifts further to the right leg.

Image 8-74

Further shift your weight to the right as you continue to push downwards, with both hands moving in unison and placed close together.

Images 8-75

Continue to shift your weight to the right and move the left leg to shoulder width from the right, and raise the left foot so that it rests on the toes as you continue to push downwards, with both hands moving to a position facing down to the floor and over the right foot.

405

5. Single Whip – 单鞭 Dān biān

This is a movement that exists in most if not all taiji styles and is repeated within all the other longer Chen forms. As always, the movement continues from the previous one seamlessly and without pause. Inhaling, turn slightly to the right and move the left palm forward, at the same time, the right palm backwards a fraction. Both palms rotate to face upwards move smoothly and together. The weight remains on the right leg, but the left pivots on the toes. As the body turns, the knees move slightly to close and the eyes fix on the palms. In a martial application, this is a defensive move, used to free the right hand from the opponent's grasp.

Now turn to the left whilst retaining the weight on the right leg. The ball of the left foot remains on the floor and, as the knee moves outwards, the fingers of the right hand fuse together in the hook position but remain relaxed with no tension, and the hand moves upwards and in a circle to shoulder height – but without inadvertently raising the shoulder. Next, lower the left hand down to the abdomen and move the eyes to fix their gaze on the right hand. This has both a defensive and offensive martial application, which is that of escaping being held and using the wrist to attack wherever there is a weakness spotted.

Now turn and shift the weight of the body to the right and then slowly raise the left leg by relaxing the hip and lifting from the knee and moving it slightly inwards. Please make sure that you remain standing erect and do not slump as you perform this move, and pay attention to ensure that the buttocks do not stick out. Also, make sure that the shoulders are relaxed and that the elbows lower slightly. Hold the position as you look to your left side. This movement will test your balance, especially the more slowly you perform it, but with practice, it will be easy to do. The movement protects the crotch and can both block the opponent's kick and thereafter respond with an attack by rendering a kick.

Continuing with the weight on the right leg, and with the toes of the left foot pointing upwards, slide the heel out to the left. The right hand remains as was and the left lowers. The eyes focus on looking forwards. Make sure that the body remains erect as you do so. Your Zhan Zhuang postures will embed this position into your subconscious.

Inhaling slowly, slightly turn to the right whilst shifting the weight to the left to form a bow stance. Pay attention to the left knee to ensure that it does not bend over the end of the toes. As always, ensure that the shoulders do not hunch up or that the elbows lift involuntarily. For martial applications, this movement can be used as a strike with the elbow.

Exhaling whilst continuously moving, turn slightly to the left, curve the body so that the left hand circles past the left knee, then open the palms and sink the body down whilst remaining both centred and with the body straight but relaxed. The head is held up and the neck stretched but relaxed, not stiff. The left foot is pivoted so that it is turned out slightly as the right foot turns in slightly.

Individual breakdown of the movement

Image 8-76

Turn the palms of both hands over as you continue the circle upwards and shift your weight to the right leg.

Image 8-77

Between 60% and 70% percent of your weight rests on the right leg. Move the right arm over in a circle and turn it over so that it moves over the hinge of the elbow of the left arm. The left arm turns inwards and slightly to the right. The right hand is open and rests on the inside of the lower left arm. The left hand is open and facing upwards. The left foot is still raised on the toes.

Image 8-78

Slide the right palm across the left inside arm and form a hook as you do so. At the same time, drop the left arm so that the palm of the hand lowers to the top of the abdomen facing upwards. The elbows are relaxed and hang down – not stiff.

Images 8-79

Relax the right hip and sink slightly as you turn to the left. The left foot is still raised from the floor with the toes, ready to step out.

Image 8-80

Lift the left leg as you relax the hip and turn the body at an angle and to the right. The right hand moves across the body to the right.

409

Image 8-81 and 8-82

Lower the left foot to the floor and drop the heel – sliding it along with the toes raised – and straighten the body so that it faces forwards. The distance between the feet is around two shoulder widths apart.

Image 8-83

Turn the body slightly to the right. The left hand remains covering the abdomen a few inches in front and the right hand remains in a hook.

Image 8-84

Continue to shift your weight to the left leg, with the left arm following across the front of the body.

Image 8-85

Now shift your weight back to the right. The left hand is still positioned covering the abdomen a few inches in front and the right hand remains in a hook.

Image 8-86

Continue to shift your weight now back to the right. The left hand moves upwards in a circle across the body level with the inside of the right elbow.

411

Image 8-87 to 8-89

Now, ready yourself to turn back slightly to the left as the left arm starts to turn outwards and to the left. The left hand turns out stretching to the left and opening out to the side, with the right hand remaining in the hook position. The eyes follow the movement of the left hand.

Image 8-90

Sink your body and turn in very slightly to face forwards. Between 60% and 70% of your weight is on your left leg. The left and right hands are level with the toes of the respective feet. Relax and look straight ahead. The body should be perfectly aligned from 会阴 – huì yīn to 百会 – bǎi huì.

6. White Goose Spreads Wings – 白鹤亮翅 Bái hè liàng chì

This movement is very deftly done and you need to co-ordinate the whole body. Continuing from the previous movement and inhaling, turn your body and shift your weight to the right leg and turn the body initially slightly and then to the right, smoothly. The right fist takes up an open palm position and lifts to a position in front of the forehead as it closes. The arms are open and the shoulders do not rise or become too tight. Ensure that the right elbow does not rise. The left hand presses down to the floor. The martial application is that of defending both the head and the groin.

Turn to the left. Pivot the left foot to the left and change the right hand from the hook into an open palm. Cross and open the arms in front of the chest and hold the left palm so it faces the right with the fingers pointing up. The right palm faces up and the fingers point to the front.

Turn right, shifting the weight there and open the arms. Press the left palm down to the left hip with the palm of the left hand facing down. Lift the right palm up and outwards to the right with the palm facing out. Move the left foot up to the left of the right foot.

Individual breakdown of the movement

Images 8-91 to 8-93

From Single Whip, open the arms and the drop the right arm as you prepare to move to the left. Then turn your body slightly towards the left knee, with the right hand over the right knee.

415

Images 8-94

This movement is that of a swift turn pivoting on the left leg as the right leg and arm move together, with the right leg lifted as the hip relaxes, the toes of the right foot pointing to the floor. The two hands come together at a distance of some six inches or so, with the palms facing each other. This needs some co-ordination to move and yet retain balance, so concentrate and relax.

Images 8-95

Closing the hands together at the wrists and moving the weight to the left, step out the right leg by placing the heel of the right foot on to the floor, with the toes pointing up.

Image 8-96

Lower the right foot flat to the floor and separate the hands slightly. The weight is on the right leg but is ready to shift to the left.

Images 8-97 to 8-99

Shift your weight to the right and turn your right arm over in a circular movement as you slide your left foot in towards the right, with the heel raised. The right hand is at temple height and the body is straight upright. The palm of the right hand faces outwards. The left palm presses down to the left hip with the palm facing down.

417

7. Diagonal Step – 斜行 Xié xíng

Continuing from the previous movement and inhaling, keep the feet stationary and turn the body to the left as you circle the left hand whilst at the same time sinking the right elbow and circling the palm of the hand to the left in a smooth circular movement. Please note again that the body leads and that the shoulders should remain relaxed and not rise. This movement is that of a defensive blocking application.

Exhaling, turn to the right by pivoting the right toes out slightly. The body turns and leads the left hand palm upwards and to the right. Move it upwards past the nose. The palm of the left hand faces down and your eyes follow the movement leftwards. Maintain the shoulders being relaxed and the body held straight. This is a continuance of the defensive application of guarding the face.

Slide the left heel in a forward and diagonal movement to the left as the body sinks down, the eyes following the movement and the palms of both hands pushing upwards. Note that the heel of the left foot should remain on the floor as it slides forwards. The martial application is that of negating an attack using the leg and shoulder.

Turn the body left, shift your weight on to the left leg and circle the palm of the left hand below the left knee. Move the right hand palm up to just below the right ear in a backwards circular motion. This is a co-ordinated movement and, when applied quickly, uses the left side of the body supported by the right.

Inhaling and maintaining the weight on the left leg, continue the movement to the left. The palm of the left hand should remain relaxed as it turns in to a hook and moves to shoulder height. The palm of the right hand is held in an upright position in front of the chest in preparation for a potential attacking position.

Exhale and turn right, and push the palm of the right hand forward and to the right, ensuring that the shoulders remain relaxed and the elbows sunk slightly. The body must remain stable and balanced, which allows for the ability to move and attack from any direction. Relax and sink as the knees bend.

Individual breakdown of the movement

Images 8-100 and 8-101

From the last posture, let the feet remain stationary and turn your body to the left, with the right arm turning inwards in a circle and the left moving in synchrony with the palm of the left hand moving backwards. The body movement turns the arms. They are not separate movements, but a co-ordinated one.

Images 8-102 to 8-103

Move the left heel down to the floor and start to turn to the right by placing the weight on to the left leg and, as you turn, raise the toes of the right foot on the heel as the arms turn, the left arm in an upward circle with the right palm level with the nose and the right palm over the knee at the height of the right thigh. Continue turning so that the right hand moves across the right thigh to the side; 60% to 70% of the weight is on the left leg.

Images 8-104 and 8-105

Raise the left leg by relaxing the hip as you raise your arms up and over to the right.

Image 8-106

Step out to the left by sliding the left heel along the floor.

421

Image 8-107

Turn back to the right, moving from the waist as you turn in a circular motion, with the right arm moving across the body and the left moving in synchrony. You sink down slightly.

Image 8-108

Now turning to the left move the arms in a circle downwards as you sink further down and place most of your weight on to the left leg.

Image 8-109

Continue turning to the left as your left arm looks to circle downwards over the left knee and the right arm follows upwards in a circle over your right ear.

Image 8-110

Continue turning to the left but raise your body as it completes the circle, with the left hand forming a hook and the right arm following the movement across the body at shoulder height.

Image 8-111

With between 60% and 70% of your weight on the right leg, start to turn the right arm over back to the right as the left hand remains in the hook position over the left toes.

Image 8-112

Turn the body slightly back to the right and open the palm as your arms move horizontally across the body to the right.

Image 8-113

Now, turning to the left, move your arms in a circle downwards as you sink further down and take most of your weight on to the left.

423

8. Grab the Knee 搂 膝 lǒu xī

Overall Movement

Inhale and keeping the body erect, sink from the act of relaxing the hips and bending the knees as you slightly lift the hands so that they are positioned over the left knee and face each other, but just slightly apart. Look to the front. This movement provides you with the opportunity to strike the opponent's head on both sides if there is an attempt to grab your leg.

Inhale as you lift the left hand in front of the right, with the palms facing upwards and in front of the chest as you move your weight to the right leg. Now move the left foot backwards, with the toes pointing to the floor, and look forwards. The posture is centred and you are ready to change direction quickly if attacked.

Individual breakdown of the movement

Image 8-114

Sink the body as you relax your hips and bend the knees. Open the arms, with the palms of each hand open facing inwards. The bulk of your weight is on the left leg. The hands are separated and open at a distance of a double shoulder width apart.

Image 8-115

Continue to sink the body as you relax your hips further, and bend the knees a little, with the weight sinking further on to the left leg. The arms come together as the palms of the hands open upwards and close towards each other about six inches apart over the left knee.

Image 8-116

Now, slowly shift your weight to the right leg in a smooth and controlled movement whilst maintaining your balance and, as you do so, simultaneously draw the left foot in and raise the heel of the left foot as you pull the arms upwards while your body rises.

425

Image 8-117

Continue to draw the left leg in towards the body, with the toes sliding across the floor to about shoulder width apart but in front of the right leg. As you do so, continue to raise the hands upwards, with the right-hand thumb facing the level of the nose facing about two inches away (depending on how big your nose is!). The left hand is in front of the right a few inches away and level with the toes of the left foot. The left foot remains with the heel off the floor.

Image 8-118

Sink the body down slightly and lower both arms so that the hands are horizontally level with each other, with the right hand palm facing the thumb of the left hand that is facing outwards. As always, relax and ensure that the shoulders do not hunch up. The toes of the left foot remain touching the floor, with the heel raised.

9. Twisted Step 前蹚拗步 Qián táng ào bù

Overall Movement

This movement is light and like the stepping of a cautious, yet confident, cat.

Breathe out and turn slightly to the right, moving the palms of both hands downwards and to the right in a circular motion. At the same time, raise the knee of the left leg with the weight on the right leg. Ensure that you remain in a straight position and do not bend your back as you perform the movement. This movement allows you to pull an attacker down and attack with the left knee.

Inhale and move forwards naturally, slowly and in co-ordination by slightly turning to the left, with the left foot forwards, placing the heel on to the floor with the toes facing upwards. Raise both the hands up in a circular motion and look forwards. The movement allows you to strike an opponent's face.

Shift the weight on to the left leg and move to the left while, at the same time, pressing the palm of the left hand down as the right hand palm pushes forwards. Raise the right foot and look forwards. Now, step forward with the right foot by moving the heel to the floor and with the toes pointed upwards. Maintain the weight on the left leg and turn the left hand backwards, with the palm pushing forwards and the eyes following.

Inhaling, pivot on the right foot and shift the weight on to the right leg and then lift the left leg and step forwards and to the left. As the body moves forwards, turn on a ninety-degree angle and lower the palm of the right hand. Now, lift the left hand, and in a circular motion, raise it to the left ear and continue until it crosses with the right arm directly in front of the chest. Maintain your gaze looking forwards. It is important that this movement is performed smoothly and not jerkily and in an uncoordinated way. Pay attention to the height of the body and ensure that you are not bobbing up and down like a cork on the water. Exhale and to the rhythm of the movement. This posture is one of anticipation, awaiting the movement of an opponent so as to be able to react appropriately.

Individual breakdown of the movement

Image 8-119

Sink the body as you relax your hips and bend the knees. Open the arms, with the palms of each hand open facing inwards. The right hand is in front of the left and both palms are raised slightly but pointing to the floor. The heel of the left foot remains raised off the floor slightly.

Image 8-120

Turn your body slightly to the right and start to lift the left leg so that the heel rests just slightly away from the right knee as you lower your hands down together to a position just above your abdomen, with the hands about shoulder width apart.

Image 8-121

Continue to turn the hands to your right in a circular motion as your left leg starts to move to the left. Sink down slightly by relaxing your hips. The left arm is inside the left leg and the right hand covers the right hip, with the palms facing down. The palm of the left hand is facing away from the left leg.

Image 8-122

Now, turn the arms further to the right, with the right arm leading and the left palm covering the front of the body at shoulder height, and with the left palm facing outwards level with the right shoulder. The left leg starts to step out to the left. All the weight is on the right leg. Keep your balance and posture.

Image 8-123

Continue to move to the right with your arms following, the right hand palm facing outwards over the right cheek (of the face) and the left hand opening out slightly. The left leg steps out to a distance of two shoulder widths away from the right and the heel of the left foot lowers to the floor.

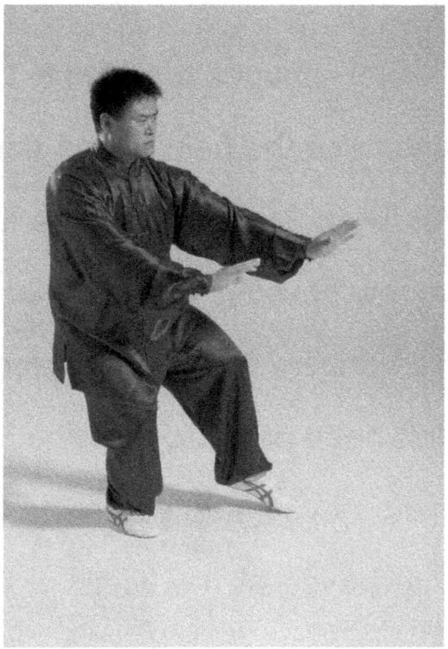

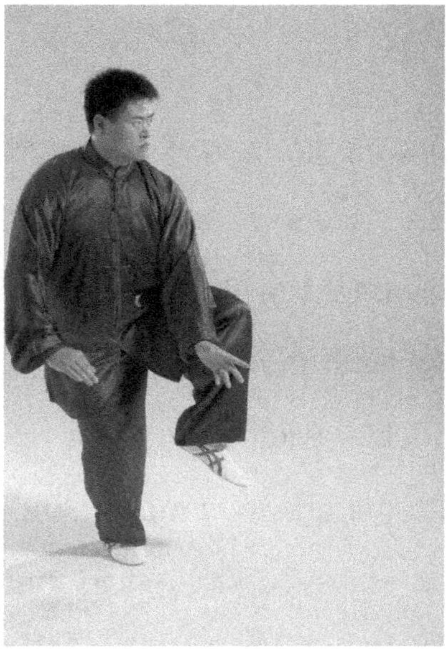

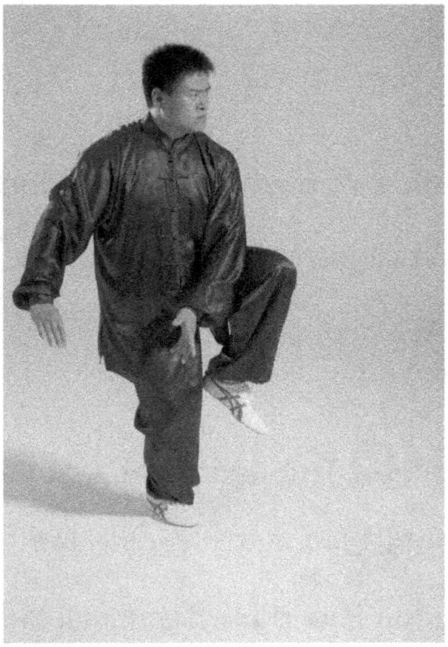

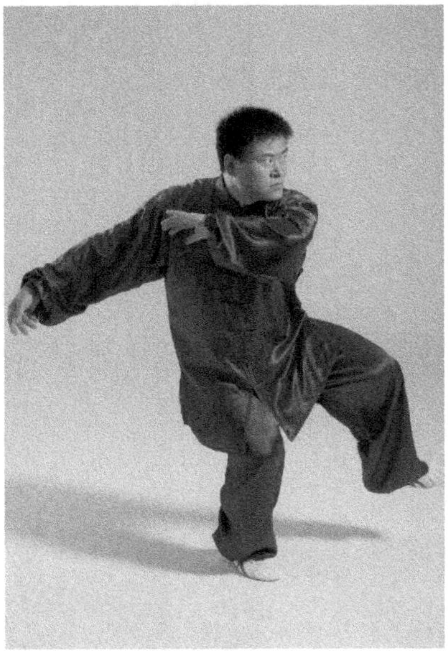

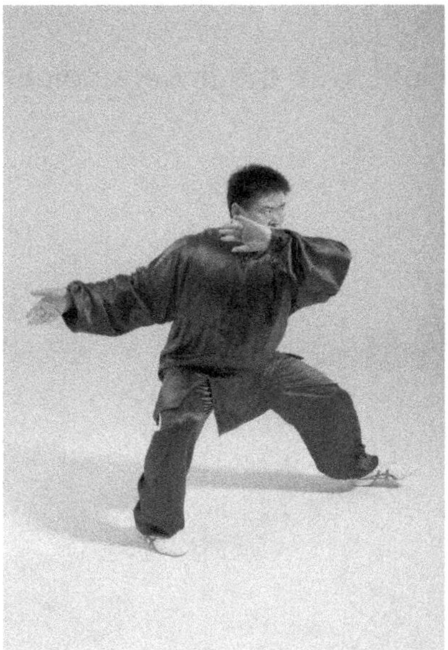

Images 8-124 to 8-125

Turn the arms over in a circle and, as you do so, pull in the left foot and raise the heel off the floor. The fingers of both hands face downwards. The feet are shoulder width apart. Continue the movement of the arms to the right in a circular downwards movement, drawing the left foot in slightly to the right. The heel of the left foot is raised and the hands are horizontally level and about shoulder width apart. The right hand palm faces to the floor and the left hand palm, to the side. As with the previous movement, turn over the hands in a circular movement as you step out to the left and shift your weight to the left leg. The arms are at the height of the eyes at the width of the shoulders, with the palms of each hand facing out.

Image 8-126

Continue your movement, with the left arm pushing out and the right arm turning at the elbow, with the right hand palm level with and just slightly to the back of the right ear.

Image 8-127

Raise your right leg and left arm together. The right heel is at the height of just below the right knee, with the toes pointing down at an angle. Sink down on the left leg. Lower your right leg and left arm together as you straighten your body. The right hand still resides close to the temple, with the fingers inwardly pointing towards it as though you were saluting an officer. The left hand is at waist height, with the fingers pointing to the floor. The heel of the left foot is raised slightly.

431

Image 8-128

Now raise your right leg again as previously, with the right leg slightly higher and positioned just above the right knee and the right hand remaining as it was, pointing at the right temple. Step out the right leg to a distance of two shoulder widths apart or thereabouts, with the right hand following and moving from the right temple outwards but at the same height, with the right elbow over the right knee.

Images 8-129 and 8-130

Move your weight to the right and stretch out the right arm, the eyes following the movement, with the palm of the right hand facing outwards.

Image 8-131

Continue to shift your weight to the right and stretch out the right arm, and sink down slightly. Both feet are turned out slightly.

Image 8-132

Start to turn over your left arm, circling from the elbow and over. As per our oft-repeated and perhaps overstated advice, do not hunch or raise the shoulders. The movement of the left arm is from the relaxation of the elbow, with the hand turning over in a circular motion. Continue to circle the left arm over towards the right arm as you move your left leg in towards the right, and shift your weight slowly on to the left leg.

433

Image 8-133

Turn your left arm over and on towards the inside of the right-inside arm. As you do so, turn your body to the right and step out with the left leg by sliding the heel of the left foot along the floor.

Image 8-134

Start to shift your weight to the left as you straighten up and start to sink slightly. The arms meet together, with the left arm over the right and both palms facing down.

10. Covering Hand Punch 掩手肱捶 Yǎn shǒu hōng chuí

Overall Movement

Exhaling and using the body to lead, turn slightly to the right as you shift your weight to the left and cross the arms outwards to open from the previous closed position. The act of separating the arms acts to neutralise an opponent.

Shift the weight and turn ever so slightly to the right, sinking the body from the hips and allowing the knees to bend as you raise the right hand and create a fist with the palm facing upwards. Raise the left hand and push the palm outwards with the fingers raised upwards yet relaxed, and aligned with the centre of the chest. This left hand covers the fist in readiness to attack.

Push the right leg into the floor and quickly turn the body to the left by relaxing the left hip and punch the fist in a quick movement, using the twisting of the waist to generate power so that the palm that was facing upwards ends up facing downwards. Simultaneously, strike with the left elbow backwards.

Individual breakdown of the movement

Image 8-135

Circle the hand over each other and sink your body. The weight is still on the right leg at the usual ratio of 60% to 70%. The hands meet in the middle and in front of the body, with the palm of the left hand covering the right wrist.

Image 8-136

Continue to separate the hands outwards, with both palms facing out and to the sides. The weight shifts to the left.

Image 8-137

Continue to circle the arms upwards, with the right hand forming into a fist and the left hand opening out. Most of your weight lies on the left leg.

Image 8-138

Shift the weight slowly to the right leg and, in so doing, simultaneously open up the left arm so that it is stretched yet relaxed, with the palm facing outwards. The right arm is moves up in a circle, with the fist at shoulder height and a few inches out. Start to slowly sink down.

Image 8-139

Continue to shift the weight slowly to the right leg and sink down further. The right arm also sinks to a position just level with the right nipple. The left arm also sinks a few inches in line with the overall movement. At this point, you are ready to punch out so you need to be sunk and relaxed. There should be no tension in the shoulders nor should they be hunched. The left arm is stretched yet relaxed. The right fist is clenched but also relaxed. You need to concentrate and look forwards. The right ankle is turned in slightly to infuse some power form the ensuing movement.

Image 8-140

This punching movement involves a shifting of the weight from the right to left leg and is directed from the 丹田 dān tián. The right fist punches across the body as though directed at an opponent. At the same time, the left elbow strikes backwards.

437

11. High Pat on Horse 高探马 Gāo tàn mǎ

Overall Movement

Continuing from the last movement, inhale as you turn the waist and change the right fist into a palm, and with a circular motion, move it downwards so that it rests adjacent to the right hip. Simultaneous to the movement of the right hand, push the left palm forwards to a position in front of the body and raise the fingers upwards so that the palm faces outwards.

Inhale and relax the hips as you transfer the weight to your right leg as you turn the body to the left. Turn in the toes of the right foot and move the palm of the right hand in a circular clockwise motion upwards to shoulder height and face it forwards. Simultaneously, bring in the left hand slightly as the arm circles in a counter-clockwise direction, with the palm facing upwards.

Exhale and retain your weight in the right leg and turn your body to the left. Move your left foot in a circle, with the toes touching and remaining in the floor. Simultaneously, move the palm of the right hand forwards in a circle. Ensure that your shoulders are relaxed and that the elbows hang down. Now, circle the left hand back with the palm facing upwards. Ensure that the eyes follow the movement and remain looking forward.

Individual breakdown of the movement

Image 8-141

Turn the left arm inwards towards the upper bicep of the right arm and, simultaneously, open the palm of the right hand from the previous fist position. Slowly start to shift your weight on to the right leg.

Image 8-142

Continue to separate the arms, with the left moving forwards in a circular upwards motion and the right arm moving downwards in a circular motion. The left hand palm faces outwards. The weight carries on shifting to the right.

Image 8-143

Further shifting the weight back on to the right leg, continue moving the right arm down and over the knee and on to the back and above in height and on to the back of the right knee. The left arm continues its trajectory forwards, with the palms further opening and pointing forwards. As always and with the danger of this becoming a mantra, please ensure that the shoulders neither become hunched nor rise.

Image 8-144

The right hand moves to a position just behind the ear, and the right toe turns in slightly to create some torque ready for turning. The left hand turns upwards and most of the weight is on the right leg.

Image 8-145

This move is one of the most difficult in terms of co-ordination (and in trying to describe it!). In actual fact, though, it is easy to perform after some little practice. Turn on the right foot with the toes turning in and shift the body so that it turns in a circle, the left foot moving on the toes in a circle. The left arm swivels around so that it rests under the elbow of the right arm. The heels are in a straight line. It is this part that needs most practice to obtain precision.

Images 8-146 and 8-148

With the toes of the left foot still resting on the floor, turn both arms out and away from each other so that the right points to the side, with the right palm at the centre of the ear and the left arm circling downwards, and the palm facing you and resting just above the abdomen. Equally, as usual we must emphasise that you need to relax the hips. In addition, the chest should also be open. Sink slightly as you complete the movement. The eyes look forward. The positions as seen from the front.

441

12. Left Heel Kick 左蹬一根 Zuǒ Dèng yī gēn

Overall Movement

Now inhale, move the right arm slightly in a circular motion and forwards in a counter clockwise direction, with the left palm facing out. Simultaneously, transfer your weight to the left leg and then step your right leg out about two shoulder widths apart.

Cross the palms over in front of the abdomen as you bend and then lift the left knee with the toes relaxed and pointing to the floor. The right leg supports the body. Ensure that you do not lean over too much and thus lose balance.

Exhaling, and with the right leg providing support and balance, lean slightly to the right without losing your balance, and kick out with the left leg at just below waist level. As you do so, punch out both fists to the side.

Individual breakdown of the movement

Image 8-149

Move the arms together, with the left hand inside of the inside of the right elbow as you lift your right leg, relaxing your hip as you do so, with the right foot rising to the level of the knee to the side and the toes pointing to the floor.

Image 8-150

Step out more than two shoulder widths apart and bring the hands together in front of the face at eye level. Both palms are facing out.

Image 8-151

Shift your weight to the right and separate the arms, with the hands moving to the level of the temples.

Images 8-152 to 8-153

Move the left arm in towards the centre of the body, with the hands overlapping and formed into fists above the abdomen. The weight remains on the right leg as the left leg is raised with the toes pointing down. Sink and relax the hips. You need be ready to kick out to the left.

Image 8-154

Sink down slightly and relax. Kick out to the side with the left leg. The power comes from the waist and hips. Do not fall into the temptation of leaning over to the right. Both arms strike out to the side simultaneously, with the palms of the hands open and facing to the sides.

13. Jade Girl Works at the Shuttle 玉女穿梭 Yù nu chuān suō

Overall Movement

Inhale and continuing from the last movement, rest the left leg to the floor and as you do, change both fists into palms crossing the left over the right by circling it counter clockwise as the right circles clockwise. These movements are undertaken simultaneously and are led by the waist.

Next, move your weight on the left leg as you slightly turn to the right, with both hands following the body and the right hand up and in front of the left, the fingers of both pointing upwards and the toes of the left foot turning in slightly. Simultaneously, turn out the right knee, with the toes remaining on the floor.

Exhale and relax the hips and sink down, allowing the knees to bend. Press down with both hands and circle the right arm counter-clockwise. Please ensure that your body remains erect as you undertake this movement.

Inhale and raise both arms, with the right moving in a circular clockwise direction and the left in the opposite circular counter-clockwise direction. Jump up so that both feet leave the floor together.

Stamp on the floor, with the left landing slightly before the right and with both hands following pressing down, the right moving in a circular clockwise direction and the left in the opposite circular counter-clockwise direction. Again, ensure that your body remains upright.

With the weight on the right leg, push both hands up at the same time, with the right moving in a circular clockwise direction and the left in the opposite circular counter-clockwise direction, and raise the right leg to waist height, with the toes of the right foot facing down to the floor.

Inhale and step forward with the right leg leading, push forwards about two shoulder widths apart and land with the right foot on the floor and exhale

Turn your body to the right at an angle of 180 degrees as your left leg leads and rests the left foot on the floor. Simultaneously, the palm of the left hand pushes forwards and that of the left to the right. The right foot moves behind the left foot, with the toes remaining on the floor during the movement. This is a tricky movement and can be broken down into stages or as you advance, it can be undertaken in one continuous movement.

The body continues to move to the right in a 180-degree movement as the weight is transferred on to the right leg, with the toes of the left foot following inwards. Simultaneously, both hands move to the right and the right arm circles in a clockwise direction.

Individual breakdown of the movement

Image 8-155

Lower the leg to the floor slowly so that you retain your balance. The majority of your weight is on your right leg. The temptation is for you to use to much tight strength in the kick and flop down. Ensure that you relax and move the left foot down slowly, maintaining your balance as you do so.

Images 8-156 and 8-157

Shift your weight to the left and as you do so, move your arms down so that both move towards each other, with the shoulders relaxed. Sink down as you do so. Continue to move to the left and, closing both arms together, above and to the side of the abdomen.

Image 8-158

Turn the body back to the right slowly and as you do so, turn the arms up and over in a circular motion, with the left hand touching the inside of the right arm at the inside of the elbow. Turn in the left foot ready to use as a pivot for turning.

Image 8-159

Shift the weight to the left leg and turn the body, using it as pivot, and move the arms over to the front of the body, with the right arm raised up level with the top if the head and the left hand moving down also in a circular motion.

Image 8-160

Press both hands down level with each other just above the abdomen. The palms face to the floor. The waist directs the shoulders which, in turn, directs the elbows, and the power resides in ten hands.

Images 8-161 to 8-163

Sink down and then jump up with the right leg, rising high, with the arms moving up and the palms facing up; then land with the right foot in front of the left and the palms facing down. The left foot lands first slightly ahead of the right. The eyes face forwards.

Image 8-164

Now, raise the right leg by relaxing the hips and, at the same time, both hands, with the palms facing upwards. Keep your balance and perform the move smoothly. Remain upright and do not hunch.

Image 8-165

With the weight firmly rooted on the left leg, kick out with the right leg to the side. Avoid over leaning to the left and losing your balance. The left elbow strikes out to the left and the right hand also punches out to the right.

449

Image 8-166

Lower the right leg to the floor, with the arms remaining in the same position, the left elbow pointing out and to the side and the right palm out to the right, with the arm stretched but relaxed.

Images 8-167 to 8-168

Move the left foot forward as you turn your body forwards and transfer the weight to the right leg, with the left toes moving simultaneously inwards. The right hand moves in a counter-clockwise direction and the left hand, clockwise.

Image 8-169

This move needs some concentration. At first it seems difficult – and it is a little if you wish to perform it correctly. The key is to ensure that the heels are aligned after you have twisted around. It is easy to end up with the feet misaligned. The key is to turn on the heel of the right foot and the toes of the left. That way the feet are level after turning. The arms simply follow the movement.

Image 8-170

This move needs some concentration. At first it seems difficult – which it is a little if you wish to perform it correctly. The key is to ensure that the heels are aligned after you have twisted around. It is easy to end up with the feet misaligned. The key is to turn on the heel of the right foot and the toes of the left. That way the feet are level after turning. The arms simply follow the movement.

451

14. Wave Hands - 运手 Yùn shǒu

Overall Movement

Continuing from the previous movement, rotate the right arm clockwise, with the palm of the right hand facing towards the left of the abdomen as you simultaneously circle the left hand clockwise in front of the body and push to the front of the left shoulder, transferring your weight on to the left leg as you sweep the right foot behind the left, with the toes pointing to the floor.

Transfer your weight on to the right leg as your body turns slightly to the right and place the right foot flat on to the floor. Raise your left foot and step out to the left with the heel grounded and the toes pointing up, and simultaneously raise the right hand counter clockwise across the front of the body and to the front of the right shoulder. The left hand circles counter-clockwise to the front of the abdomen. The eyes follow the movement and rest looking forwards. This can be a tricky movement. So, ensure that when your legs cross, there is a space so that you do not become entangled.

Slightly turning the body to the right, move the right hand clockwise in front of the abdomen, with the palm facing to the left. Simultaneously, circle the left hand clockwise, with the palm resting and facing to the left. Shift your weight on to the left leg as you move your right foot along the floor to cross behind the left foot.

You can repeat this movement a few times. It is good practice in its own right as a single movement.

Individual breakdown of the movement

Images 8-171 to 8-172

This move is also one that you need to concentrate on in order to achieve the smooth flow and co-ordination necessary. Once you have practised a little, though, the movement will flow and become natural. Move the palm of the right hand on a downwards circle and the palm of the left hand across the body to the front of the left shoulder. Shift the weight to the right hand side and move both arms simultaneously in a circle as you step out to the left about two shoulder widths apart.

Images 8-173 to 8-187

Turn the body to the right with the hands moving in a circular motion as was explained in the Clouds Hands in the previous chapter. When the arms have moved in a circle then move back to the left and at the same time shift the weight to the left and move the right leg leftwards behind the left leg so that the left foot and the hands are level. Then step out to the left continuing the movement of the arms. And repeat this sequence three times – moving to the left. On the last movement when the right leg has been positioned behind the left turn on the left leg and swivel the body so that it is facing forward with the arms following the body. As mentioned this is difficult to describe in words without making it seem too mechanical or convoluted. Please refer to a video demonstration for further clarity. It is easier watching than looking at the photos.

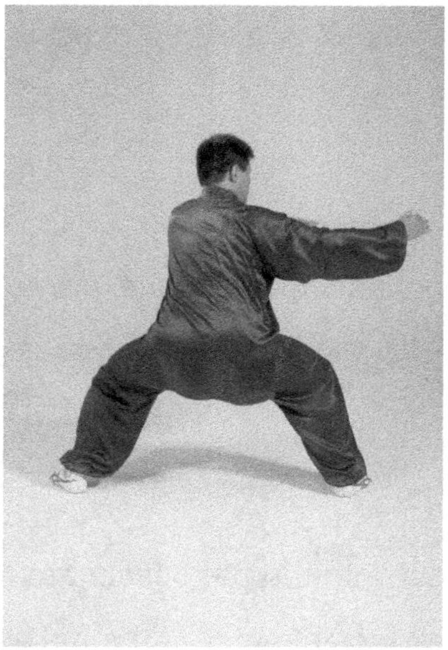
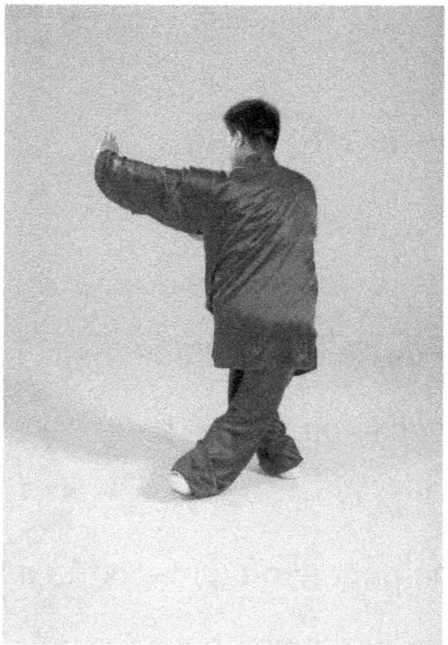
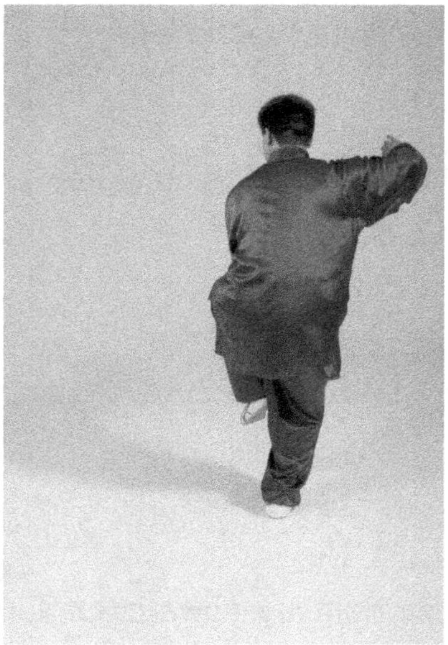
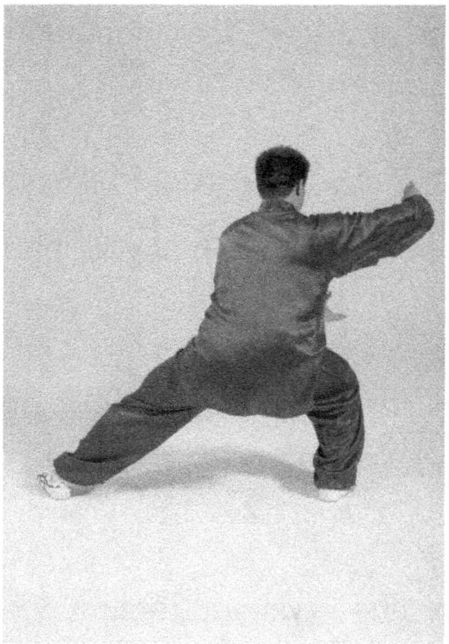
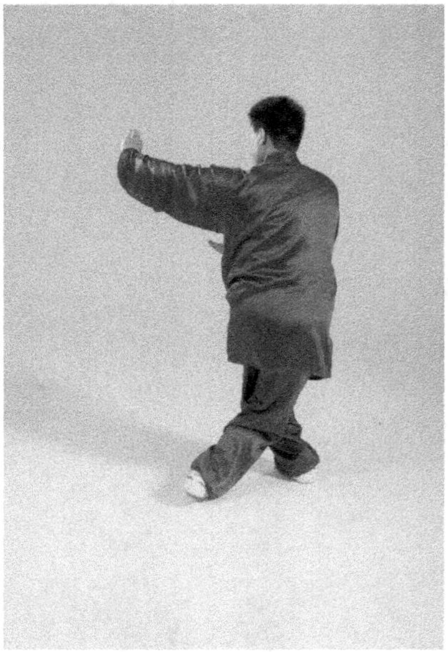
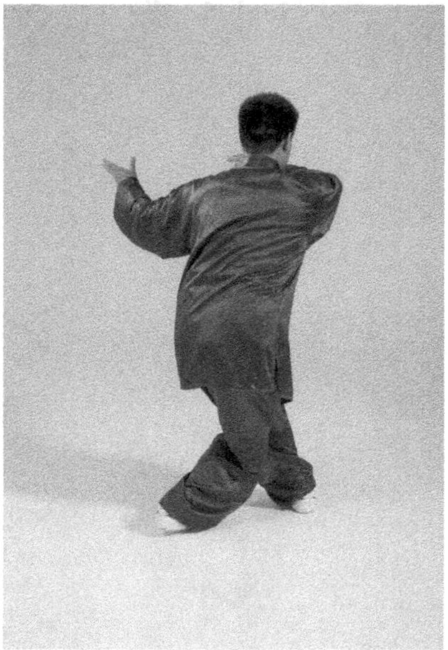
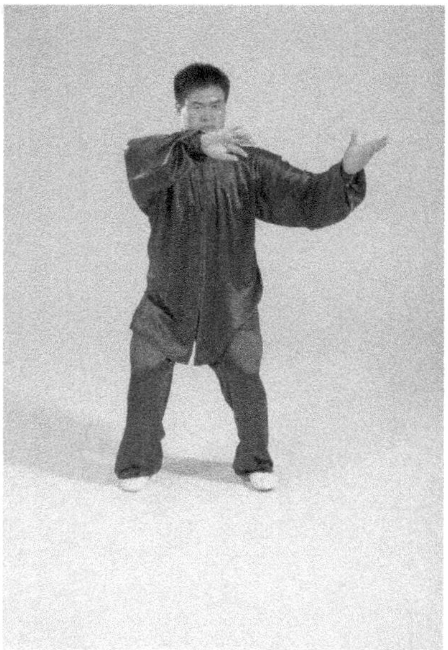

15. Turn the Body with a Double Lotus Kick - 转身双 摆莲 Zhuǎn shēn shuāng bǎi lián

Overall Movement

This is a tricky one as it requires a good deal of flexibility and co-ordination. From the previous movement, circle both hands counter clockwise and turn the body to the right a full 180 degrees by pivoting on the left heel, using the right heel to turn.

The right hand turns to the extent that it rests in the middle of the chest with the palm facing to the right and both palms facing outwards. The left hand rests in front of the left shoulder and the palms face upwards. The shoulders are relaxed and the eyes face forward and slightly to the left.

Now take this next part slowly. Shift your weight on to your right leg. Lift your left leg and keep your knee bent as both of your hands simultaneously move backwards and to the right.

Turn your body to the right and shift your weight on to your left leg whilst the hands circle clockwise and rest adjacent to the right waist.

Breathe in as you retain your weight on the left leg and now swing your right leg upwards and in a circle, and sweep both hands together across the left foot and breathe out as you complete the move.

Individual breakdown of the movement

Image 8-188

This is another twist around where the feet should be square on and level upon completion. Turn on the heel of the right foot and the toes of the left foot so that you are now facing forwards, with the feet level. The arms have followed the movement and are in their relative positions, with the right palm facing out and level with the relaxed shoulder and the left hand level with that of the right hand. Continue the turning movement to 45 degrees on the right heel and end with the toes of the right foot pointing upwards slightly.

Images 8-189 and 8-190

Push out to the right and lower the toes of the right foot. The arms are relaxed and follow the movement.

Images 8-191 and 8-192

Shift your weight on to the right leg and slowly lift the left foot up by relaxing the hips and lowering down slightly. The eyes are following the movement looking forward. You are ready to step out so you are looking at where the left foot will land.

Images 8-193 and 8-194

Slide the left foot forward on the heel of the left foot and then turn to the right, with the arms moving in a circular upwards and backwards direction. The weight has shifted to the right leg.

Image 8-195

The arms complete the circle in unison and start to push back to the left and the weight shifts back to the front left leg.

Image 8-196

Continue to shift your weight to the front leg and as you do so, slide the toes of the right foot forward to close the gap between both feet to a just a few inches. The hands turn over and meet level with the right foot.

Image 8-197

This is arguably the most difficult single movement in the set. It is fast and dynamic and requires both balance and co-ordination. You are going to raise the right foot up and over in a circular movement and, at the same time, hit the foot with both hands. This requires flexibility of the hips and co-ordination. Relaxing the right hip, swing it round in an upwards circle and as you do so, bring both hands down so that each hand slaps the foot. Keep looking forwards.

Image 8-198

This is another twist around where the feet should be square on and level upon completion. Turn on the heel of the right foot and the toes of the left foot so that you are now facing forwards with the feet level. The arms have followed the movement and are in their relative positions, with the right palm facing out and level with the relaxed shoulder and the left hand level with that of the right hand. Keep the body upright and do not collapse the upper body

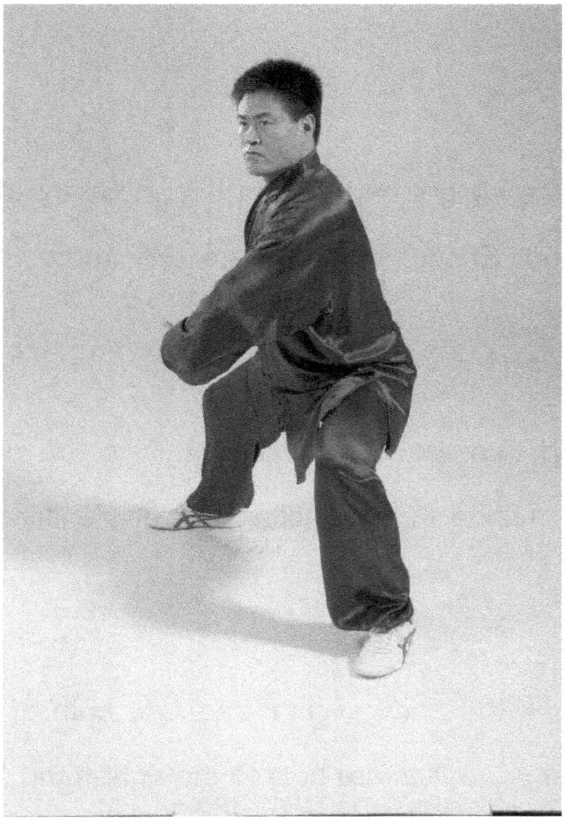

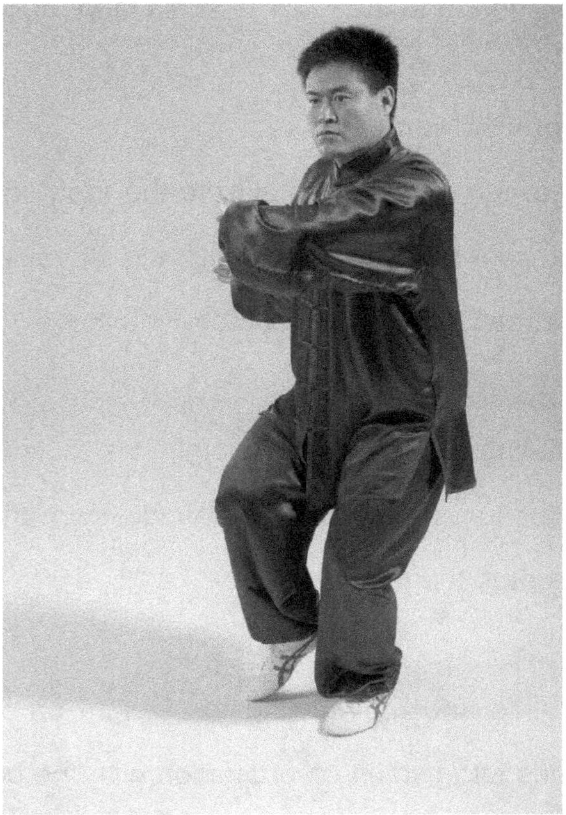

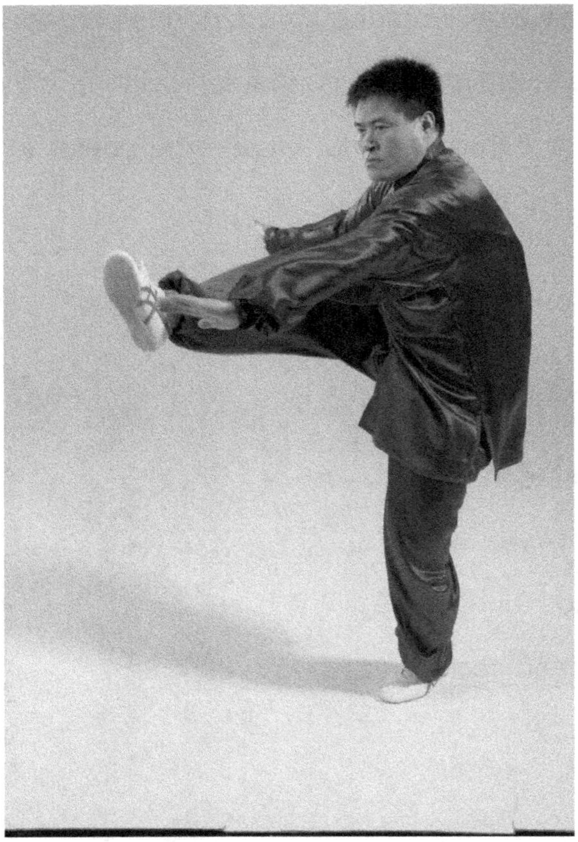

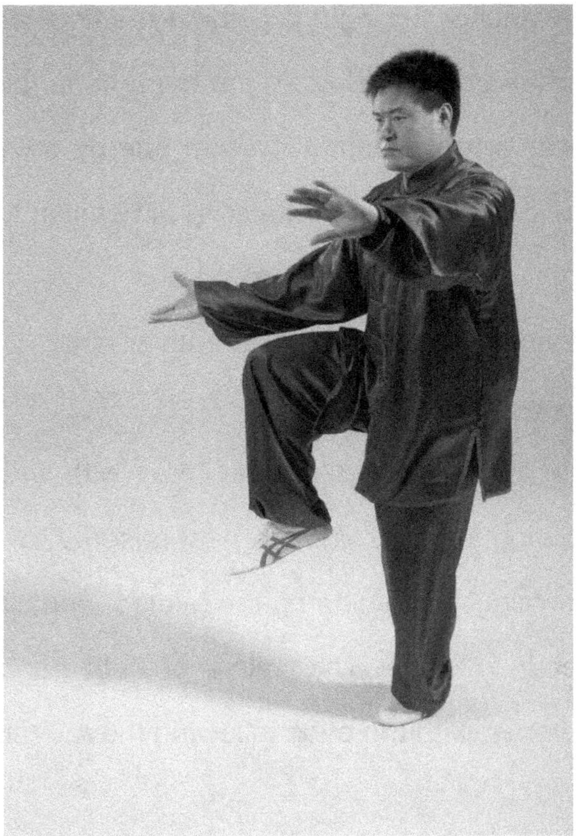

16. Cannon to the Face 当头炮 Dāng tóu pào

Overall Movement

Following on from the slap to the foot, and maintaining the weight on the right leg, move the right foot backwards whilst continuing to raise both hands up. Ensure that you are fully rooted and balanced after the kick before you move.

Exhaling next shift the weight on to your right leg and turn the body just slightly to the right. Simultaneously, move the hand down and in a circular motion in conjunction with the shifting of weight to the right leg.

Inhaling and maintaining the weight on the right leg, sink the body slightly and turn both of the hands into fists in co-ordination with the body's movement. The fists are held to the right-hand side of the chest.

Now anchor the right floor onto the floor and move your weight from the right to the left leg as you turn the body quickly. At the same time, punch out with both fists. This is an expression of internal power with concentration and energy driven to the fists. Exhale as you release the power to the fists. This has a very attacking dimension to an attack by an opponent.

Individual breakdown of the movement

Image 8-199

Drop the right leg to the floor, with the toes of the right foot touching the floor and the heel raised. Move the body forward, with the weight remaining on the left leg, and circle both hands together over in a circle. The palms of both hands are open and facing forwards, with the eyes following the movement.

Image 8-200

Drop the right heel to the floor and place the palms of both hands into fists and start to pull back to the left leg backwards.

Image 8-201

Using the waist, continue to turn the arms backwards and shift your weight to the right leg and sink slightly. You are setting up the body to punch so you need to be relaxed, balanced and sunk sufficiently to raise some power.

Images 8-202 and 8-203

Now release your energy into the punch by turning the waist and shifting the weight forward, using the right foot, which is slightly turned in. Both arms move together. The fists are clenched and face each other but are not tense, and the eyes look forward. Do not bend the waist, and remain in a straight posture albeit with the body sunk naturally from the relaxing of the hips.

17. Buddha's Warrior Pounds Mortar 金刚捣碓 Jīn gāng dǎo duì

Overall Movement

Now change the fists back to open palms and pull back and to the right as both hands work together as you shift your weight on to the right leg from your left and look forwards as you do so. Then move the weight back from the right leg on to the left leg, pivoting the foot and stepping it down so that you feel balanced. Shift the weight to the left at an angle of 45 degrees, circle both hands forwards and downwards and hold the left hand in front of the chest with the palm facing downwards.

Now move the left hand forwards and upwards and back in a circular motion and let it rest against the inside of the right forearm. Simultaneously, lift the right hand so that the palm rests in front of the chest to the right. The right hand palm is held facing upwards and the left hand palm faces downwards. The right foot steps out in line with the inside of the left foot and the toes remain at rest on the floor.

Now, move the left hand palm outwards and then lower it to a point in front of the abdomen, facing upwards. Make the right hand into a fist and lower it on to the palm of the left hand, facing upwards.

Raise the right fist to shoulder level in a circular movement but do not raise the shoulders as you do so. Simultaneously, raise the right leg, with the knee bent and the toes pointing downwards. Drive down the right foot to a position of shoulder width to the left leg at the same time as you land the right fist into the palm of the left hand.

Individual breakdown of the movement

Images 8-204 and 8-205

Shift the weight from the front left leg to the back right leg using the waist to drive the arms in an upwards circular direction.

Image 8-206

Continue to turn towards the back, with the arms moving together and circling over the right knee at shoulder height. Once the arms have turned shoulder height in their backwards circular movement, start to turn back to the front, shifting your weight to the front left leg.

Image 8-207

In one motion, shift the right leg on the toes towards the front, with the ankle almost touching the left ankle. The arms sweep forward, led by the left arm, with the fingers of the left hand pointing to the floor and at shoulder height. The right hand is open.

Image 8-208

Continue to slide the right foot forward on the toes and move it in front of the body and at shoulder width apart. The arms meet in front of the body, with the right arm moving forward and the right hand aligning with the right foot and the palm facing up. The left hand moves down to meet the right arm and rests on it, with the palm facing up and the fingers pointing forwards.

Images 8-209 to 8-211

The right hand forms a fist and the left hand slips down the level of the abdomen, turning over so that the palm faces upwards. The right leg rises at the same time as the right fist.

Images 8-212 to 8-213

Now, drop the right leg at the same time as you drop the right fist into the open palm of the left hand at the level of the abdomen in front of the body. This is a co-ordinated movement and should be done with power but smoothly. The arms are rounded and relaxed, with the usual gap kept between the arms and the torso. The knees and hips should be relaxed.

467

18. Closing Posture 收势 Shōu shì

Overall Movement

Breathe in and now open the palm of the right hand to change from a fist and separate both arms so that they hang down either side of the body. Sink the body down by relaxing the hips and allowing the knees to bend as you exhale. This movement is one of co-ordination of the whole body.

Inhale and circle both hands upwards together whilst maintaining relaxed and loose shoulders and sunk elbows. Now exhaling, press both hands down together slowly and with the same downwards momentum until each hand stops either side of each respective leg. Relax and let the 气 qì settle in the 丹田 dān tián. Now, raise the body to its natural height when it is relaxed but stretched from the head as though a hook was pulling it upwards. Move the left foot to the right so that the feet are together. The palms of the hands face in to the sides of the legs. Look forwards. You have completed the 18 form set.

Individual breakdown of the movement

Images 8-214 and 8-215

Separate the arms by opening out with the hands, with the palms open and facing upwards. Raise the arms upwards, and the body rises slowly and effortlessly until the hands reach over the head. Keep the body straight. It does not bend.

Images 8-216 and 8-217

Bring the arms down together, with the hands coming together as the body lowers itself. Keep the body straight. It does not bend. The lowering of the body is achieved by relaxing the hips and bending the knees. Relax and allow the 气 qì to settle in the 丹田 dān tián.

Images 8-218 and 8-220

Continue to sink down and then rise and relax. Well done you have finished one complete set of the 18-movement form.

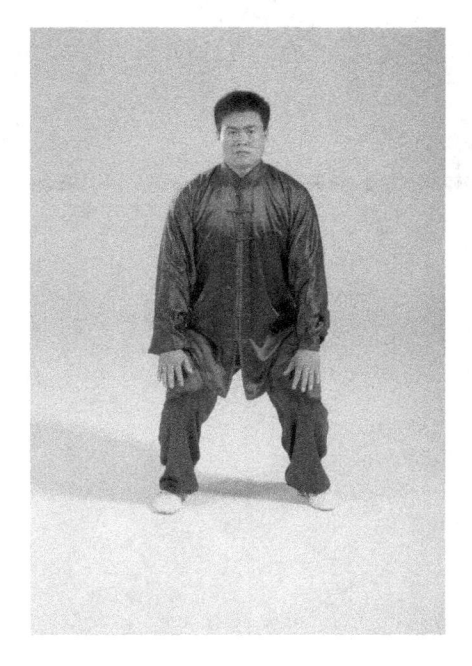

We hope that you have enjoyed the literal journey through this book. We started off with some, at times, pretty heavy theory and ended as we have done with the practical application of the forms. As stated at the outset, learning the 18-movement form incorporates all the basics that you need in order to advance in Chen style 太极拳 tàijíquán.

The ensuing forms, for example, 老架一路 lǎo jiā yì lu, 老架二路 lǎo jiā er lu and the 新架 xīn jià frames, as well as the weapons and push hands. These all make use of the basic elements that we have explained in this book. We advise you, the reader, to find a good teacher, one who has a high level of skill and with the right credentials, and thereafter to practise as much as you can.

You will benefit from the practice of Chen style 太极拳 tàijíquán, no matter what age you are. It will help your physical body by loosening your joints and stretching your tendons. It will also improve your posture. Your breathing will be fuller and deeper. You will be better balanced and co-ordinated.

Energy will flow through your meridian system more smoothly and powerfully. Your mind will become quiet. These are all just the simple associated benefits that will accrue to everyone. In addition to all of this, you will be able to learn a sublime and very sophisticated martial art that is very effective and will not injure your mind, body or spirit in so doing. We both encourage and welcome you to try this for yourself, and you will be most definitely pleased with the results.

The 站桩 Zhàn Zhuāng standing practice is easy to assimilate into your daily routine as it requires very little physical space and will energise you so that you require less sleep, thus creating some extra time available. Indeed, it becomes something that you will look forward to once you have learned to relax both mind and muscle.

Anyway, we hope that you have enjoyed reading the book and find some value in its contents.

CPSIA information can be obtained
at www.ICGtesting.com
Printed in the USA
LVHW02s1033211018
594317LV00005B/152/P